General Turchin

Where Will the Blacksmith Settle Down

ALEXANDER BORSHCHAGOVSKY

Translated by John Mandelberg

ETT IMPRINT
Exile Bay

ETT IMPRINT
PO Box R1906
Royal Exchange NSW 1225 Australia

ISBN 978-1-923205-41-3 (paper)
ISBN 978-1-923205-42-0 (ebook)

First published as Где поселился кузнец, Александр Борщаговский by Советский писатель - Soviet Writer, Moscow, 1978

Contents

General Ivan Turchaninov, better known as General
John Basil Turchin, Union Brigade Leader during
the American Civil War, and author of the book
Military Rambles.

INTRODUCTION

Alexander Borshchagovsky was born on 1st October 1913 in the city of Belaya Tserkov (Kyiv province), into a Jewish family. His father was a lawyer and journalist, his mother was a midwife. After his basic education he graduated from the Kiev Theatre Institute (1935), When he completed graduate school he went to the front of the Great Patriotic War 1941-45. He also became head of the literary department of the front-line theatre, a civilian employee, and was awarded the medal "For the Defense of Stalingrad."

Borshchagovsky was a popular writer who also wrote patriotic novels including *The Russian Flag* (1953) and *Milky Way* (1968). After the war, he headed the literary department of the Soviet Army Theatre (1945-1949); In 1949, as part of an ideological campaign against "rootless cosmopolitans," — ie. Being a Jew he was fired from his job, expelled from the Communist Party and deprived of the opportunity to publish for participating in an "anti-patriotic group of theatre critics."

He found himself in the first class of "rootless cosmopolitans", whom Pravda subjected to crushing, so-called "party" criticism —which was tantamount to a judicial verdict. I learned that this was followed by draconian "organizational measures" : Borshchagovsky was removed from the editorial board of the "New World", expelled from the Central Theater of the Red Army, deprived of not only work, not only the opportunity to publish — the path to literary and theatrical criticism was closed to him - but also housing...'

Lazar Lazarev,Soviet writer & literary critic
From "LECHAIM" JUNE 2006, SIVAN 5766–6 (170)

Later he returned primarily as a prose writer. His greatest fame came from the story 'Three Poplars on Shabolovka,' he revised into the script for a famous film *Three Poplars on Plyushchikah* (1967). Further films based on his stories were *Only Three Nights'* (1969), *Train to Tomorrow* (1970), *Door without a Lock* (1973), *Glass Beads* (1978), the play 'The Ladies' Tailor' (1990) became a film that highlights the tragic mass murder of Ukrainian Jews by the Nazis at Babi Yar in the Ukraine in 1941.

In 1991, he published his memoirs, Notes from the Minion of Fate (1991) which are replete with a lot of inaccuracies and distortions. The works Charged with Blood and The Hollow Monolith are also dedicated to the period of "the fight against cosmopolitanism."

In 1993 he signed the "Letter of the 42" which was an open letter signed by forty-two Russian literati, and aimed at Russian society, the president and government, in reaction to the 1993 Russian constitutional crisis. In recent years, he lived and worked in the writers' village of "Krasnovidovo".

Alexander Borshchagovsky was a literate moral man and a realist who told stories about human beings under stress from life, even under the thumb of Communism he was able to enlighten and entertain his Russian audience over many years.

Alexander Mikhailovich Borshchagovsky died on May 4, 2006 and was buried in Moscow at the Vostryakovsky cemetery.

Borshchagovsky's novel *Where Will the Blacksmith Settle Down*

After the 1949 ideological campaign against "rootless cosmopolitans" where the writer was stripped of any rights to publish, depriving him of work and access to housing, Borshchagovsky was able to start to write acceptable stories, articles and novels that were reviewed and approved for publication by the Soviet Writers Guild. So to curry favour with the United States in 1976, the year of the 200th anniversary of the founding of the United States, Alexander Borshchagovsky wrote the novel *Where Will the Blacksmith Settle Down* that is set in the US before the start of the American Civil War and uses the life and times of John Basil Turchin and his wife Nadine Turchin to tell a story of an attempt of Russians and Americans to work together caught up in the turmoil of the four year fight for the Republic between the Union and the Confederacy, which had been formed by states that had seceded from the Union.

Based on the true story of Russian Colonel Ivan Turchaninov – known as "the Russian Thunderbolt", an historical novel that tells about the life and work of Ivan Vasilyevich Turchaninov, better known in history under the name John Basil Turchin. This name was especially popular in the United States during the war between the Civil War between the North and South. Historians called Turchin "Lincoln's Russian General," a lot has been written about him both in Russia and in the USA, but this is the first time the reader will encounter such a wide and complete artistic canvas dedicated to this outstanding man.

Ivan Vasilyevich Turchaninov was a young intellectual, who espoused democracy. He left the oppressive system of Czarist Russia with his wife Nadine to settle in America to start a new life there.

When the American Civil War broke out, he raised a regiment and went off to war. He was among those who radicalized the war – transforming it from a purely military effort to preserve the Union into a crusade to free the slaves.

In this way, Turchin helped to free American democracy of the hypocrisy of human bondage.

Colonel John Basil Turchin was a leader of men and a brilliant tactician, whose important role in the Civil War was overlooked.

What was remembered of John Basil Turchin was the result of the libel and slander of Southerners as part of the 'Lost Cause' myth.

Freedom is not the right to do what we want, but what we ought.
Let us have faith that right makes might and in that faith let us to
the end dare to do our duty as we understand it.
Abraham Lincoln

John Mandelberg
Hamilton, New Zealand

PROLOGUE

They walked side by side - a general in a black felt hat, shifted from the maned nape to his forehead, and a woman with a light, sliding gait, dressed as simply as generals had not been dressed for a long time on the Potomac, buried in lace, scallops and frills, crowned with hats that an abundance of fruits, flowers and feathers argued with the Flemish 'still lives'. They walked squeezed together by the crowd, walked so closely that it would have been easier for a woman to pass her right hand, free from a chequered bag, under her husband's elbow, but, walking hand in hand, they could not manoeuvre ahead of a gang of soldiers in blue uniforms, officers jangling with spurs, gentlemen in short spring coats, in coloured frock coats, in brown top hats that shone copper at sunset, efficient servants and messengers with white-toothed grins on their black faces, intoxicated Irish artisans and incessant Germans, as if the imminent victory over the rebellious South had made them drunk, immigrants, more than the Washington natives.

"Let's take a break! You drove me, Vanya! For ten years now, English has become their everyday language, but in anxiety, in love, in forbidden pity for each other, Russian words came to them."

"God! You didn't hide the tickets: if you lose them, you'll have to buy new ones!"

As was the custom here, he tucked his discounted Chicago tickets, Chicago-Washington-Chicago, behind the moiré band of his hat. The woman took off his hat, revealing a high bald head.

When they got off the train, they did not even look in the direction of the hired cabs, but moved into the city through the station square, among the locomotive screams, the clatter of hooves, the inviting cries of cabmen, the ringing of bridles. It was Good Friday, April 14, 1865, a wind from the ocean, from the Chesapeake Bay, drove rare clouds above and promised bad weather.

The general needed President Lincoln. In his wife's bag, wrapped in a sheet of newspaper, lay a small book, just printed in the Chicago printing house of John R. Walsh, named by the general in such a way that he himself would have found it difficult to translate the title into Russian. "Military Rambles": military reflections, a kind of prowl of angry thought, a sarcastic review of everything that caught the eye of a military traveller, a mixture of a merciless memoir and fables, thickly peppered pamphlet escapades. About everything passionately, resolutely, passionately, completely in the Russian spirit.

The wind carries the voices of trumpets from the naval barracks and the shipyard, from the Arsenal at the arrow that cuts the Potomac into two arms. Orchestras are playing at the White House, at the War Department, at the Treasury, at the Ford Theatre and the Grover Theatre, the triumphant voice of copper calls into the sky of the capital more and more thick, viscous, ever lower floating clouds. The doors of taverns, saloons and beer halls are open, there is no way to push through: a drunken yell, the rumble of forged rims, the creak of dusty chariots and wagons, the Babylonian mixture of languages and dialects, the flickering of wheel spokes, the proud swaying of horsemen, the rustling of silk dresses, the intoxicated, restless tap dance to the whistle and the clapping of onlookers—Washington cares neither about them nor about Mr. Turchin's snooty little book.

"Colonel!"

Turchin did not respond to the voice. Although worn, he is wearing a general's uniform, with a large star on each shoulder strap sewn across his shoulder.

"Colonel!" In front of them, leaning over, danced a stocky man, covered from neck to knees with a theatre poster.

"Madam! He pulled his hat off his tousled red head and bowed ceremoniously, showing the second poster on his back. - Happy to see you, Mrs. Nadine!"

His puffy, freckled face, with eyes as blue as the general's, vaguely disturbed his memory.

The general moved forward, a fellow soldier settled down next to him, crouching on his leg.

"Yes, you have to keep up with this herd, otherwise they will trample it ... I am Paddy!" Sergeant Paddy! Weasel Paddy! - He amused himself with the general's unconsciousness and made charades, naming not his own name, but the nickname Paddy, common to all Irish immigrants, from the patron saint of Ireland, St. Patrick.

"Come on, I'll throw you a bait on a hook! In the autumn of the sixty-first, I got into your batmen. I did not like Madame's orders, and I said: let the lady not command me, and you? Sergeant, you said, Mrs. Turchina can command a whole regiment of people like you. Go!"

"Sergeant Mullen!" Nadine exclaimed.

"Barney O'Mullen!" All that's left of him! The former sergeant's voice trembled. Barney pulled up his left leg and lifted the prosthesis off the ground. "Many have lost more: especially those who didn't take care of their heads.

"How you have changed, sergeant!"

"And you are not the same either: perhaps the uniform will not fasten on you. And you, madam... Ah, how we loved to repeat aloud: madam, madam, our madam! [1] As if there were no other women in the whole world."

Throwing back her head, Nadine laughed happily, as if she were not in the crowd, but high in the saddle, next to her husband, and the sun blinded her eyes, and things were going as well as possible.

They walked quickly, Barney limped, he forgot that he was released into the streets of the capital to, tearing his throat, call out the name of the play and the name of the beneficiary, star Laura Keane.

"You have nothing to fear, madam: the world will grow old, but you will not."

Nadine leaned close to her husband and whispered: "And you hesitated whether to go to Washington!"

"Saint Patrick! I have never regretted so much that I had not a cent in my pocket; Here I would make you a drink!"

"We don't drink, sergeant."

"How are you still not drinking?" Barney exclaimed sadly. "But you will regret it, but it will be too late."

"Today is a holiday, today I, perhaps, will dance so that all my pieces of iron will crumble. Of all the recent battles we've won, do you know what's the smartest one?"

Barney was silent for a moment so that Turchin wouldn't think that the former sergeant was really examining him. "When we elected Lincoln for a second term! Abby won't let them raise their heads! He will order them all to be taken to Fort Sumter, given to them with holes, the lousiest boats we have, and let them sail across the ocean to their friends, to hell, forgive me, madam!"

"Blessed are those who believe, sergeant. We've been praying for you, at every opportunity. I just don't know if the prayer of Catholics helps Russians? ..."

The crowd pushed them back to the huge van, which drove right onto the sidewalk, under the green, transparent canopy of spring foliage. There was road dust on the wagon, heavy farm horses unharnessed, on the front end a fat black woman who had come to such a big city for the first time; she rolled her eyes at the crowd and muttered loudly:

"Lord! Jesus! Do each of them have their own name?"

"Yes, auntie! Barney replied. - And not one, but two names, or even three, and the Germans, it happens, have more."

"What are you doing in the capital, Barney? " the general asked.

"Children! I can't do anything better than children, excuse me, madam." He did not understand why the face of Turchin had changed. "I was lying here in the hospital, well, the local girl picked me up; now she's tearing her hair out, but it's too late. We live at the shipyard." He gave the address. "She rarely lets me on the threshold, and when she blunders, it costs her dearly. We have three, the first time she gave birth to twins."

4

Turchin closed himself in a heavy and, as Barney imagined, contemptuous aloofness, as if there were no former sergeant here, no whites, no blacks, no darting servants, no fuming torches, but only them, a noble stray couple in the middle of the desert. But he approaches them with an open heart, he tells them about legitimate children, if only they would look at his twin boys! And Barney began to sort out the anger.

"Good town!" he shouted. "The stable smelled through and through, wherever you spit, the stable. Horses are a feast for the eyes, and in saddles every second is a bastard. They say the midwives take them ready, with silver spurs on their heels! Today, these sons of bitches would not be shown their noses on the street, but you look how many there are, and the pockets of purses sagged. And I have nothing to drink! Here!" He tore the poster aside, revealing a worn waistcoat and a tattered shirt open at the chest. "Laura Keane keeps me warm! Do you know who gave me this paper coat and a dollar a day? Remember William Christian, he was all around the printer when you started printing the Zouave Gazette? Now he's an important bird in Ford's theatre, a bouncer or something, I don't quite understand."

From a side street poured out a crowd of newsboys in uniform caps and jackets, in the new-fashioned New York style, with heavy bundles of evening newspapers. They shouted loudly that President Lincoln and General Grant and their spouses would attend the Ford Theatre play 'Our American Cousin', starring Laura Keane.

A cab stopped a few paces from the Turchins, and the general shouted to the cabman that he was taking it, although there was not much left to the White House; Turchin was in a hurry. Having already seated his wife, he slipped a five-dollar bill to the sergeant.

"Borrowed! Only on loan!" said Barney O'Mullen hotly. "Otherwise I won't take it, otherwise I'll be offended. You know my unbridled temper! "

However, the money was already in his pocket, and he spoke more for onlookers than for Turchin. "Otherwise, it's the devil knows what: alms! You and madam must come to visit me!" He repeated his address.

"Not alms, sergeant, but a duty, a duty of honour," said the general, helping the Irishman out in the eyes of onlookers.

Leaning back into the depths of the cab, they felt separate from the crowd; life rumbled outside like shrapnel and rapid rifle fire, pounded the leather tops of the carriage with brass bands, explosions of holiday firecrackers, and the hiss of firecrackers. Nadine buttoned her husband's uniform, her fingers soothingly and gratefully feeling under the uniform cloth his body, not heavy, not sluggish, but impetuous and strong. She feared that her husband, through drooping eyelids, still saw in front of him the unknown wife of Sergeant Barney and her red-haired children. They, the Turchins, are not given this happiness: it was decided in Poland. Now it is even more late: in a year she is forty, he is four years older. For years they did not leave their saddles, rough on the thighs; hardy, patient, their flesh was stale, churning ever tighter. Now you can't even think about children, and therefore you think, think, think with anguish clouding your consciousness. Nadine clung to her husband and kissed him, breathless, impatient, young, as if not in a momentary shelter, but at the very beginning of a long night.

The horses turned to the left, it was already lilac dusk in the alleys, the hooves gently stepped on the trampled earth. The driver pulled on the reins, doubting whether to go further, and Turchin shouted to take him to the very entrance of the White House. And again, for a moment, but sharply, hurtingly, Turchin thought that his undertaking was hopeless, he was late and croaked amidst general jubilation; that he, an immigrant, cannot rise to the people's sense of triumph and the joy of victory; that he grows old and grumbles, reads sermons when the hour of the jig and psalms of thanks has come.

They were separated from the White House by a vast lawn and thinned trees.

A policeman and someone in a dark frock coat, with a hat in his hand raised, were already hurrying towards him, either calling for a cab, or threatening him not to move on, to a two-story white building with an Ionic portico. The cabman preferred to turn around and, having received a coin from the general, rolled back.

Turchin did not have time to ask the policeman about Lincoln, as a swarthy handsome man stepped forward. The general recognized him as a young Chicago businessman, a major who had left the army too soon but had succeeded in politics.

"Are you with the president? What a pity! He's just gone to the theatre... Laura Keane... a farewell tour," the retired major said frequently. He was shaking with excitement, with secret gloating and a premonition of failure, which must, without fail, befall the reckless, impudent general; that this proud posture, these long, immaculate arms and the amazing perfection of a pale, half-open mouth, her gray eyes, all her beautiful, natural flesh, he could not outbid for all the wealth of the railway company, not to bargain with God, from fate, from a bearded general with speech of a well-born Tory. "The president is at Ford's theatre, but not with Grant, no, Mr. Turchin, their wives don't get along. Ms. Grant believes that from now on all eyes should be turned only to her husband. But you and I know how much Abie did for the victory ..."

Turchin grunted something in parting and turned away from the evening street, lit by gas lamps, the flame of torches, the bright light of taverns and beer halls. The Pittsburgh-Chicago train left after midnight. After circling the park, the Turchins made their way to Tenth Street, to the illuminated facade of Ford's Theatre. In front of them rose a three-story, brick masonry building, the lower floor was plastered, white-washed and highlighted with lights, against which the carnival action played with all colours: varnished carriages slowly playing with the red knitting needles of landau, silhouettes of black, bay and ash-dark horses under the saddle, bulky women's hats, dresses cinched at the waist and wide at the bottom, the uniforms of senior officers stitched with two sparkling rows of buttons, Parisian outfits that even the war did not prevent from crossing the ocean. Apparently, Lincoln had just appeared in the box: the orchestra behind the brick wall played "Hail to the Chief", the crowd on Tenth Street responded with shouts, roar, top hats and hats flew up, flew so high that few returned to their owners. Suddenly, the crowd rushed back to the sidewalks, opening the way for a carriage with four horses. Enthusiastic cries: "Grant! Grant! Long live Grant! - drowned out the sound of hooves; the crowd recognized their favourite in the tired, preoccupied, forcedly straightened man. He was sitting in an unbuttoned uniform, as with Turchin, next to his wife.

Pressed against the wall, Turchin tilted his head as if Grant could see him in the crowd. A feeling of pity for his wife, worn down by someone's indifferent backs, surged so furiously, so repentantly, that he wanted to plough her way through the crowd with his fists.

And she seemed to guess his confusion and said gently: "We're on target, my friend. Let's wait for the president - and let everything be as it will be."

"I won't let you wait outside! We could go to the theatre together."

"Fire me! Turchin exclaimed. "I'd rather take a job as a groom for these tricksters, clean up manure, anything, but not watch an idiotic farce! ..."

Behind the theatre, in the twilight of the Theatre Alley with black elm trunks, Turchin was surrounded by silence. The rear wall of the theatre was vaguely dark with a gate laid with wooden beams, with a service entrance to the stage, with an anvil not far from the door and a hitching post aside. You could barely hear the laughter of the onlookers here as the service door was opened to throw out a cigarette butt or look outside for rain that was brewing in the Washington sky. Turchin sat down with his back to the theatre, on a stone pedestal. He did not notice that a horseman rode up to the theatre, deftly jumped off a high saddle and impatiently walked along the firewall, as if he was waiting to meet someone here and, to his annoyance, was deceived. Turchin jumped to his feet when behind him someone said very close: "That bastard Spangler is always late!" and the whip slapped the general on the back.

"Sorry."

The man reached for his hat, but did not reach the dark, half-mast brim and stared pugnaciously at the general.

Turchin had never before seen this stately fellow with a nervous, arrogant face, with a dark mustache bent downwards and with an unkind squint of unyielding eyes. And he, having seen enough of the uniform of a northerner general, looked at Turchin not only with a challenge, but also with personal, conscious, vindictive hostility.

The service door opened, the faint light blurred against the wall; barely touching the anvil, someone hesitated on the threshold, getting used to the darkness, and moved towards them.

"Ah, Spangler!" said the man with the whip to himself.

A long-armed, wiry, tall-necked man approached. An unclean beige hat, drooping on the left, shaded his face, and under his work jacket he wore a coarse, collarless shirt. Now they were both looking at the general: one tensely and suspiciously, the other with lazy swagger.

"To hell! " Turchin said without an address, referring to the accident that brought strangers together in the backyard of the Theatre, and went along the alley towards the stables. The service door behind him opened and closed, then everything was quiet, Turchin was left alone. And suddenly strange sounds rustled in the darkness behind the trunks of elms: "White mud"

Turchin listened: had he not misheard? Why would it be in the capital, at the walls of the theatre, where the president himself is, the words so well known to the soldiers sounded, with which the rebels expressed their hatred of both the poor whites of the South and the white opponents of slavery.

"White dirt! Are the spirits following him?

"Russian mud! Here's how: the spirits know him!" The voice comes from below, as if someone, hiding, crawled past.

"Russian mud! " was heard behind the elms, this time from the other side, and another voice, drunk, hoarse.

The spirits started a game with him. They appear here and there, cunning, teasing, deceiving. Whispering, choking breath, idiotic, clucking laughter, cautious steps and quick, shallow stomping, the step of a child and the throw of a badger rushing in the grass.

" Russian mud!"

Turchin rushed to the voice; his hat fell off his head, he bent down, groped for it, and a whisper slashed it, burned him on the right, and, blinded by rage, he rushed there, knowing that this could not be done. Turchin ran into the trunks, hurt his shoulder, rushed about, hearing mocking laughter behind and from the side, and blindly stumbled upon someone crouched near the ground. He did not escape from the hands of the general, but dutifully wandered to the theatre, where it was lighter.

A big-headed freak, almost a dwarf, with a dull, grown-up face. He stood with his head turned to one side and looked at the general without fear. And at the same moment when Turchin saw him, someone, quickly rounding the theatre building, shouted into the darkness:

"Mister Turchin! Mr. Turchin! Where are you?"

As soon as the tall, skinny figure appeared before the general, he recognized William Christian, a young Chicago journalist who had volunteered for the 19th Illinois.

" Oh, General! Dear General! " Christian was delighted and immediately saw the freak. "What are you doing here, you bastard?"

We played hide-and-seek with the general. Turchin let go of his hand, and the dwarf realized that there would be no beating. "Mr. John Booth told me to watch the mare. Get her to the hitching post and get out."

He hobbled towards the theatre, shaking his big head on his wrestler's neck; the service door opened, and two men slid backstage: Spangler and someone else, thin and fast.

"Come on, Christian, I'll take a look at you."

They went around the corner of the theatre, under the gas lamps. Despite his extreme thinness, everything in and on Christian was proportionate, well-built and fitted, as if in a minute he himself would go on stage. The frock coat is long-skirted, with small lapels trimmed with braid, under it is a bright waistcoat, a starched shirt with a collar propping up the head, a skilfully knitted bow, sideburns, a pale, bloodless mouth and an evasive, embarrassed by its insight look.

" Who are these people?" Turchin nodded into the darkness.

" Scum. The owner of the horse is John Booth, an artist, a handsome nonentity. Depicts on the stage of life an aristocrat of the spirit, but in essence, rubbish, a tavern hero." He waved his hand, saying, do not talk about them. " I met madam! Do you understand what this event is for me?! You and madam! General Turchin and our confessor mother!"

"Easy, Christian!" I endured flattery harder than enemy fire.

"I have excellent seats for you and madam in the theatre. Baron Stackl ordered a box, but something has changed with him, he will not be in the theatre. The box is on the left, opposite the President's box: let Abie puzzle over who this ferocious bearded man is, who has taken the place of the Russian envoy, Baron Stackl!

"Have you got a family, William?"

"Never, or at any rate not before I become a famous writer. This is incompatible, General - wife and literature!

"Do you think a wife and war are easier?"

"Madame does not count: she is the only one of her kind, at least on two continents!...

"Yes, you will be a scribbler!" Turchin said. " Famously you learned to flatter, and that's half the battle."

" I am ambitious, Mr. Turchin, I want you and Madam to say to someone in ten years: we knew him! He was a good guy even then, but who could have expected?! " All this he said without a smile, only a soft voice expressed both mockery of himself and the disturbing aftertaste of a dream. They walked through the thinned crowd at the theatre, past the clerk who checked the tickets, up the gentle steps and the deserted corridor along the balcony and the boxes. "In the meantime, General, I'm talking about you and Madame, and so far, they don't believe me, but I swear that everything is true, the true truth; He is such a person, and he treated me well."

" All right, Christian; and if without jokes, how do you find our affairs? I won't go to bed alone."

Christian nodded in understanding.

"Our affairs are complicated, General. The nation has won - the nation is sick."

Christian moved forward to fetch Turchina, but the general stopped him:

"I need to see the president. I am not looking for positions or favours from him: what I want to talk about concerns us all. Is he alone?"

"There's also a wife and Major Henry Rathbone and his fiancée.

"Maybe an officer I know is assigned to Lincoln?"

"With him today drunken Parker - I would not trust him to be the coachman of the presidential carriage." Parker sat down in the theatrical buffet, and it was easy to enter the president: "kick open the door, then the second, click his heels and - hello, Abi!"

"Something didn't please you, Christian."

"I pray for him every time I start praying out of forgetfulness. I'm counting on Lincoln, General," he continued dryly, "but I wouldn't look for anything from him. As soon as he smelled of success, he was surrounded by people who would not give you an inch. You will be a stranger to them, an immigrant, a troublemaker, even in this uniform."

Turchin laughed in relief. "He is there with his eccentric woman," said Christian, "what a longing to talk about business in front of her!"

Turchin opened his bag and took out a book published by Walsh.

"I need to convey this to Lincoln: my military and political plays are collected here, there is no role for Laura Keane in them. I want to give the book to the president, that's all. Don't look at me with such regret: I'm not the Marquis of Pose, my dear!"

But the connection was already broken: Christian looked past Turchin, glimpsed absent-mindedly along the wall with dimmed lamps, at the dark figure that flickered in the corridor and went to fetch Turchin's wife, and the general, upset by his sudden coldness, absent-mindedly entered the box. He wanted to find Nadine below the barrier, to see how Christian would beckon her and she would slip between the chairs with her head bowed, and the men would look after her as if she had not interfered with them but had shown mercy by passing by. And women will look - squinting, anxiously, not immediately finding sharp, angry words.

The hall was in bursts of laughter, in a hubbub, in excited gestures, in joyful nonchalance. It was like a different world than the one from which the general had just emerged: where a big-headed freak rushed behind the elms, and the first drops of rain fell, and the mare of the handsome actor beat with his hooves on the hard-packed ground, and the blank brick wall of the Ford Theatre went up into darkness, like a huge primed canvas on which the creator had yet to paint a fresco about a guilty and bloody world. Here the air is stuffed with perfumes, smells of creams and powders, electrified by rubbing silk, envy, enmity, silenced for the time allotted for laughter. A stage in lights and two figurines on it - the famous Laura Keane and comedian Harry Hawke; he famously hooked his thumbs into his suspenders and shouted out lines in an unnaturally loud voice. And above all this, across the hall, the president's box. Spacious, of two connected lodges, with a removed partition, as on old frigates, when cabin bulkheads are removed for the convenience of gun servants; softly illuminated by a bronze chandelier with frosted balls, the president's box seemed to float above the hall, and the folds of three striped flags and wavy curtains seemed to sway, filling with wind and picking up speed.

Lincoln sat in the chair. He is so tall that, leaning back, he towers over everyone in the box. And it is impossible to understand what connects this emaciated man with the profile of a sage and a satyr and a vain, puffy woman in lurid flowers on her head.

Maybe the reason for this is the distance and diffused dull light, but it seemed to Turchin that the president was tired and suffering, that he could not bear the forced, prim silence of the box amid the general fun and the inability to concentrate, to think about his own. Turchin even got up at the sudden thought that Lincoln would be glad to send theatrical tinsel to hell, meet him, shake his hand, settle down in the dark lobby, talk about business, breeze something like that, about which the newspapermen will make noise for a long time, laugh, finally, over the general, who gave him a lot of trouble, and think, think - stroke the book published by Walsh with long fingers, slightly twitching his sunken cheek and touching his plebeian cartilaginous ear with his palm ...

But Nadine had already opened the door of the box, and Mr. Trenchard, the American blacksmith, portrayed by the first comedian of the Ford Theatre, Harry Hawke, desperately, in the manner of a true Yankee, pulled his suspenders with his thumbs and shouted so cockily after Laura Keane as she left the stage that the hall burst into laughter.

A dark figure appeared in the box behind the president, a hand raised with a pistol, Turchin recognized John Booth, screamed, waved his arms, but no one heard him; even the sound of the shot was covered with laughter. And when Booth threw Major Rathbone aside, wounding him with a knife, when, jumping on the barrier, he jumped down, catching his spur on the flag, when, shouting: "This is how tyrants end!" - he rushed across the stage into the wings, dragging his broken leg, - even then, few in the hall realized that a misfortune had happened. But Turchin saw where the lead fell, how Abraham Lincoln leaned forward and then, supported by his wife, unconsciously leaned back on the back of the rocking chair.

The Turchins sat in the box while the excited crowd squeezed out of the Theatre, and someone wept loudly, someone shouted about the pursuit, and then about the fact that the murderer had galloped away on a horse that a little freak kept ready for him. Medics and officers crowded into the president's box, and soon he was carried away to the loud cry of his wife.

...

They walked out onto Tenth Street, packed with people despite the rain. They strayed along the small streets, past saloons and billiard rooms at night, where they still did not know about the murder. We walked, soaking through, under the April, squally rain. At the bridge over the Anacostia River, not far from the shipyard, they stopped: there was not much time before the departure of the Chicago train. Here street lamps are rare, and the streets are almost deserted. Turchin read the name of the street; Barney O'Mullen told them about her. Some fifty steps separated them from the house of the former sergeant.

"I want to see Barney," Turchin said. He will take us to the station.

14

The light was on in the one-story, shabby house, the door flew open at once, as if a small, busty Irish woman was guarding at the door, but she was waiting not for this strange couple, but for someone else with whom it would not hurt to settle scores. She watched silently and expectantly.

"Does Barney O'Mullen live here?" Turchin asked.

"Perhaps here, when he crawls to the house."

" Could we see him?"

"Still could! He's somewhere in Washington: it's Good Friday, some slacker slipped his hubby dollars, a war debt. Now you can't wait for it!"

"I'll ask you a favour, Mrs. O'Mullen." And he, for the last time in this endless evening, took a book out of the bag. " I'll leave a book for Barney, here he will find something interesting for himself. I don't remember anything he read. Apparently, it spares the eyes."

" Then here's what ... " Turchin leafed through the book, turned down the top corner of the thirty-fourth page. "There are only a few lines here."

Barney's wife bolted the door, turned the wick of the lamp brighter, and read:

"Valiant and devoted patriots! The bones of many of your comrades turn white in the hills of the South, the cemeteries of your slain spring up here and there in the territories of the rebellious states; you undermine your health, lose your arms and legs, you die bravely for your country and freedom, and if we fail, no one will dare to say that this happened through your fault, and if we win, then impartial history will someday knock us off our pedestals many of our generals from politics and will write in his annals as an eternal edification to future generations: "The politicians brought this Republic to the brink of destruction, the volunteers saved it!" [2]

The woman stared at the yellow, smoky light. Something alarmed her, but vexation with her husband soon got the better of her. "A-ah-ah!" she thought. "All of them are one Zapyantsovskaya company ... Let them just talk about the war! ..."

BOOK ONE

Chapter One

On the back stairs of the boarding house—long unswept, narrow, crushed by iron railings and walls—there was a stagnant smell of a beggarly kitchen. The dusty glass panes were reluctant to let in the light from the yard, which was as vast as cattle, but even there, there was little light; old men and women, whom Vladimirov had just met in the courtyard, loomed gray shadows through the staircase window, they leaned towards the wind, surrendered to the mercy of bad weather. As soon as he entered the gate and passed the old people he met, Vladimirov sharply and cold-bloodedly felt that all of them, still alive, did not belong to life, but were departing, were the past century, the sinful nineteenth century, with which, as Vladimirov believed, humanity parted without groaning and bitterness.

The year 1900 was coming to an end, and in people who thought often and on various occasions, there was a feeling of a borderline time, a boundary, and even a dividing abyss. It was different, young Vladimirov had nothing to look back unnecessarily; from his father, a gymnasium teacher and an unknown writer; at the poorly sheltered, poverty stricken in the yard of the boarding house, and, presumably, at the old men on the third floor, where the cast-iron flights of stairs rested against the attic door.

The father considered the old man a celebrity, a general, deserved in the States no less than Skobelev in Russia, complained that the mail between them was interrupted, that he so easily missed friendship, moved away from him and from his wife Nadezhda Lvova, princess, "beauty", which does not exist now, and smart, smart so that one conversation with her was worth a book. Vladimirov would no longer look for the old man, justifying himself by saying that he found no trace of the celebrity either on Washington Street or in Kenwood, now captured in the electric orbit of the city - in the American world everything is quickly mixed, divided and multiplied, everything is boiling, as in a drop of raw water under a microscope lens, they don't live here for thirty years in one place, and if they do, it's beggars, not celebrities - he wouldn't even think about this Russian anymore, he would take a sin on his soul: that he travelled, they say, and in Radom and found two crosses on the grave there, but his father's will overtook Vladimirov in Chicago. Here money and letters from St. Petersburg were waiting for him; my father remembered the names of the Chicago publishers Walsh and Fergus: they had once published famous books and had been on friendly terms with him. His father was waiting in Petersburg for news, worried and grieving, and it was not difficult for Vladimirov Jr. to find out that both cases - Fergus and Walsh - were still in progress.

The young Walsh's did not even remember such a name, but in Fergus's bookshop Vladimirov found the owner, gray-haired, fast, busy, overjoyed that a young gentleman from Russia was looking for their friend. Well, for thirty-five years now, they have known everything about each other that friends can know, the general visited them often, he came to her husband's funeral with Nadine, took the violin from Nadine's hands and played, played until the coffin was lowered, and the dry, August earth fell into the grave. Of course, then she did not hear the music, she was later told how divine the general's violin sounded. The widow Fergus wrote down the address, explained how to get to the general, and all with such an air of importance, as if it were a rich estate. Having already said goodbye to Vladimirov, she ran after him in slippers on dry, bone-protruding feet and handed over the paper-wrapped books to the general.

The whim of his father took away a short December day from Vladimirov, drove him to the publisher's widow, to a bookshop, with signs of decay on everything: from a hoarse doorbell to dull shelves with old book spines; first there, and then to a small village, to a boarding house, where several dozen old people beat their loneliness.

His father did not supply him with a letter to a celebrity; he sat down at the table, tried to write, and would certainly write if he himself were mere mortals, but he is a writer and was looking for style; but how will you find him after a quarter of a century of silence, when you don't know what your addressee is now, you don't even know if he is alive ?!

Here is a ribbed cast-iron platform, a dark door and a steep iron ladder to the attic on the side. Who is he, the old man who settled behind the brown, unpainted door? A homegrown fiddler playing at a funeral? A lucky, long-forgotten general? A student of Fourier and Saint-Simon, as could be understood from his father's oral memoirs? Former American businessman, railroad builder and city builder, now bankrupt?

After wandering down the corridor unanswered, Vladimirov asked in English, in a low voice: " Is there anyone here?"

No chair moved behind a single door, no floorboard creaked, no old man's preparatory cough was heard. And, annoyed with himself, Vladimirov moved to the front door, angrily, in Russian, exclaimed: "There's a living soul here, damn it!"

The front door suddenly opened inward. Against the light, Vladimirov could not distinguish the old man's face, he only noticed that it was wide, and his forehead was huge, he saw a white halo above his head and a gray, shaggy beard.

"Excuse me, I'm looking for Mr. Turchin."

The old man stepped aside; not vainly, with slow dignity. Entering the room and looking at its owner, Vladimirov was amazed at how the old man's unhurried, inviting gesture did not agree with the intense gleam of blue, fading eyes under severe eyebrows.

"You are Russian?" Turchin held out his hand, slightly trembling in weight. "Well, of course Russian! I would recognize you as a hare even without these "hell" and "living soul"!" - He repeated Vladimirov's words in Russian, then chewed, rolled English idioms in his tongue and chuckled with satisfaction. "It happened during the war: you swear to exhaustion in English, but the Russian devil still sits on your tongue. He remembered how they started. - And your English is perfect, Petersburg, and if I met you in Chicago, I would pledge my soul to the same hell that I saw you once."

"You really saw me, Ivan Vasilyevich. By the way, the name and patronymic of the old man came to mind. He's almost thirty years old."

"You weren't around then. He looked tenaciously, evaluating." Are you fresh, maybe also a student?"

"Finished the course this spring. And you saw my father, in appearance he was the spitting image of me."

Turchin suddenly appeared to be an ingenuous old man; a mouth covered with gray, neglected hair, a protectively raised, fencing off hand, a homemade, blue velvet jacket and under it a white shirt pulled together with a scarf.

"I am Vladimirov, Nikolai Mikhailovich, and my father travelled around America, feeding himself with his own hands."

And again, the movement of a plump, unhealthy hand, an imperious gesture, a mute order to sit down, be silent, let him remember himself, run back into the depths of years. Vladimirov took off his overcoat, placed the package from Mrs. Fergus on the table littered with papers, and sat down on a chair.

"Vladimirov! Misha!" whispered Turchin, trying on the half-forgotten name to the larynx, to the tongue and to the ear. He asked: "Are you alive? " And, without waiting for an answer, he exclaimed: " Alive! He was not many years old, and he was a burlatsky in health, why not live. People in the war die before their time, and Mikhail, as far as I remember, did not keep up with the Turkish from the States."

"Father managed to go to war," Vladimirov said. "America prompted him to work: when he returned, he got married, published a book about America, and then he was wounded, his left foot was amputated. - The military injury of his father - a leg frostbitten in the Balkans - as far as Vladimirov could remember himself, was called a wound, and here, in front of the general, he did not want to deviate from the family tradition."

"By mobilization or volunteer?" The old man looked inquisitively, as if it was about today. Lay the head for the brothers of the Slavs! "By good will," Vladimirov replied dryly. "I know Turkish from the newspapers, at that time I had my own war, it was not before that.

20

But the brain has apparently already surveyed the Turkish campaign, separate from his life, and dictated his assessments. "A lot of things were wrong back then. Mediocre and bad, covered in blood. The kings always have it in abundance - the people are great. The third Plevna was not needed at all, criminally not needed. And the Balkan crossing does honour to the Russian soldier. I don't know if it was good for the Romanians under the Turks, but the Serbs and Bulgarians without them, it's true, are better. You say Michael released a book?" "Yes. Russian among Americans".

Turchin advanced on him, brought his head, bare to the crown of the head, sown with old man's freckles, and looked angrily.

"Many write," he muttered aloofly. "One of the deceivers of mankind carelessly said that a lie has short legs - nonsense: slander has a quick, evasive step, and printing presses, and locomotives, and soon there will be wings."

He spoke recklessly, crimson appeared both on his cheeks and under his white hair, still whitening it, he spoke without considering Vladimirov, who could be offended by his father. - Why is there not only printing lead and paint and helpful publishers! But, - he pointed to the papers with which the table was littered, - priceless creations of the mind and talent, and for forty years they did not move things: forty years! It's time to despair, to curse the hand that dared to write it all down. Burn! Burn! - He turned to the table with such determination to act that Vladimirov felt madness, obsession, and everything fell into place: people knocked down under a heavy wind, an iron staircase, a kitchen stench, the silence of a corridor where old people hid behind numbered doors. Do you think everything is on the table? No! - He crossed the room in three quick steps, tapped a leather chest with a stick - one of those old, indestructible products that competed with iron in strength and long service life - and, recollecting himself, as if apologizing, leaned over and touched the traveling chest with his palm.

"And it's full! Tales, stories - they could make a glorious page in Russian literature, but it is written in French, yes ... more in French ... I sit here, busy with my own, but occasionally, for relaxation, I translate into Russian. I don't know much anymore, I forgot ... in old age, Don, Cossack words especially come to mind. Bad..."

It's time to say goodbye before the old man is driven away: Vladimirov had no doubt that Turchin's manuscripts were lying on the table and in the traveling leather chest - junk, rags - and now, in the desert of the old man's obsession, he also began to rewrite the pages that had lain for half a century.

Vladimirov was not immediately found: his father told him more than once in instructive conversations that he was happy, happy both with his fate and with the general progress, but his look at these words was not very prosperous.

"Is he happy with life?" the old man asked. "Not a table, no prosperity, no wholesome cooking, but life?"

The old man waited, and the young Petersburger did not dare to answer him with a short 'yes' or 'no.' "You can't tell about your father separately from Russia, and the old man didn't know the new Russia, the serf remembered it, with the heroes of Senate Square still scattered in the remote corners of Siberia, with Gatchina and with the old-fashioned, proud that did not allow Admiral Nepira to Petersburg, Kronstadt. How can I tell him about the Zemstvo, about the new-fangled parties, about the brave, noble people who, in their impatient pride, plotted to change the government with one bomb, an assassination attempt, and after the assassination, with success or without success, laid their heads on the scaffold? How to describe the bitter, humiliating need of the peasant, for which there are no words, and at the same time thousands of miles of new railways?

"And how are you to us: in the paternal manner? Without a penny in your pocket, all with your hump?"

Vladimirov, a teacher from Russia, in 1872 got off at the pier in New York with a few dollars in his pocket and lived in the United States for four years, traveling the country from ocean to ocean, from the Mexican border to Canada, did not disdain any work and a penny payment, honouring and his luck during the years of the financial crisis that gripped the country.

So, they decided at first, but the father could not stand the character. He sent me to repeat his lesson and lost his peace: why am I here and how? Provided me with everything. He raised his hands like a tailor's fitting. "Money has already arrived in Chicago."

"And you would have them back!" The old man's eyes lit up. "Here's a lesson for the old man! Are you the only one with him?"

"One. And there is not enough character," Vladimirov admitted, he felt freer and simpler, and thought that they, the Vladimirovs, in the whole world do not have close relatives, and he grew up, developed without relatives, as if it should be so, and this was imprinted on him with coldness, stoic calm even at the sight of someone else's misfortune - already as a student he came up with the idea of being proud of this calmness, attributed it to medicine, to the conscious preparation of himself for the career of a doctor. - Man is weak, especially against care; so convenient to be secured. No one could help my father, he is an orphan, so he floundered ...

"Lived!" said the old man thoughtfully. "Michael lived!"

He was lifted from his flattened pillow; leaning heavily on the table with his stomach, he sprang to the window. In silence, one could hear the boring tapping of the clock laid on the table, the onslaught of the wind, even, gloomy, through the whole window, the pressure, when the wind does not look for loopholes, but wants to move all the housing - and walls, and doors, and windows, and a roof over the loneliness of old people . The old man quickly got out from behind the table and leaned against the window. Plush trousers tucked into warm boots hung down at the back, from the back it was better to see how short his neck was, however, in a squat and coarse figure, not only the former strength was visible, but also an avid interest in this moment, such an aspiration for outdoor life, in front of which and the dry window, strong in the wind, and the brick wall, and that second, seemingly prison-like fence, had neither strength nor reality. What connects him with the life behind the windows? He is alone, right, and his wife is no longer alive, otherwise why would he be here at the boarding house?

On the walls there is only one ancient daguerreotype, faded from time to time: a half-length portrait of a young woman, a black Amazon pulled her chest, over a high collar with a velvet bow-bow, a gentle, one-touch, oval face, a sensual mouth that warns with mockery and bright, persistent, crazy eyes ; they laugh both at themselves and at the one who stands at the photographic camera.

Vladimirov said that he was not looking for physical labour, modern specialization leaves no time for chimeras; he will still have time to earn his living, an apartment, and books; he proposes to live in Chicago, to practice in a hospital, he will receive an allowance and then make a gesture, return the money to his father ... He coolly checked the old man: "if he hears him, he will certainly respond to such words as "chimera" and "gesture".

The old man was silent, his eyes fixed on a lone figure looming at the intersection of two roads: now the short-sighted Vladimirov also noticed it. In silence, they waited for the traveller to approach, and the old man was the first to recognize the postman in a state cap, in a cape, with a bag hidden under it. Turchin turned liberated from the window, he again belonged to the guest, the room, the papers, but the expression of tender and excited expectation did not immediately leave his face.

"Are you waiting for letters?" Vladimirov asked. "That time has passed."

The daguerreotype on the wall is old, perhaps even from Russia. There is no longer a wife, there are only papers and crazy persistence to impose on humanity, which is choking on books, some new ones, or rather, old, home-grown ones, for some reason translated from French into Russian.

"The last letters to me were three months ago from the War Department and Congress." He nodded at the table, explaining that the letters were somewhere here, among the chaos of papers. "They informed me about the denial of my pension."

"What an injustice!" Vladimirov was politely indignant.

Turchin casually waved it off and stared at his fingers, trembling, pinched in pain from the sudden movement:

"What do you see as injustice? The old soldier was left without money, gray hair was not spared? And what, how is this the highest state wisdom?! Damn them! Damn! Previously, we used newspaper greyhounds on him, but now, why? He himself crawled into the hole, and his fingers cramped, now his hand won't take a sword, and won't hold a light bow ... The old man was all in an impatient movement, like an ox shaking his head, puffy fists angrily hit the armrests. " If you please, I will introduce you to a sort of steward of human destinies." "How?! Distribute the money that our good people get by the sweat of their brow? Throw them to the vagabonds, the detractors of the Republic! Assign money to those who did not have it at the best of times! For what other merits, dear sirs? Why do they need dollars? What did they incite, set up, thirsted for blood and shed it? Why worry about the destroyers, about someone else's, alien blood! I, Nikolai Mikhailovich, am proud of this refusal, yes, I am proud and consoled, although my bread is bitter and already the last one," he said muffled.

"In refusal, I also find recognition: it means that I am alive, and the malice towards me of those who should experience it in a world where a living jackal is better than a dead lion, is alive. They are angry, they remember, and with this memory of theirs - I am alive, I am in the saddle. I'm not all, not all yet," he whispered inflamed, with a hoarse obsession. "I still have enough to complete, finish, and no one can prevent this." He was ready to jump up, move the semi-armchair to the papers, set to work right there, in front of the guest.

The old man got up, blood rushed - from the neck to the face, to the temples, the gray hairs around the reddened head fluttered with bright prominences. Vladimirov was afraid of trouble, a blow, he jumped up from his chair and, grabbing Turchin by the arm, sat him on the bed.

"And so, all my life!" Vladimirov was amazed. "Why did you choose this execution for yourself?"

Turchin did not immediately raise his lowly lowered head, fashioned from the back of the head in the way that nature manages once for millions of heads.

Shuffling, slowing footsteps were heard outside the door, and, throwing up his head, the old man grinned:

"I chose justice, or rather, I have been looking for it all my life. I searched in the republic, among the ocean of despotic monarchies." As if there had not just been an explosion, a bloodshot face, weakness, he spoke almost evenly. "There are many false idols in the world: for example, blood. Blood! Blessing and Curse! Service to personalities or speculative teachings, not excluding Christianity. All these are convenient roads to the mutual extermination of people". A sudden thought raised the old man from the bed, he put his hand on Vladimirov's shoulder, and he heard his laboured breathing closely. "If I heard right, are you staying in Chicago? Oh, how I need a faithful Russian man!"

- He looked into the gray eyes of the guest, slightly warmer behind the crystal-strict glasses of pince-nez, looked with a request, as if he himself was ready to serve, to surrender to someone else's will.

"Darkness is coming, an eclipse." He parted his eyelid with his fingers, exposing the eyeball and slimy redness, and at the same time revealing the bitter meaning of his prediction. "I want to complete, finish, and it takes a little time: a year, a year and a half, but will he be released? Don't say anything!" he warned Vladimirov's impulse. Picking up a stick leaning against the wall, he walked around the room: from the table to the leather travel chest, and back to the table, and to the bed, and to the closet, blindly pushing against the doors. And he spoke - quietly, quickly, in a hurry: "All this must be preserved, handed over to people. It is difficult to say to whom, if I could answer, I would have flown there. To pass it on to those who come forward, that's probably for sure; who will come forward, not in money, not in cars, but in justice. The living closed the eyes of the dead, but it happens that the dead also open the eyes of the living. And this was written by two,

two, and all their lives, and what was in French was mute, did not come to life, and little English came out. Look!"

Turchin finally lifted the light lid of the chest - on top of the papers with which it was stuffed, lay an old notebook, burnt from the corner. "I saved it from the fire, from the fireplace in Radom, but a lot of things burned, letters, stories. He pressed the notebook to his chest, felt sorry for its injury, belatedly feared her insecurity.

"I translated this story in Russian, secretly, into my half-forgotten Russian, and, believe me, I suffered while copying. The story is more than forty years old, but everything is alive, has not grown old: thoughts live longer than us."

The old man bent over by the bed, shifted the violin, groped for something under the mattress. "Take it!" Turchin pulled out a home-made, sewn notebook from under the mattress. The hand itself knew the place, the fingers chose the right one and handed it to the guest, fearing both his sudden act and a possible insulting refusal. He pressed his finger to his lips, squinted at the door, made a sign to be silent, would not footsteps be heard again under the door? "It would be nice for you to have the original, but you yourself saw it, it was burned, I translated the top lines from memory: "I read it so many times! Her name cannot be lost forever; the story is brutal, but not to the same extent. There must be logic, otherwise - why? Why is a man moulded from a handful of earth, from dust, and so exalted by the mind?! How quickly he changed, how easily he fell into a frenzy, into angry and proud despair."

"Excuse me, Ivan Vasilyevich..." Vladimirov kept his hand on the flyleaf, as if he hadn't accepted the notebook yet, but took it so that it wouldn't fall. "Who are you talking about?" The old man stared disheveledly, hostilely at the guest and already moved his hand to take away the notebook. This lasted for a fraction of a second, the broad, defence-less in its openness face immediately softened into a guilty smile.

"Read it, my dear, and everything will be determined: both measure, and honour, and meaning. You don't see that my hand is trembling, my handwriting is good. No! No! he was ahead of Vladimirov. And don't look at the title, only later, in solitude, so that you start and to the very end, without breathing, without brushing away your tears!"

Without taking the notebook out of Vladimirov's hands, he folded it lengthwise into a narrow rectangle, and squeezed it tightly, silently persuading him to hold it like that. He ran his hand over his forehead, as if he had finished a difficult job and was happy with it. The desert of abandonment, the thirst for friendship were behind this joy. If in his youth his face did not know the mask of concealment, did not develop the nervous and muscular apparatus of pretense, then how painful his whole life must have been! With such nakedness, one can live a tyrant on a throne or a condemned man ten fathoms from the chopping block.

The gray space outside the window was shaded by rain, the onslaught of the wind became louder and harsher, the voice of the church bell that sailed from afar brought not peace, but alarm. The old man noticed a dark figure under an umbrella about thirty paces from the gate, rushed about the room, began to fuss, getting lost among familiar objects, shifted the violin case to its original place, suddenly handed Vladimirov's coat and escorted him out, pushed him out the door, speaking excitedly and loudly, apologizing, assuring the guest that he needs it, it is necessary to leave, and quickly, and so that the local people do not notice him; they will certainly see each other again, they can no longer do otherwise, they were bound by this, this ... a notebook hidden from the rain in a coat pocket; and not only that, but Vladimirov Sr., and the past, the past too; "Russian among Americans!" - how well he said it, perfectly said: he would have taken it and stolen, he would certainly have stolen for his memoir, if he had stolen before, at least once in his life, but it's too late to start ...

On the landing stood a small man in a gray dressing gown and a gray starched cap, from under which liquid strands fell on his forehead and temples, and below dark and dreary, filled with tears, like those of the Kuril Ainu, eyes and thin tense nostrils catching the air. Vladimirov ran down the cast-iron stairs, buttoning himself up as he went, and on his way out, pushing with his hand, he realized that the old man had managed to slip the widow Fergus' gift, two small format books, into his coat pocket.

A woman was limping across the deserted courtyard. An umbrella covered her face. Vladimirov saw a fur coat - dark, trimmed with miserable mustelin fur stuck together in the rain, high lace-up boots, a free, confident step, despite the injury. When they approached, the woman raised her umbrella, and Vladimirov was amazed, it turned out so unexpectedly that under the umbrella he found an old woman, with gray hair from under her hat, with a wrinkled mouth, but also with such bright and unfading eyes that it was impossible to connect them in one portrait.

He quickened his step, feeling the slippery bumpiness of the cobblestones with his new boots; he walked, leaning into the wind, like old people recently in the yard of a boarding house; he was walking to the dreary church alarm, for some reason certain that he would return here, and when, due to the distance, he finally decided to turn around, he saw a dark brick building and an even array of windows, the same as the numbered doors in the corridor, at which he exclaimed in his heart: "There is a living soul here, damn it ?!"

Chapter Two

"History of Lieutenant T."36
Op. Nadezhda Lvova

If it were not for the fires of the bivouac fires and the sleepless neighing of regimental horses, it would be hard to believe that a bloody war
is going on among these green mountains and picturesque valleys.
In a Magyar town, at the bottom of a valley indistinguishable from
the windows of the castle, the sexton announced midnight with bells,
declaring to the living that the sorrows and blood that had fallen on his
land could not interfere with the eternal order of things. But, having
struck twelve times, the sexton did not leave the belfry to light a candle
and pray for the salvation of his people: rare, single blows did not
stop - not a bell, but the voice of a sinking brig. "Ba-am!" - came from
the mountain river, - today the infantry and cannons forded it, easily
crossed it, ba-am - in a voice not of a church, but of a worldly warning.

Lieutenant T. could not sleep: as usual, he did not snore as soon as
his ear touched the saddle, although he boasted that he needed only
two things for sleep - an ear and a pillow. Today the lieutenant threw a
Cossack mityuk on the carpet and put a saddle at the head. He walked
for a long time with his horse-artillery battery from near Warsaw,
from the banks of the Vistula, he walked without meeting the enemy,
without engaging in action, except for the chatter at daylight meetings,
receptions in the living rooms of the Galician landowners, or two or
three volleys of his guns at random, through dense dark oak forests.

His name was lieutenant Ivan, the Don gunners also added to
his name, so common in Russia, a patronymic - Vasilyevich, but we
will call him lieutenant T., according to the first letter of his surname
- Turchaninov. He had a reason to write letters to Warsaw in the still-

ness of the night, when the thought is sublime and detached from vain affairs, and most importantly, when Prince L., the commander of the regiment, was assigned to the artillery battery of the lieutenant, in the past, 1848, transferred from the Don to the Vistula, along with other batteries and cavalry regiments of the Don Army, totalled forty-six thousand Cossacks.

Prince L. sleeps, terribly pleased with how easily his regiment entered the country of the Magyars - for this ease did not put up barriers to obtaining orders and proper production; after all, it happened that a close rival of the prince, a certain Kolzakov, made the career of a general exclusively and only by skilfull marching. The prince sleeps, not knowing that the lieutenant is writing letters to his twenty-three-year-old daughter Nadezhda. And how could he know about their friendship, which flared up like dry gunpowder; the young people met recently, in a secret circle of Warsaw freethinkers.

Ba-a-am! This time the blow was stronger. Isn't this a signal, isn't it a sign that allows someone on passes and mountain paths to find a goal and a way, just as a ship finds its course by the light of a beacon in the dangerous proximity of rocky shores? The barefoot lieutenant, in an undershirt that shone white in the twilight of the room, went up to the open window. To the left, the wing of the castle crowned with a tower darkened gloomily, in front lay a park, regular at first, but soon turned into a dense forest, to the right - a mountainside in fires, stables, barns, stone buildings of a farm. The horses crossed with slowness of the night, the lieutenant guessed they were on the lawn by their soft step and by the way they covered, as if swimming, the fires.

Ba-a-am!

Only yesterday, the lieutenant would have missed the bell chimes: at that hour he would have slept peacefully and dreamed of Novocherkassk, the green reeds of Aksai, the Nakhichevan Armenian shop next to his parents' house, or the Warsaw Barbican, where the lieutenant said goodbye to Nadezhda L. Yesterday it still seemed that no force can slow down their victorious progress. The words that bothered the lieutenant were already erased from the memory, with which the prince admonished the regiment at the Austrian border, from where the land of Galicia opened, black and yellow railings on bridges and the same, snake-like, alien to the eye, colour of barriers and roadside poles. Wise providence led their regiment through a foreign land so circumspectly that only the mountainous echo of artillery thunder reached their ears. It happened that a mounted officer or a courier passed through the location of Prince L. in search of an army headquarters, and by his appearance, by skimping on some details of officer equipment, by gambling or longing eyes, one could feel the near front. Until this morning, the lieutenant had not met either the arrogant Magyar nobles who had disturbed this virgin corner of Europe, or, by the way, the allied Austrians; the last time he saw white Austrian uniforms was in Galicia, in the garrisons. The Austrians joyfully shouted "Vivat!" to the Russians, who went with their blood to rescue their emperor, they, out of love and hospitality, inserted green branches into their gun muzzles and shakos.

Until this morning, the lieutenant had met many Magyar peasants, shopkeepers, artisans on the threshold of their houses and workshops, saw silent, not obsequious inhabitants of villages and Carpathian towns. If anyone rushed to the officers with courtesy and service, it was a rich merchant, or an innkeeper, or a money changer in a yarmulke, forced by unfortunate experience to seek and change any gentleman. Going over in his memory hundreds of peasants faces - for they were the same peasants, only in the Magyar manner - the lieutenant was perplexed and upset that these faces were gloomy and gloomy that the needy Magyars looked at the manner of the Caucasian highlanders. Although the lieutenant was not a young man, having lived in the world for 27 years, and read more books than any other nobleman, kindly born and enlightened environment, although he singled out justice as the main goal of his life, understanding it passionately, but not clearly, partly from French books, but more of all his observations of Russian life - in one thing he remained a youth: he had never yet been thrown like a chip by a hostile stream, he still believed that he himself, and his battery, and all his guns exist, as it were, on their own and they will do only what he himself wishes, and without fail serving good and humanity.

And in the morning their regiment was baptized by the war. It was baptized, you can't pick a better word here: it's true, the two miserable little guns firing small cannon balls immediately fell silent, but the muskets of the rebels pierced the regiment's position with lead in all directions, as if the Magyars stood in a circle behind each rock, under the cover of oaks and beeches. With muskets, with pikes and sabres, they went on a close suicidal attack, determined to fall or win. The guns fired cannonballs, then shrapnel, but it happened that the artillerymen repulsed the enemy with a bayonet. Everything was mixed up, even the most serviceable map by noon would not have helped to establish where someone was standing, to top it all, among the trees and bushes, Austrian recruits in clumsy linen frock coats and caps were rushing about. In vain tried to stop them, mounted officers and a thin, bony, dark-faced Austrian general, more like a Tartar than a European.

More than once during the long day the lieutenant saw the rebels up close, their hastily altered Austrian uniforms, hunting caftans and jackets, homespun peasant clothes, their flintlock muskets, sabers and pitchforks. It seemed that such an army should not have held out against a trained enemy, but meanwhile the lieutenant was convinced how stubborn and strong they were.

However, isn't it time to tell about the lieutenant's youth in order to explain some of his unusual actions during this night?

He was born in the Don Cossack Region, three years before Emperor Nicholas I ascended the throne, and, by his most merciful decree, the heroes of Senate Square were executed in order to mercifully save them from the inevitable execution of the people or long martyrdom in casemates and mines. The history of this reign has more than two decades, but a genius has not yet been born in Russia to fully describe the wisdom of his reign and the extent of the suffering of the submissive people. One thing must be said: just as a skillful and kind doctor bleeds an ill person for any reason, so he never found it difficult to bleed, shed not rivers, seas of blood, and in this vigilant healing he assigned the role of assistants or assistants to the brave Don Cossacks. Their gray overcoats with blue collars saw the recalcitrant Caucasus, and Shumla, and Brailov, and Zhurzha, and Silistria, and many other fortresses.

The ancestors of the lieutenant in the male line, as worthy of the Don people, served the sovereign with weapons in their hands, but only those Turchaninovs who left for long-distance service, closer to the court and Gatchina, were promoted to the ranks - the Don Turchaninovs, faithful to the steppes, did not go more than seconds - and prime majors. So, the father of lieutenant T. deserved the rank of major, having determined the future military career of his blue-eyed son with his life.

Gone are the days of the Cossack freemen, when the Cossacks took care of Poland and Ukraine, and the Don people - Moscow, when wolves roamed the steppe unrestrictedly and flocks of bustards darkened the sky, when the Cossacks took wives from Polonyanka and lived while they lived, often without consecration and rites - the old Cossack boats made of linden logs rotted in the reeds of the Don, Donets and Aksai, rich grain villages tamed the feather grass steppe, and damask knives and gold-rimmed Turkish guns hung on carpets, like a memory of long-standing campaigns. Now the rich Don brides were dressed up by the Kuznetsky Most, and jewellery - diamonds and pearls - were presented by the shop of Zilberman, a subject of the Austrian emperor, a generous crown, who also kept his efficient agent in Novocherkassk.

Cossack children often played war - marching out of the city and building reed fortresses - and although our hero more often trailed behind the winners, in clouds of dust, with guilty tears in his bright eyes, nothing could turn him away from service. In 1836, at the age of fourteen, having already tasted two foreign languages and playing the violin, he showed up in St. Petersburg, in the Imperial Artillery Cadet Corps.

He spent five years in corps classes and came out of them a new person - skillful not only in mathematics, drawings and ballistics, but also in disputes that were not related to military affairs. He improved in languages not according to old novels, both philosophers and notorious Jacobins were in his hands, and the writings of daring authors became the gospel of our hero, who cared not about increasing wealth, but about healing society, about its healing from tyranny and slavery. God was forgotten forever, and if he was not subjected to blasphemy, it was only because the young man, brought up in an Orthodox and God-obedient family, never crossed this last, blasphemous line.

Returning to the Don after five enlightened years in St. Petersburg was like a violent fall. Approaching Novocherkassk along the old Voronezh road, passing the depressing Aksay steppe in autumn, the floating bridge and the triumphal gates, the lieutenant felt anguish in his heart and a premonition of an empty life. The guest yard seemed miserable, knocked down from boards, the gymnasium, enclosed with a dilapidated fence, was not pleasing to the eye, in the neighbourhood of the barracks there were houses covered with reeds. Amidst this poverty, stone two-story buildings of eminent and wealthy people, an arsenal for light tools and two triumphal gates, built at the request of the ataman Count Platov in the reign of Emperor Alexander Pavlovich, so merciful to the Don people, towered. But even more meagre, after St. Petersburg, was the emotional life of the Don capital. The spiritual embarrassment in which the lieutenant found himself, drove him to despair, and he made a vow to himself not to drink wine - even the best, not to struggle against all officer traditions, not to sacrifice nights to cards (a hryvnia to Boston, a small one to whist). And although he was approaching thirty and his blood was pursuing his - he kept his word of vow, served without complaints, and gave long hours of freedom from service not to the fair sex, - according to the rumour, unusual in those parts, but to painting portraits and landscapes with colours and play on a violin. So, his life could have ended, with a reputation as a misogynist.

But here Europe began to get excited, Paris and the Germans were indignant, followed by Italy and Vienna, fertile drafts stretched into Russia, so dangerous to the health of the emperor: there were not enough windows and doors to close tightly, bayonets and guns were also needed, and shakos

(NB. a stiff military hat with a high crown and plume), and camping cloaks, and kitchens, and cups of aniseed vodka. So, with European drafts, the battery of Lieutenant T. was pulled out near Warsaw, along with a violin and a leather traveling chest that had seen Berezina, Borodino, Poland and German cities, with a chest bequeathed to a young cadet by his cousin uncle, an associate of Kutuzov, Lieutenant General Pavel Petrovich Turchaninov , who died in the autumn of 1839.

Having found new friends in the Warsaw circle of freethinkers, warmed up by the close heat of European passions, the lieutenant perked up, and only the shortness of the terms did not allow him to fully understand the direction of the times. The circle, which agreed in condemning tyranny, monarchy and serfdom, was torn apart by disputes and cruel disagreements. Cursing the tyrant, the lieutenant could not hide his love for Russia and more than once caught the ardent, dissenting glances of young Poles. And on the contrary, denying God, going further than fashionable German philosophers, he met with the same young Poles such devotion to Catholicism that he could neither understand nor accept, not finding the advantages of this faith over the Orthodox and any other. The Poles saw in Catholicism the possibility of uniting the nation, even at the cost of the abyss in the Slavs: Catholicism, they believed, was inevitable for all Slavs, for in Orthodoxy they suspected a special kind of paganism not outlived by history.

So, in the confusion of the emotional substance, aggravated by the first friendship in his life with a woman, the daughter of Prince L., the lieutenant fell under the orders of the Hungarian war. Like the entire Russian army, he was under the high hand of the Warsaw Prince, Field Marshal Count Paskevich, walked through the Uniates, went to their churches, (Uniates: [noun] Christians of a church adhering to an Eastern rite and discipline but submitting to papal authority.) where they sang in Slavic and without an organ, but not from the kliros music, but from the choirs; I walked past the huts, where the icons hung not inside, not in the red corner, but outside, nailed over the doors; marveled at the wealth alone and extreme, even on the Don unprecedented, the poverty of the life of the mountaineers-Rusyns(Ethnic Group: Rusyns, also known as Carpatho-Rusyns, or Rusnaks, are an East Slavic ethnic group from the Eastern Carpathians in Central Europe); he also saw the captured Hungarian Landsturmists (trans. 'Young Guard'), marvelling at the white "Hungarians" unsuitable for war with light green hussar cords, small half-headed caps and half-boots laced on the side. He dragged guns with the Cossacks to the passes, inhaled the smoke of the artel boiler; on the passes I met watchmen from the bottom, with a lance in one hand and the reins of a saddled horse in the other; laughing, he snatched out potatoes baked in the ashes of the fire, and thought about when the enemy would finally open up, when and how he, Lieutenant T., would find in his heart enough dislike for him to honestly fight and kill, and whether there would finally arise the affection of the Russians for the Austrians, who are so briskly fleeing from the Polish general Bem and the Hungarian Gergei; affection for the officers, whose names were so reminiscent of many Russian generals - for the Millers, Kaufmans and Pistolkors?

Now he stood at gunpoint, in front of the old Magyar, but the duty of an officer brought the lieutenant here, and not malice towards the people whom he did not know. This was his weakness and sad advantage: the Magyar, if he was somehow smart, could not help but notice the absence of hostility in the open look of the Russian.

"Give me your word," the old man said, "that you will not resort to weapons and will not call your soldiers."

"Willingly," replied the lieutenant. "There is no military enemy here, and my lead is not dangerous for civilians." Then the man lying on the bench spoke in delirium, spoke not in Magyar, but in Polish. "Are you the owner of this estate?" asked the lieutenant. Furious, unrestrained Magyar curses escaped the old man's lips. "The owner is a traitor, he is with the Austrians!" Finally, the old man replied. "Both he and his sons are unworthy of the name of the Hungarians. We found a short refuge here: my family, daughter and son. The son fell ill, he does not regain consciousness ..."

"Why is he lying, did I clearly hear the Polish language?" thought the lieutenant, but did not incriminate the old man. The lieutenant introduced himself, the Magyar identified himself: Istvan Kodaly, and they went under the stairs.

"Maria Kodai," the old man said, touching his daughter's dark hair, which fell on the green velvet of the cloak, but did not name either the decrepit, bent monk, or his son, whose deathly pallor most likely spoke of a dangerous loss of blood. The head is tied with a bloody scarf, the body is covered up to the chest with a peasant cloak made of sheepskins. István's daughter did not look at the lieutenant, even hatred was now powerless before her suffering, before the feeling that united her with the dying Pole. No matter how much she loved her brother, such a look of a sister is impossible, it showed love, great love, which one day makes a person born of another woman the closest to you.

"Who is this Pole?" Lieutenant T asked. "I will tell you everything as a nobleman and a man of honour," said the old man. " Let's leave them."

He took the lieutenant to the far corner of the hall. The lieutenant managed to make out on the bench next to the lying Pole both a downy, low-crowned hat girded with black crepe, and a bloody dolman, and a wide sabre with a hilt in a sheath, thrown on the edge of the table, and realized that he had met with not entirely peaceful Magyars. The old man did not hide this: age did not allow him to join the regular Hungarian army, and he indulged in the activities of the Landsturm, made his way from village to village, from manor to manor, skillfully bypassing Austrian spies, mountain outposts and patrols of lancers. Hundreds of militias owe him the fact that they chose the path of honour and service to the great republican idea. Gergely and Bem and General Dembinsky drove the Habsburg soldiers, in Transylvania the Austrian troops openly fled, the republic was close to victory, but the irreparable happened: Paskevich's corps entered the business. At the first approach, the Hungarians instinctively, deceiving themselves, mistook the Russian Cossacks for disguised Austrian lancers, believing that the tsar gave the Austrian emperor only costumes, a wardrobe of a European tragedy, but not performers!

Alas, the awakening was terrible: against several thousand Poles, whom the heart had brought under the banner of the republic, the Austrians received 200,000 borrowed bayonets, artillery and Cossack hundreds. A young Pole, an officer of Dembinsky, was sent to help the Landsturm in training and in battle, events cut him off from Dembinsky's headquarters, and he remained here; he fought honestly, won the hearts of the militias and, more than others, the heart of Maria Kodaly. At her insistence, today they got married in the castle, and the bells in the night are God's consent to their holy union ...

From under gray, blackened brows, the old man watched the lieutenant. Taken by surprise, forgetting even to fasten his uniform, the lieutenant thought about what to do? The old man hates the Austrians - well, let him, these are their household accounts; after all, he himself has no more sympathy for the Habsburgs than for the Russian autocrat.

Once, at a halt, the lieutenant asked the leader if he knew why he had come to a foreign land? "How not to know!" cried the resourceful Cossack. "Drag the Germans out of the mud by the ears!" God be with her, with Austria, the favour that St. Petersburg did to her today will still respond with treason and treachery, for there are favours that humiliate, calling for revenge, and the current generosity of the Russian court is of this kind. But these people - a stern old man and an oblivious Pole - did not shoot at Austrians alone yesterday; and why are the Poles here, what good is it to them that the Hungarian magnates, dressed up in chikchirs embroidered with gold, in beaver hats, in hussar boots, with expensive stones planted wherever possible and impossible, boss around the poor people, competing in splendour with the French nobility and German princes?!"

After listening to this objection of the lieutenant, the old man sadly dropped his head on his chest. What a delusion! What a hopeless confusion of European affairs, if an educated and noble Russian officer understands events so wrongly. And, getting more and more excited, swearing to God and threatening to fall on his knees before the lieutenant (from which he warned him twice, not letting him fall), the old man told the lieutenant the true story of the destruction of the Habsburgs in Hungary and the creation of the republic. So, the lieutenant learned the truth about the Sejm of Pressburg, about the abolition of humiliating political privileges by the Hungarians, about the abolition of corvée I, and tithes, about the unrestricted freedom of movement of peasants, about universal suffrage and many other things that the lieutenant would consider happiness, the fulfillment of his most daring dreams to see in Russia.

"What do you expect from me?" the lieutenant asked dejectedly.

"Help us get out of here. A Pole must not die without help. We were going to get out alone when everyone was asleep ..."

"Ba-a-am!" came the distant chime of a bell. The sexton did not stay in the belfry in vain - perhaps these blows mean more than the wedding bell?

"But will you leave your weapon to me?"

The old man chuckled, his closed, stern face expressing bitter regret at how little this man understood him.

"As long as I breathe, until God calls me," he exclaimed with an oath, "until the executioners of Gainau cut off my hands, I will not lay down my arms!" The lieutenant was silent, and the old man was also silent, no longer daring to ask for help, but then his daughter's voice rang out, hoarse from shed tears.

"He woke up," Isstvan said, "we must not delay.

The lieutenant trudged after the old man, despairing, not daring to raise his eyes to a man who was driven to a feat not by service and golden semi-imperials, but by the conscience of a republican.

The lieutenant would have more easily accepted the sound of a close shot, a bullet whizzing past his very ear, than the quiet, contemptuous and so familiar voice of a Pole:

"Ah! Turchaninov! Lieutenant Jean! So, you are convinced how difficult it is for a person not to be a slave if he is ... a slave!

Ludwig! Soft, rasping speech, bulging gray eyes, arrogant in whitish eyelashes, a large hooked nose - how did he not recognize him, barely seeing him ?! They converged in the Warsaw circle, converged closely, and suddenly tore, flared up with disagreement, and most of all in how to turn a slave into a man. Lieutenant T. insisted that first the structure of society and the methods of obtaining daily bread, even its price, and then people, would change; Ludwig, getting excited, foaming at the corners of his mouth, argued that a person is obliged to expel the slave from himself, and if he does not do this, no new structure of society will become possible.

Met so humiliatingly, the lieutenant loyally rushed to his friend - this was his whole character: impulsive, clean, stubborn, but not persistent in error.

"Ludwig!" He put his hand on the Pole's trembling shoulder. "Oh, Ludwig! How could you!" Ludwig seemed to grieve that they were no longer together, as they were together in Warsaw.

"Don't you know the inevitability of the military mechanism!"

"Break the sword... break the treacherous sword and... both halves away... Let them execute!" He seemed to be returning to an incoherent mutter. "Better death... than betray your mind..."

"If only the Poles had stood aside! Oh, Ludwig, if only they hadn't entered.

" This made him furious.

"The king would have found dozens of other reasons. It wasn't the Poles that frightened him, but the revolution... it chose equality... it allowed the Slovenes, the Croats... to use their mother tongue... it destroyed the Habsburgs... it printed its banknotes... He wants to tame you, Turchaninov..."

Ludwig's head threw back again, the Pole had put on a cloak of homespun cloth to be carried along the mountainside, in the direction of the church, where the old man hoped to find friends. The incessant strikes of the bell gave him a sign that the road to salvation was not completely blocked. The priest, clasping his hands in prayer, said:

"In nomine Patris, et Filii, et Spiritus Sancti!" and they lifted Ludwig.

I will not describe in detail their dangerous descent through the location of the regiment of Prince L. and the mental anguish of the lieutenant, who was led by friendship and indulgence, and the duty of an officer tormented from the inside. In one, he found a shaky excuse for himself that he was leading a priest, a young woman and a deep old man, and the only officer was lying unconscious. And shouldn't a mortally wounded enemy be an object of mercy? Had it not happened before that they gave bread and red Erlau wine to the wounded Magyars, who lay in huge, sluggish vines? Passing by smouldering fires and dark tents made of cloaks and cloaks on tall goats made of lances, past saddled, in fetters, horses and drums composed by a group with a banner laid on them, responding to posts, sentries, night watchmen, the lieutenant deeply suffered from the need for deception.

Suddenly, the hits of church copper, rare up to that moment, turned into an alarm, rifle shots rained down, and the wooded slopes with mountain paths, along which they moved closely during the day, on one horse, again became a military Theatre, with all the illegibility of a night battle. Duty called the lieutenant to the battery, to the Cossacks, and he said goodbye to the Magyars, to Ludwig, who had regained consciousness; no longer needed a lieutenant. But he didn't have to leave either; Magyars ran up from the darkness of the forest and, seeing the lieutenant's epaulettes, raised their guns at him.

Istvan Kodai defended the lieutenant and ordered him to be released to the Russians. Kneeling, the lieutenant embraced Ludwig and was shocked to feel the Pole's tears on his face. "Our mutual blood," Ludwig whispered, "is not salvation... the triumph of tyrants... Farewell!"

The lieutenant rushed up the slope, in the direction of the castle, from where the voices of his cannons were already heard. He ran straight ahead, into the gunshots, into the thunder of guns, as if he was looking not for salvation, but for death. It often happens that when a brave man is looking for death, the bullet flies by, the bomb shell does not burst close enough and the pike strikes easily, tearing the cloth.

The lieutenant had almost reached his guns when he met the regimental commander, Prince L.

"Where were you, lieutenant?" The lieutenant was silent: the night adventure cannot be told in a few words, and for many there is no time in the midst of battle. His appearance did not inspire disposition; a guilty look, an uncovered head - the cap is lost - a face scratched by branches, a tattered uniform.

"I will dismiss you before the end of the battle," the regimental commander said and held out his hand for a pistol. "Go under house arrest."

A battle flared up all around, and an unarmed lieutenant walked past his cannons and gun servants, who noticed how Prince L. took a pistol from their officer. He reached the castle, entered the room, where he left a sweatshirt, a saddle, and a chest made of leather, as if wrought from rusty iron; the chest heard the thunders of the Kutuzov campaigns, but apparently fell into the wrong hands ...

On this night, the lieutenant was formed over more than seven years of vegetation at the Novocherkassk arsenal. He thought not about a severe punishment, when he would not conceal anything from the night incident before the prince, - and not about his future fate, the features of which were eclipsed deftly, the lieutenant reflected on the highest appointment of a person. Why are people so eager to subjugate others? How many legends did he hear about the Cossack freemen, about the elective principles of the Don, about the natural disobedience of the true Cossack and patriarchal freedoms: but where are they on the Don? As soon as Alexander I welcomed the Cossack elders to the nobility in payment for the jealousy and service of the Don Cossacks, it was worth the royal grace to remake brigadiers into colonels, military foremen into majors, captains into captains, centurions into lieutenants, and cornets into cornets, give Don noble institutions and the leaders of the nobility, - and on the land of the Don reigned not a new truth and understanding of the needs of one's neighbour, but shameful slavery.

Five years had not passed, and already in the previously free steppes, serf souls were mourning grief, with a bill of under a hundred thousand, and all because what kind of master are you if you don't completely sit on someone else's back? There was a time, they used to say on the Don - they put water on the father, but don't approach your son with a collar! But it turned out differently: they put a collar on many tens of thousands, iron bits in their mouths and a ring in their noses, like working cattle; the whip freely walks along the slashed backs so that others can ride in expensive carriages, send Cossacks to foreign pensions and eat on silver. But in their veins - and in those whose breath is clogged with iron bits, and in those whose tables are breaking from silver - one blood flows; Russia is their beginning and mother, and they are forever enemies. And is it blowing away from the fall of tyranny, or is it an eternal curse in humanity, a pestilence, the healing of which will be sought for centuries and will they find it? In Warsaw, he fancied himself the brother of Ludwig and other Poles, and it was not the Slavs that made them related, although there were many speeches about the Slavs. What is it, the common Slavism of Ludwig and him? In the patriarchal Russian church? But he rejected the church along with God, and even before God, and Ludwig is a faithful Catholic, a destructive enemy of the Old Slavonic worship.

And the Hungarian bells rang out, the republic called out to Europe, its tricolour banners and red banners, with a coat of arms embroidered in the middle, and Ludwig was there, against him. And the lieutenant himself, executing his unfulfilled duty as an officer, did not repent of anything. He did not go so far in his thoughts as to see himself as a Hungarian insurgent, although he was already standing between the armies, not knowing to whom to give his warm heart.

Prince L. entered the room with a quick step, but stopped suddenly, because volleys struck in the valley, near the mountain stream. He remembered the Caucasus and the Circassians and distinguished an honest military exchange of fire from punitive volleys, from shooting. An hour ago, having handed over the captured Landsturmists to the Austrians, the prince received an oath assurance that no one would

44

be punished without a fair trial. The echo of a false oath has not yet resounded in the mountains, but the execution has come to pass.

Here it is appropriate to say that little about Prince L., without which not everything will be clear in the dramatic development of his duel with the lieutenant.

He has a right to this both in himself, as an unfamiliar person, but also as the father of Nadezhda L., who won the tender friendship of the lieutenant. The regimental commander would have been retired long ago if it were not for his most influential relative, Alexei Fedorovich Lvov, the son of the late director of the court choir and the musician himself, whose fame went beyond Russia. Within the limits of the Russian kingdom — and beyond its borders, where the union did not supplant Orthodoxy — not only "Like Cherubim" and "Thy Secret Supper," but also many other psalms and prayers, almost the entire annual calendar of Orthodox services, were sung from his notes.

And, proving how inseparable are the high peaks of faith and the greatness of God's anointed on earth, our musician, author of symphonies and operas, the high priest of divine church prosody, a successful concert performer in Europe, was also a zealous courtier. After graduating from the Institute of Railway Engineers, he took his first successful steps in the military settlements of Arakcheev and moved up steadily, more than once being in the retinue of Emperor Nicholas I, already with the stars and the rank of adjutant wing. Well, what does it matter to a dignitary so exalted to a quick-tempered relative, an early widower who raised his only daughter in the garrisons, with nannies & batmen! But the adjutant wing was kind to the orphan and to her father, who in the memorable year 1825 lacked guilt in Siberia, however, even the one that was on it, even with high intercession, drove the prince first near Orsk, and then, in a special way. mercy, to the Circassian saklya.

The small, restless colonel did not lighten the soul of the adjutant with gratitude, and, as it happens, it was his incorrigibility, the heavy cross that he placed on the back of the benefactor, the crown of thorns, that inspired the musician to the feat of mercy. So, they competed - the colonel in ungrateful indifference, the dignitary - in unruly nobility. And between them - a child, a girl, then - a girl - over the years, she leaned more and more towards her father; his views were without system and without direction, but his sharp, even stunning, actions tended towards goodness and justice, which was more than enough for daughter love.

The morning volleys would not have attracted the detached attention of the lieutenant, if it were not for the change in Prince L., he stopped in front of Lieutenant T. with an open expression of resentment and insulted honour. A dry, reddish face, chasing graying temples and short sideburns, with clean brown pupil enamel and quick angered nostrils, with an effort returned to the lieutenant and the inevitable interrogation.

"Well, I'm listening to you, lieutenant!" If the young officer had not been so shocked, he would have noticed that the regimental officer had brought his pistol with him, and this, of course, was a good sign and a possible step towards reconciliation. The lieutenant trusted nature and did not hide a single detail from the colonel, even the fact that he was honoured with the name of a slave.

"Well, you are brought up on the Don! exclaimed the Colonel, after listening to everything. Don is great, there is goods for every merchant. So, you give honour to your native land!"

"The accidents of birth have no value in my eyes."

"What has it? Honour! Activities for the benefit of people and mankind."

"Ba-ba-ba! The Colonel pulled up a chair, sat down, and prepared to speak. What words: humanity! Honour! How much honour to secretly, villainously lead through their scouts!"

"I did not lead but carried - this is my justification."

"Do not rush with indulgence! - Having spoken the sacred words about honour and humanity, his sacrifice, without hesitation, will give an answer even to those questions that lead to the scaffold. The Pole was wounded, and you carried him; and if he were intact, would you deliver him into my hands?"

"A healthy person wouldn't need me."

"Will you need it? Like in a cover, in a password?"

"Circumstances would show that.

"How?! The regiment did not sit on a chair. There's a war going on, what the hell are the circumstances?! Did they give you the word of the nobles that they would lay down their arms? That would excuse the lieutenant a lot; the prince almost wished that the artilleryman would confirm, nodded in response."

"I asked the old man about this, but he replied that he would not stop until Gainau cut off his hands!"

"Here is the firmness of the spirit! Here is a lesson for you!"

"When in such a duel one person shows more firmness of spirit," said the lieutenant, and his cheeks turned white, "it means that the cause for which he stands is higher!" "So, you are giving primacy to the rebellious Hungarian!"

"He's a model Republican!"

"Nobody measured this republic and did not take it by the tooth; your cannons are her first test, and, you see, she collapsed. From your free-thinking hand and fell! - He tested the lieutenant, took out his soul.

"This is my grief," the lieutenant said proudly. "I recognized him last night.

"So, the championship - the old man?"

"He is protecting his." "He, then, has trump cards, and we have meagre ones; do not hold above seven in our hands?!"

"I did not graduate from the Lomber Universities."

From the first days of a quick and stupid campaign, the prince liked the young officer, devoid of servility, a big-headed Don Buell, with intelligent eyes that did not take anything for granted; but this morning the lieutenant took on an extreme tone, unsuitable for a delinquent.

"Cards, my dear sir, are also an academy of honour: not everyone can do it. The Hungarians, it turns out, - the matter, the Poles - they will not even sip a cup of coffee without a conciliar cause; I don't even speak about the British and French, their enterprises are certainly universal, the hungry Irishman has the task of planting potatoes, and we are living without meaning?! So, we pull bast and bast shoes with all the people, and we prick a torch! Without a higher, so to speak, appointment?

"The purpose of my homeland is enormous," said the lieutenant with perfect and bitter simplicity, "but not a single mind in Russia now knows the dates and the way, and if it does, then it is either unfortunate or it is not given to it to say."

"Yes, you have a whole philosophy ready: everyone is simple-haired, but you are wise and see in advance. Not a regiment was thrown into the cause, but as many as two hundred thousand bayonets ...

"That's why there are a lot of them," the lieutenant decided to interrupt, "because the campaign must be ended quickly, as other dirty deeds are sure to be done before dawn. This war cannot be explained to the people, and, thank God, we are already at its end."

"Do you know, lieutenant, that the Hungarians are also ready to help them with every Polish uprising—both with a purse and a sabre?"

"I am intimately acquainted with the Russians, who are not alien to Polish freedom!

The lieutenant challenged, his openness was based on the conviction that in the head of the regimental judgment was over, and the case would not be held by a formal court."

"I see one of the defenders of Poland in front of me," the prince said slowly, anticipating something bad, pacing around the room. "Where are the others?"

"I'll give you a name. But first, remember that, having received a Russian officer as a prisoner, the Hungarians released him in the middle of the battle."

"I'm waiting, lieutenant."

"This name is familiar enough to you." And the lieutenant recklessly named the name of Princess L., without hiding that the wounded Pole is their mutual friend.

If he even named Field Marshal Paskevich himself, it would not have produced the explosion that the name of the princess did. And not because for the regimental her views were a secret; she had already worn him out with her criticism." But the prince flattered himself that this was their family, from all the closed lists.

"I'm with you about the officer's honour, and what are you talking about! He covered himself with a grimace of contempt. Where are you looking for allies? Where the hair is long, but the mind is short! Do not dare to object, I read Schiller and did not master more than one monologue at a time. What came up, young ladies to count!" "Defending himself, he offended the honour of his beloved daughter, wondering why this causes so much suffering in the lieutenant. "Take your pistol." He put the weapon on the table, showing that he did not want to come into contact with the lieutenant, even through iron. "I would have brought you to justice, but the Austrians relieved me of this duty. It is the Austrians who are so unkind to your heart...

The lieutenant took the pistol.

"I gave them all the prisoners," the regimental continued. "That is the requirement of our treaties with the Austrians. The prisoners were promised a fair trial, both the old man and the woman, but you heard, lieutenant, volleys; their blood washed away your guilt.

"So, no! exclaimed the lieutenant, almost losing his mind. "I will execute the sentence of justice myself!" He put the pistol to his temple, and if the colonel had not stood by, the lieutenant would have been dead after the shot. The bullet scratched his forehead, gunpowder burned his hair. The shell shock nevertheless threw the lieutenant to the floor, the colonel pushed the saddle under his head, poured a little wine into his mouth from a small Hungarian bottle, as if forged, and muttered in confusion:

"That's how it is, that's how brave! ... And all the hotheads, all the extremes ... Tell me, why did the Hungarians allow the Poles into the army ?! An insignificant little legion, some five thousand, but how much noise ... and what a reason for him, a reason ..."

The colonel unbuttoned the uniform and shirt on the lieutenant, watched how heavily his chest heaved, fussed over the recent criminal, rejoicing that, in the midst of universal blood and treachery, he had saved a man's life.

Chapter Three

How long have I been watching from the Sevastopol forts, the evolution of Dondas frigates, how long ago did I go out on rowboats to the Gulf of Finland to count the idle sails of Admiral Napier, and already an English private ship was drawing us from Portsmouth across the Atlantic, and with each turn of the propeller the promised land, called American states.

The ship, even in the first class, with mirrors and polished wood, shook violently when the machine stubbornly started a squabble with the weather; what was it like for the Irish and Germans, who filled the passenger hold as closely as you don't often see on military ferries.

Who drove them across the ocean? Hunger? But for many years, potato tubers have grown without interference in Irish soil. Blood and fear?

But the blood shed by monarchical Europe has gone into the ground; it seems that the tyrants were fed up, so much so that our Unforgettable himself choked on it.

So, it was not hunger that drove it and not the fear of new blood, but hope? And if there is hope, it would be better to say: "It didn't drive it, but called, called to the Republic, to free, unoccupied lands." She called us too: me, a stout gentleman in a dark English-style frock coat, and my Nadine, yesterday's Nadezhda Lvova, the daughter of my regimental officer. Our cabin heeled, we clutched our hands, happy at any occasion, and if we threw them stronger, into hugs, then even better. Happiness is complete, as a moralist would say, egoic. What is it for us? - we would ask heaven, God, if with every mile we did not move away on an English ship from the old continent, from the white stone churches of Russia and the psalms of the adjutant wing of Lvov, the favourite of two reigns and with the hope of a third, for these days of summer 1856 and the solemn coronation of Alexander II was being prepared.

However, there was not much English in that ship: I can only vouch for the captain, navigator and flag - England conducts her business so successfully that her product comes from all four countries of the world and here, on the islands, receives a British label.

Apart from Nadia and me, on the ship, even if you searched it to the bottom, there was no third Russian soul to be found. Just now they flashed in abundance in Europe; it seemed that if you pulled up a tablecloth in a fashionable Alpine resort or in a Parisian restaurant, a dignitary would be found there, and not just anyone, but under the stars, and with a portly wife covered in stones. It was as if the whole Crimea was waiting, crouching and indignant that a Russian soldier was dying so slowly near Sevastopol, interfering with their friendship with Europe, their spiritual communication. Now we have to make up for the missed years - let the soldiers' bones die to themselves! - life does not stand still, and high-born wives certainly need to know how Parisian tailors cut cloth and silk today. Once in Paris at the Hôtel de Ville[4] I was approached by a general dressed in civilian clothes - defiantly and expensively - but the Petersburg general lay on everything from his bearing and mustache to his intimidating look.

He did me a favour by speaking in an undertone, in French and confidentially:

"What is it you, colonel, decided to flaunt with a uniform?"

"I'm not ashamed of my uniform," I answered in Russian, causing him to seethe with bile. After all, our banner has been defeated, but we have not renounced it.

"But, for what's the sake of... embarrassing yourself?"

- Out of gratitude: "not all honour to the Dondas and Mackenzies, it is necessary for the inhabitant of Europe to enjoy the defeated uniform."

He hated me, and that's all I need, and Nadia too. Nadia - doubly, we are both evil, impudent, and these feelings were also part of the diabolical composition of our happiness.

And now the Russians, so conspicuous in the crowd of Paris or London, cut off on the Atlantic coast, following the steamers and sailboats with their eyes; only one in a thousand risked trusting the ocean.

The family of the Virginian landowner crowded in the cabin nearby; he, a wife, as short and fussy as he, and three resourceful boys who did not know the word of prohibition. There was something glorious, simian about the landowner: crooked, short legs in tight chequered cloth, felt sideburns, deep eye sockets and a cheerful, fanged mouth with the lower jaw moved forward. And his wife, a swarthy beauty, I took for a Creole, although she turned out to be the keeper of the old, seasoned French blood of the first settlers of America.

The Virginian seemed to have taken a vow not to stop the useful activities of the landowner on the packet boat; like a weaver's boat, he scurried among the immigrants, avoiding the Germans and entering into negotiations with the Irish.

He explained something to them - he wrote on pieces of paper, multiplied some numbers, laughed at their suspiciousness and treated them to tobacco - not a fragrant cigar, which I often got, but crappy tobacco, with the then unknown name "chu".

One starry evening we were standing on the deck, and I decided to ask about the nature of his activities. "I'm looking for workers!" he wondered. "Why don't you approach the Germans?" "The Germans are another matter. Coming down to our shore, the German will not rush somewhere headlong, he is looking for his German fereyns. The German loves to start his own business, even if it is small, but his own: a bakery, a pub, a printing house, a pharmacy. This is one grade, and the second is rebels running from the law ...

"To be honest, it became strange from his words, it smelled like a kennel, but he believed that we, first-class cabins, would understand each other.

"What makes you so closer to the Irish?" I groped for the truth.

"I need a cheap worker, the cheapest one," he laughed, revealing enviable strong teeth and all his active cigar mouth. - I would say - gratuitous, but a gratuitous worker is bad, everything should have a price.

"What to look for cheaper than a Negro? I asked cautiously. "He's already without rights, like working cattle." The Virginian even closed his eyes from resentment, stood as if praying to his offended god, and then said quietly and sadly: "You haven't even made it halfway to my homeland, but you are ready to believe the slander of the vile pastoral offspring!" I understood whom he had cursed: the woman, the pastor's daughter, known to the whole world as Harriet Beecher Stowe. I freed my sleeve from under his suffering palm but said nothing.

"If you want to know, we are our blacks and fathers, and mentors, and guardians, and healers, and shepherds ... And there are such jobs that you can't send a black man, even though he would be better than any white rubbish on it." "What kind of work is this?" "But at least on my estate: draining swamps or loading cotton onto riverboats." - "Why is the Negro out of place there?" — "Have mercy!" he was amazed. "I'll lose half the people in the swamps, but you know how it is when loading: freight roads, the captain hurries, work without rest, even at night, the bales are such that on a rare day someone doesn't fly over-board, but with a broken ridge ... "- "Is it better to break Irish backs?" "The Irish back is not my concern, but Saint Patrick, the patron saint of the Irish. After all, I pay an Irishman for a job, but a Negro needs to be bought, and now he costs so much that risking his life is sinful and ruinous ... "

From that evening, the fate of the Irish, condemned to eclipse the need of the unfortunate Negro slave, was revealed to me. The Virgin-ian did not understand the change that had taken place in us, called us to his place, but I answered that we were going for a short time, to the wedding, and thus cooled his ardour.

The ocean was long, so long that one day Nadia, following the wave with her eyes, which made her eyes borrow delicate greenery, asked me: "And what, how is this road without return? If we are destined for this, then so be it!" I replied with ease.

Nadia's face is harmony and a miracle of proportion, even a great portrait painter would not have thought to change anything in it; for me, any detail of her face was a separate world. Lips - large, mobile, elastic and so expressive in contempt or mockery; her straight and regular nose, with distinct, speaking nostrils, slightly shortened, which gave the impression of enthusiasm and lightness; her tender, fair skin, her heavy, blond hair, was a separate world. What can be said about the eyes, about two steppe lakes in marten - both in colour and softness - reeds! And this rare luck of nature contained not an ordinary mind, but that special composition of thought and morality, which is even rarer than the most perfect beauty. That's how rich I was!

A night in the Carpathian Mountains took away my youth but gave me love.

The military campaign soon ended, it was dangerous for young Russian officers to stay at the Habsburgs: among the streams of Tokay and sabre rattling, quarrels flared up more and more often, shots of duellist's clapped - Russians and Austrians came out to the barrier. The generals were free to ascribe extremes to the action of Tokay and Erlau red, to interpret impudent speeches in reports as drunken cries of revelling youth, and these were groans of a strangled conscience and belated repentance. The regimental officer became attentive to me; often called to him on the return days. At first, I suspected pity in him and went wild as soon as the people of the Don know how, to the point of gnashing of teeth, reddening of the cheekbones and trembling in the legs. Then it passed: I discovered in him an interlocutor, he had seen a lot and leafed through hundreds of books - the first impression of a gambler and a graying brace turned out to be deceptive. What we didn't talk about at the fires and on the way, when we rode side by side, stirrup to stirrup, on the side of the road broken by regiments and autumn bad weather. In one thing, we were not frank: none of us mentioned the name of Nadia. It hovered between us, was heard in the dry rustling of leaves, in the sad chirping of cranes, for me her name came to life in everything - in the ringing of well chains, in the sudden nightly laughter of a strange woman, in the quivering rise of a bird from under the very hooves, in the song of the Rusyns so close to the tunes of my homeland. Even at night, in the moonlight and in the glow of the fire, I saw her with us, heard the quiet neighing of the third horse, the silver ringing of his bridle.

In Warsaw, parting with the regimental, I did not receive an invitation to the house and was not too upset - I knew that Nadia and I would see each other, even if there were fortress ditches or a monastery charter between us. I was waiting for this meeting, like awards and executions. Nadia's father considered Ludwig a traitor to Russia; and what was Ludwig to him? - a wax profile with life flying away, a youth thrashing about in delirium? Nadia knew Ludwig intimately, his ability to tremble at the sounds of the organ, but she also knew his devotion to freedom. Ludwig is dead, and Lieutenant Turchaninov has come to brag about how, together with others, he disposed of someone else's freedom. The long road was cut off, not a simple explosion of a shrapnel shell awaited me - the powder magazine had to rise into the air.

And he got up, so much so that not everyone could see me from the sinful earth. Himself not tea, the regimental brought us together with Nadia, and brought us together forever; it happens that excessive precaution has the opposite effect. He himself taught his daughter the whole lesson of the Carpathian campaign. And when the lesson was coming to an end, the prince inadvertently mentioned the name of the artillery lieutenant, told about the night act and about the imprudent bullet.

It was then that the powder magazine was raised to the Warsaw skies! Nadia was thrown towards me across the entire garrison settlement; in a house dress and a thrown over Amazon, with her hair falling on her back, she ran without noticing the passers-by. It was not an execution that fell on me—even though I was worth it—but forgiveness and love. In an instant, we stepped over months of rapprochement, indistinct whispers, naive signs of love, generously described by novelists. Separation, the danger of my death and other people's crimes - everything was mixed up, preparing an explosion, and my father's inexperience, like a set on fire, brought fire to this mixture. When we woke up, there was no hostess or orderly in the room. We sat down on the couch. Nadia touched my forehead with her fingers and, bending down my head, searched for the scar with her lips, regretted that the hair had grown back, that it was not she who had to heal me; she behaved like an older woman—this has remained with us for the rest of our lives—and she repeated only one thing, shaking her head:

"How could you?! How did you decide? How dare you do this to yourself?! I'll go to your father at once!" - Such determination rose in me, after many years of betting, such energy awakened that I was ready to run to the regimental one.

"First, I need to prepare him ... He will not listen to you." I did not immediately understand what lay behind her caution: my noble nobility or my own small rank? It turned out - the jealousy of the father, who raised his daughter without a mother, the fear of loneliness.

"He is kind and loves me like no one else will. It seems that he would agree to see me as the bride of Christ, if only not to give it to another. "Then I'll definitely go to him!"

She laughed a mocking, overbearing laugh.

"And I love him no less, and I don't see life apart from him. So, I suddenly came to you ... I don't know, maybe you will condemn it, but it's impossible to leave him all of a sudden, it's murder."

"And I no longer care about my father: I kissed her hands and swore that I would never judge her for this step."

Have you ever taken the face of your beloved in the palm of your hand so that it lies calmly, everything, like the head of a baby in the hands of a mother; look and look, remembering the features, not daring to kiss, so as not to destroy the beautiful and almost immaterial world, and at the same time taking its alluring, only materiality, the breath of living flesh; take it in the palm of your hand and look, look until the rest of the world blurs, sinks into fog, leaving you only this secret, this precious vessel? ..

My father was rude to me. If I had found myself in a dugout Don canoe in the middle of a stormy ocean, I would have felt more comfortable than in the living room of a former regimental officer.

"How can you be like that, - suddenly, headlong? What about your mother and father? Or are they nothing to you?"

"They will not interfere with my happiness."

"Deftly, you settle down ... in your backwoods! He chose more painful words. - Well, how can I not believe?"

"Excuse me! I will ride to Azov, drive more than one horse, and back.

I restrained myself, humiliated myself, and he saw in my readiness only haste, greed to get the princess as his wife.

"Why torture horses?" he scolded me. "It's not like that in our circle, Lieutenant.

I am not a retrograde, but matchmaking has its own rules and customs."

"My circle is really poor!" No serfs, no tax, not even a decent estate!

And what, you decided to fix things? - It seemed that he had a sword in his hands, he was inflicting bloody injections on me, and I was standing naked, not protected by cloth. Do you expect a dowry?

What strength it cost me not to offend him, but Nadia is upstairs, in the mezzanine, Nadia is waiting; as I approached, I noticed her in the window.

"Our opinions about life - mine and your daughter - exclude self-interest!"

Put her on bread and water? Turn into poverty, into need! Take her away to Aksai, into the reeds, doom her to a meaningless life! What is your honour? Long months next to me, and not a word about your collusion, the game ... the game and the Cossack tricks. You were shooting, my dear, cunningly, so that the regimental commander had time to take his hand away! Oh right, wrong!

How he suffered, deceived by jealousy, how he believed that in front of him was a rogue, a seeker of money. But it's not easier for me either; anger was dizzying, and I did not recognize my own voice when, hoarsely, breaking loose, I shouted out only two words:

"Your Excellency! ..."

"What a simpleton I am, merciful Lord! I'd like to let you shoot. You wouldn't, would you! That night - all lies, only truths, that your betrayal! I was already leaving, ready to kick open the door, kick it out with my shoulder, vent my despair on it.

"I would like to put you on trial... as you deserve... Put you on trial before it's too late. "I urge you to carry out your threat. And I promise to confirm everything, your every word, even though the witnesses are dead," I said, leaving. He rushed after me into the hallway, blind jealousy made him deaf to reason.

"I would give you to the Austrians! he shouted. - Austrian speedy court!"

"I hope ours doesn't blunder, Your Excellency." Enough and agility and paragraphs!

My old-world father and mother, among worries about jams and pickles, about the correction of the house near the Novocherkassk prison, in the usual circle among family holidays, official affairs, madeira, homemade mead, the famous herbal casserole and archipelago wines, they did not even suspect that their son was in trouble.

Now Nadia has entered the war again. Her campaign turned out to be shorter than my Carpathian campaign, but the victory in its ambiguity resembled ours. Cruel condition! Our union was recognized, but the marriage was postponed for years.

I had to go to St. Petersburg, to the Academy of the General Staff. Peace came, the generals, even smart ones, did not foresee the imminent war, and without artillery thunder I could stay in the subalterns for too long, representing an insignificant interest for the princely family. Instead of a wedding table, I received a secondment to distant St. Petersburg, to the classes of the Academy. My father's calculation was simple: separated by a plain of thousands of miles, Nadia and I would be cured of the whim. And we only lacked the obstacle, the abyss dug by others, the scaffolding for the feat of feelings and spirit. I suffered severely from separation, Nadia had a more difficult test: she did not receive Petersburg, novelty, unexpected acquaintances and discoveries, and it was necessary to have her will in order to live in Warsaw, pass her classes, her science, not inferior to mine. She succeeded in languages, in history, in ancient and modern literature, and in the maturity of thought she surpassed many pupils of the Academy; the female mind, not burdened with service, forms, official chains, turned out to be freer and more unleashed in flight. We wrote to each other, wrote furiously; my father kept his word: happy with my departure, he freed our correspondence from domestic censorship.

We longed for heroic deeds and trials, and we received them in abundance: not months, but years passed. When I graduated from the Academy with a small silver medal, Nadia was no longer twenty-two, but twenty-six years old, the age, in the minds of the mothers of profitable Warsaw suitors (and not only Warsaw ones!), is reprehensible. So, the jealous regimental inadvertently set his other leg to his own trap: over the years, his paternal gaze involuntarily turned to Petersburg, to an unworthy groom.

I felt a change when I suddenly received an invitation to Alexei Fyodorovich Lvov - not to a home concert, for which his house was famous, but to the daytime twilight and silence of the rooms, to carefully familiarize myself with the restive Don tarpan, which had encroached on the prince's stables. Lvov the musician immediately realized that neither the luxury of his carpets, nor the historical canvases flickering in the semi-darkness, nor the gilding of the frames made the necessary impression on me and turned the conversation to scientific subjects: at the Academy I chose fortification, fortress building as a specialty, the prince graduated in his youth Institute of Railway Engineers. He was carried away by the conversation, and when he found out that I was not yet a stranger to the violin, he forced me to play an ancient instrument that is more than two hundred years old: here it is in front of you, in the same case, and it was made by the famous Italian Gasparo da Salo. The prince praised me: forgetting that he himself was married and the father of a family, he began to say that if a musician had to make a choice between the muses and earthly love, preference should be given to the muses. I did not argue with the old man and, it seems, left him convinced that the family business was going well.

My current Petersburg was not the same for me when, as a simple-hearted Don Cossack, I entered the classes of an artillery school. It's not about me, - time has grown wiser, moved forward, although with leaden gravity, and my eyes were washed bitterly with Carpathian blood, They say about the esprit de corps [5] of the students of the Academy, - this is pure nonsense; we were disunited, divided into clans, and if some dreamed of epaulettes and orders, then for others, freedom and the daredevils who fearlessly served it were sacred. Free poems, forbidden comedies, fresh issues of Sovremennik - that's what we lived in those days. I made acquaintance with Kolbasin, a well-known publisher in St. Petersburg, through him I obtained new books.

The academy is behind me, I am a second major, I was distinguished by being assigned to the guard, in the retinue of Tsarevich Alexander - my new epaulettes could satisfy Nadia's father's vanity. My path again lies in Poland, Nadia and I are already counting days and versts, but the generals were mistaken when they spoke of a long peace: saved by Gatchina from republican destruction, the French and British took advantage of the very first pretext to take revenge on their benefactor with lead. Austria did not throw down her gloves, but she did not stand aside either; sensitive Europeans unitedly rose to the defense of the hated Ottoman port. The war began in the Crimea and - a new, for three long years, our separation.

And in the early spring of 1856, I was again in Warsaw, in the position of chief of staff of the guard's corps: under me were fifty thousand bayonets, above me was the will of Tsarevich Alexander, the emperor of all Russia, not yet crowned. And now I am a colonel, I have equalled the rank of my future father-in-law, even surpassed him: he is an ordinary colonel - I am a guardsman, he was resigned - I am in an important service, he is old - I am young, I just turned thirty-four. We were separated by years and blood, and we were related by love for Nadia; he was tired of being jealous.

I entered the house with a mezzanine, where I had been rejected seven years ago, and found an old and glorious prince. Of course, he noticed a change in me: heavier shoulders, a dark blond beard, in addition to the old mustache, slightly swollen eyelids, as if the sleepless nights of war and the long spectacle of death were forever cast with heaviness.

" Nadia! he shouted upstairs. "Go see who's here!"

He stood white-headed, withered, sharp-cheeked, in an old dressing gown open on his graying chest, and I felt sorry for him, the cruel words froze in my throat, and thank God: I wanted to announce to him that I was absent from the Academy and, catching fear in the pit service , rushed south, through Tosna and Kresttsy, through Tver and Moscow, to Mtsensk, to Belgorod and Chuguev, to Rostov, to the Zmievskaya village, and there to the Novocherkassk parental home, for permission to marry the princess, the maiden Nadezhda Lvova, - I wanted to reproach him with the past. I again saw this house beautiful, tastefully furnished, although opposite the Petersburg mansion of the aide-de-camp, it was a shack with prefabricated furniture, Parisian chairs and armchairs side by side with the walnut office of Peter the Great and the mahogany of the reign of Elizabeth.

"That's what you are, Ivan Vasilyevich! the prince flattered me; we were sitting at the table, and Nadia was pouring tea into cups. - We got our way.

I did not know what to attribute his words to, to an upcoming wedding or to the colonel's guard's uniform, and answered accommodatingly:

"Your Excellency, the Cossack is cunning, he will pass everywhere; crawling, where it is slippery; quiet, where low!" He offered me wine, I refused. "Right!" he remembered. "You didn't drink it in the Carpathians either... But I'll pour myself some."

He drank and chewed his lips. Pity pricked our hearts: he sat defenceless before us, looking forward to a family, and not a lonely old age, and we saw too well his future loneliness. Everything is decided for us: in less than two weeks, and we will leave for the wedding, to the waters, improve our health, look at Europe, marvel, study, remember ... Everyone will think so, not excluding father, who takes us to St. Petersburg for a farewell visit to the adjutant wing - and we were leaving forever. Nadia's eyes now and then were covered with tears, she closed her eyes, stroking his sluggish hand with a quick, warm hand.

"Fifty thousand bayonets in one of your guard's corps! - the father was perplexed; he was still interested in the Russian army. "Why such a breakthrough of troops? The Poles have calmed down, I do not see leaders among them."

"We don't always see leaders an hour before an uprising," I said.

"You were in Sevastopol ... well, is everything true? One heroism and random mistakes that ruined the business?"

The heroism is special, complete, I could not even imagine such a thing! But the war in Crimea was lost before the first shot. Even in St. Petersburg, mediocrely lost, in the bud.

He raised his head, looked shrewdly, as if he sensed that I was not saying even half of the insulting words.

"I was not in Sevastopol during the entire campaign. You know this: by the will of the Tsarevich, I was sent by Totleben to Petersburg to build forts on the shore of the bay."

Nadia lit the candles, their fire fluttered in the brass of the pot-bellied samovar, in the glass on the table, in the colonel's brown eyes; the room became cosier.

" And what about Totleben? the prince asked jealously. "Before the war, no one had heard of him."

"He is one of those to whom everything goes for the future: malaise, heart disease, whom he hurts at the right time, and exactly where it is needed,

"Aha!" the old man rejoiced; he greeted the appearance of a new name grumblingly, finding in everything the consequence of favouritism. "I see you don't take pity on him?"

"Say 'you' to him, father!" Nadia asked. "It's time."

"I will gather courage after the wedding, and I will also shout."

"We are sinners against God, we cannot get married."

I spoke lightly, falling into his joking tone, but I was talking about a matter that had long been decided between us. And he clutched the table with his hands, as if he had lost his footing, looked from me to Nadia and back to me, and said, moving his whitened lips:

"Marriage without a wedding is a sin ... cohabitation. I ask you one thing: do not confuse your judgment on worldly affairs with the customs of the church ..."

"In the old days, here on the Don, this is how it was done," I still joked.

'The bride and groom went out to the square, the groom called out the name of the bride and told her: "Be my wife!" And she fell at her feet: "And you be my husband!" 'That's how they got married, without priests, without a church.'

The prince got up, in anger more like the former regimental commander than he had been all that evening. "And who did you marry? Circassian! Kalmychek! Captured Turkish women! How can you equal?!"

I could tell a lot about the beauties of the Circassians, about the faithful Tatar wives, about the Kalmyks; they gave rise to more than one glorious family, not excluding ours, Turchaninov's. The idea of equality of blood already then possessed me, but I was not so blind as to enlighten my offended father.

"Our alliance with Nadia is strong," I hastened to correct myself. "It would be enough for two lives, you could be convinced of this."

He calmed down under the gentle palms of his daughter, postponed the conversation about the church, in his heart not believing that we would not give in.

"How about Alexander?" he asked suddenly. "How did you find him, serving close?"

He gave my contempt to Nicholas, yesterday's ruler, in the hope that the new one would be better.

"A lost war will oblige him to a lot," I said, - here even the strongest will not be able to hold on to his former position. And then? Don't know. If there are no changes, without constitutionally correct institutions, then a new butchery is inevitable. Tyrants are made more by circumstance than by natural character."

In St. Petersburg, we were given an affectionate reception - our patron did not know about a civil marriage, without a wedding. The father felt an imminent misfortune: everything he wanted - a quiet church sacrament, albeit without guests and in an empty church - slipped away from him, the daughter, precisely the daughter, turned out to be harder than he could have imagined; he closed himself off, watched aloofly how quickly we got ready for the trip, and secretly hoped that perhaps Europe would bring us to reason, calm us down, take away our young bitterness, and, upon returning, we would perform the ceremony. But above all there was a heavy, vague premonition of trouble.

And the benefactor was perplexed: we decidedly did not take a dowry, did not take money, lying that there was nowhere to put our own, - of all his generosity, we appropriated only blue stones for Nadia - on the chest and in the ears - and the old violin of Gasparo da Salo. If he knew that we are in a hurry not to European society, but to freedom, how many wise tirades we would hear from him, how incendiary he would draw a new liberal reign, although the two previous ones gave him glory, cast orders and wealth. A man of undoubted talent, he was ridiculous and low with his vigilant Byzantineism - only we saw our father, sitting in the carriage, his head, tightly covered with short gray hair, eyebrows raised in surprise to hold back tears, a thin figure, as if striving after us.

We rarely spoke of him, but I felt that the old man followed Nadia everywhere; and how could it not be if I, offended by him, in the sleepless hours of the night more often saw not my father and mother, but a short military old man who looks at us, pursing his lips so as not to shout, not to stoop to a plea.

Chapter Four

From a letter from N. Vladimirov to his father.

"...Turchin told me: I was defenceless before life because I was looking for a higher goal, instead of just living like others. But that made me strong and I won." Much around him is a mystery, although in front of me is an ordinary old man who is wary of hidden irony - he is smart and touchy - an old man in velvet jacket and tight ankle-length trousers, in houseboots, an old man tied to papers that are not even removed from the table.

I owe my protracted acquaintance with the general to you and the publisher's widow. They have a special family, daughters and mother live by the same interests, as sisters could: they are equally sensitive, reckless, equally in need and equally light in this need.

The eldest lives in her home, the life of the youngest, Virginia, and Mrs. Fergus, I watch and see how they protect the dollar - not out of stinginess, but because it is important. Strangely, both women, in the rooms above the bookshop, live in the general's interests more than their own. The widow is ready to spend any amount available to her on a gift to the general, on the purchase of the thing he needs - the best paper, a dressing gown or a neckerchief. Where does these kindred feelings of Illinois American women for a native of Russia come from? I don't have the key to this riddle, the old dragonfly - that's how I mentally call the thin and light as a feather old woman, with a set of imitated flowers and berries on a wicker hat - is silent. Nor do I know whether Madame Turchina, nee Lvova, is still alive; here only her spirit hovers, the Fergus's are silent. Do not ask about her and the general; something happened between them, perhaps when she began to burn her writings - the French manuscript about Lieutenant Turchaninov in the Carpathians was burned along the edge.

The Fergus' are busy with his pension; they did not lay down their arms and after two refusals of the Congress, they are looking for his old officers, who are influential in the country today, they are trying to push the governor of the state to participate, for whom the deeds of that war are a legend of deep antiquity. It turns out that Turchin forbade them to write about their pensions, offended by previous refusals. He is a proud beggar, to whom it is not easy to help: he is truly defenceless before life - what is his victory?

I began to worry about the fate of his papers. The story of Nadezhda Lvova about Lieutenant T. was written by her in her youth, in 1851, and taken out of Russia. In the States, she wrote a lot and, according to Turkey, everything is better. Grandfather's travel chest is full of papers. Will they be needed here?

The general tells me his life, tells me about 'quatitatim' - drop by drop, in small doses. And yesterday I heard his violin and I'm still under its influence. I heard the violin as I climbed the iron stairs; at first it was hard to believe that only one instrument was playing, there was so much in the sounds of depth and voices. I moved from step to step inaudibly, afraid to disturb the music. The old man did not stick to one piece, at first, I fancied phrases from Alara's "Memories of Mozart", then Paganini entered, with such force, which did not often happen after the death of a great musician. You could listen to this music for hours, even on the stinking stairs, without taking off your wet coat. But soon it ended, the bow played badly, the musician lost his breath, it seemed he was losing his mind. A small man in a gray coat jumped out into the corridor, he began pounding on the general's door and shouting resentfully: "Pig! Pig! Why spoil it!" I ran downstairs, fearing to meet Turchin at such an inopportune moment, and after waiting, got up, knocked and went in to him. Turchin stood by the bed, squeezing the fingers of his right with his left hand. He hastily pulled up the edge of the blanket, covering the violin, and looked at me, wondering if I heard the music. I didn't even show it.

Do you remember the old copy of your book "A Russian Among Americans" that fell apart into notebooks? If you can, send a notebook where Radom is described and your acquaintance with the general. A convenient moment may happen, I will show him, but most likely for me alone.

General Turchin.

Chapter Five

I have already mentioned that before the Sevastopol war I made acquaintance with the St. Petersburg publisher Kolbasin. Having learned that I was going to the waters and would be in Berlin, he gave me a package for his German colleague Ferdinand Schneider - Berlin, Unter den Linden, 19 - hinting that it was better not to show this package to anyone, but, however, he added with a smirk: "Who will ask, even at the border, from a Guards colonel traveling to the waters with his young wife, the princess!" The fate of the package worried Kolbasin, and he came to see me again. At this meeting, Kolbasin decided to open up, to make me his accomplice: after all, it could happen that I did not find Ferdinand Schneider on Unter den Linden and suddenly decided to return the package to Petersburg by mail. He said that the package had been assigned to Alexander Ivanovich Herzen, and gave the London addresses of the publisher Trubner and the bookseller Tkhorzhevsky—I could send them a package from Berlin.

The name of Herzen was, if not a banner, in the absence of a political party in Russia at that time, then a secret password for everyone who despised the bygone reign and was not too deceived by the new one. We read some of his writings as believers of the Bible, and one of his heroes, Vladimir Beltov, served our friendship with Nadia: she confessed later that at the first meetings in Warsaw, when I was shy and kept quiet, something Beltov was revealed to her in me, his loneliness and weariness of the heart. We'll talk about Beltov ahead - we read him more than once, at first, we raved about him, we believed that his life would not fade away uselessly, but when we grew up, we were afraid of his fate, a vain, irreparable waste of strength. Chance opened the way for us to Herzen, and we decided to ask him before the Atlantic; a better oracle could not have been dreamed of.

Historical memory is needed no less than history itself—they are inseparable. In half a century of American life, I have fully comprehended this - her life is still without history, it is in its infancy, it is still nothing more than a family tradition, an oral legend, and legend is not asked for truth, but beauty. Historical memory is indispensable, it will soon come to Russia too, hatching out of its undoubted past, despising dynasties—but it is precisely memory, and not that living, tormenting force of nerves, torture of the heart and passion, by which past people lived. If this force did not decrease, did not overflow into the cooling forms of memory, but with all the heat was transferred to the future, which has its own passion and new tortures, life on earth would become impossible, burning into ashes and smoke.

Understand me: in London there lived a man whose word weighed more for us than all other words, and I could come to him not idle, but on business. After Berlin, Paris, and Ostend, and the foggy strait to the prophetic bells, and the white cliffs of Dover, and London with the first days of the device - everything was filled with a new meaning: a visit to Herzen is ahead.

We started not with Trubner, but with Tkhorzhevsky, hoping to find in him a person familiar with Russia, and we were not mistaken: he met my epaulettes with apprehension, was dry, then softened - we also found common acquaintances - Tkhorzhevsky knew the elder brother of the unfortunate Ludwig. A Pole by birth, Tkhorzhevsky loved the Russian exile with brotherly love. In all my life I have not collected collections - I am too inclined to give away being a collector, but my memory keeps one collection sacred - examples of brotherhood besides and even in spite of blood. If great and small tribes only populated the earth in order to fence themselves off with mountains, rivers or seas, to quietly forge weapons of enmity, to cherish their own blood, finding its composition higher than other people's blood, then general extermination is a matter of time and the destiny of mankind.

After meeting Stanislav Tkhorzhevsky, for a long time we did not dare to return to the cool room of the boarding house with icy water in a white jug with a blue pattern, linen fresh every day and soothing silence. To the point of extreme fatigue, we wandered along the vast Regent Street, now hiding in the darkness of the night street, now emerging under the light of gas lamps. I held in my hand the book I had received from Tkhorzhevsky—the second Pole Star, published a month ago—while our thoughts and hearts were consumed by the misfortunes of the great exile. Rumours of deaths around Herzen reached us before the Crimean War, but they were so obscure, so sympathy and anger, fright of the public and the black croaking of saints were mixed in them, that there was nothing to even think about the truth. And then we heard the sorrowful speech of a friend, together with him we experienced the death of Herzen's mother and son Kolya in the deep sea, somewhere between the mainland of Europe and the island of Jer, and the agony of his wife, and with her the breath of the newly-born Herzen. We saw the mournful funeral of a Russian on Italian soil, on a high mountain near the sea, a procession that went on for a long time, passing the suburbs and groups of onlookers, marvelling at the huge wreath of scarlet roses on the coffin, but most of all at the absence of the priest. There were moments when it seemed to me that I was losing Nadia, that it was not Natali who was fading away, but Nadine, broken by a foreign land, that in the consonance of two non-Russian names that lovers called Russian women, there is a secret and intent of fate.

We did not sleep that short June night. It was full morning, sunny and clear, when we closed the book, felt hungry and had breakfast in the first tavern that we came across. At noon we hired a cab to get to Finchley Road in time. Nadia gazed silently at the thinning houses of the London suburbs; The closer the well-fed horse moved us to Herzen's house, the more obvious was Nadia's confusion.

At the beginning of Finchley Road I dismissed the driver. We slowly moved along the even side, watching the houses, hedges and gates on the odd side. Behind the 19th number, the outskirts greenery broke off, a gloomy brick wall stretched out, high, with points of broken glass sparkling in the sun on top. And greenery was buried behind the wall, trying to look out at Finchley Road, but the main impression was the wall and the same stone, austere, with a touch of bureaucratic house and a cold gray gate with the number 21.

"You will go alone," Nadia stopped on the pavement.

I was taken aback: she passionately wanted to see him - is he really the one who seemed to us when we read him? After all, she herself wrote, you know her first test about Lieutenant T., and which of the conscientious writers of Russia did not dream of shaking his hand.

"Did this guardhouse frighten you?" I pointed to the stone wall.

- I can't! - Her gaze seemed to have already penetrated the brickwork and retreated in front of something. – "You do not understand?"

I looked around in bewilderment. I was going to talk about Nadia too, asking him if it would be possible to send to London the best of what she had written. It is more convenient to talk about this without Nadia, but Tkhorzhevsky dispelled our doubts, said that when meeting with Herzen, there was nothing to think about bad conventions, and we decided to go together. "What happened to you?"

He is unhappy, but we have happiness on our faces, our separate, selfish happiness.

She is right! Even in confusion, exhausted by a sleepless night, her eyes retained an expression of tenderness and happiness.

"Here we are: two happy compatriots, two safe Montagnards. It is not the power of the moment," she said firmly, "I must not go.

Stepping on the cobblestones of Finchley Road, she made a stern review of our past, measured a single act by the yardstick of a lifetime: this is how she always did.

"I'll take a walk and wait. If the conversation drags on, I'll go to the hotel. Take your time," she whispered, pushing me forward, "and try to remember everything."

The English gate was locked, I rang it, and soon it was unlocked from the inside by a servant, either the porter or the cook, who spoke foul French. He led me into the house and pointed to the stairs, at the top of which I did not immediately make out a stout bearded man. He stood, leaning his hands on the railing, in a gray frock coat, fastened with one top button, so that the skirts parted freely on a plump figure.

As soon as we drew level, not a word was said between us yet, as both of us, I think that both of us, a strange, paused thought occurred to us: I vouch for myself, but in Herzen's lively, mocking eyes I read the same surprise . When he opened the door and it became light on the landing, the mirage disappeared: his eyes were brown, mine are light, and his hair was darker and thicker, longer at the back of the head, so that it lay on the collar; there were many differences in us, but at the first moment, when we stood on a level, in the twilight of our home, it was precisely the similarity that struck us. Both are short, stout, large-browed, bearded, with persistent liveliness of their eyes, with energy that cannot be hidden even in immobility. If I were in a gray frock coat, with a tie tied with a bow, the owner of the house would have been surprised for a moment: why did his double appear on Finchley Rod, but now it was not a double, rather a caricature, for Herzen was in a guard's uniform and with a cap on the crook of his arm - the image belongs exclusively to satire.

But, alas, I had grown together with the uniform, squeezed into it, one might say, straight from the Don linen trousers, and now I was wearing the uniform - like a sackcloth.

I introduced myself, apologized that I interrupted him from work, to which Herzen very simply said that it was empty, that he would have time to work, especially since he had a serious illness - as he put it - "a serious illness: I can't sleep at night" .

"Do you often see so much Russian gold in your place at once?" I asked, stepping back and inviting me to see the uniform.

"Russia knew glorious uniforms," Herzen looked sharply at me.

"They ripped off their epaulettes, and then they became glorious," I replied. "And I got mine to Albion safe and sound; the shopkeepers are flattered to see it."

- Although uniforms and uniformity are long-standing passions of despotism, today's Europe is more and more seduced by them. He seated me on a chair, while he paced not wide at the table, on which lay books, fresh newspapers and sheets of the manuscript, interrupted by my arrival. "Wouldn't it be hot for you: the summer in London is stuffy.

I easily, with a mockery of myself, said that I had appointed myself this test - not to take off my uniform in victorious Europe, not to make life easier for myself with a civilized masquerade. "Omnia mea mecum porto"[6], I concluded my repentance.

- Completely in the Russian spirit! He smiled for the first time. "Impose a voluntary penance on yourself! And visiting my house you put yourself as a penitent?"

- I'm with you on business. — I got up. The owner did not sit down, but he was older than me and famous. - I have mail from Mr. Kolbasin. I have already taken the package out of my briefcase. — Kolbasin addressed the package to Ferdinand Schneider, but I did not find Schneider in Berlin. I received your address from Mr. Tkhorzhevsky...

He no longer heard me: the Petersburg papers took possession of him entirely. He shook the package in his palm, then laid it among the papers, looked like him, squinting at him, I took the scissors from the table and thinly cut off the edge of the package.

"I really need to open it!" he apologized to me, habitually working with his fingers, removing a pack of paper sheets. - I am a curious gentleman, greedy for news ... Here is the last book of the Polar Star, you have not seen it in Russia yet ... I will not be long, I will not read.

He spoke, and his eyes were already running over the lines of the letter, and then the manuscript. The expression on his face became impudent, now no one would say that he was ten years older than me, but by the severity of the loss, by the zealousness of labour - by a bitter age. Herzen walked around the study, now slowing down, pausing on some sheet, then throwing them quickly, circling around the table and shaking his dark mane.

I stood at the window, looked out, over the wall with bottle fragments, and was rewarded: Nadia walked along the sidewalk, her head tilted, forcing herself not to look into the windows of the house. In the whole of Europe, at that hour, there was no modern man for me, revered higher than the one who breathed behind my back, but for a moment I forgot about him too; I saw only Nadia, from her small boots, her unoccupied hand in a gray glove, to her high neck, on which a vein always beats so nakedly and defencelessly, and her open blond hair: she held her hat in her left hand. My heart sank with anxiety: she was a woman and a child whom I led to a strange crossroads, and how much longer was she destined to walk along unknown front gardens, gates, fences, strange entrances?

"Are you afraid of surveillance?" Herzen's voice sobered me up. I didn't understand it right away. He looked strangely at me, at the package, unsuccessfully trying to tie us, or at least me and Kolbasin.

"I am an ordinary courier, Alexander Ivanovich, and not a hunter for other people's laurels."

He felt the bitterness of my words, and in me there was a need for honesty and young pride: however, pride is one of those feelings that hardly grow old.

"We need this letter for our Russian cause." - He said that with the death of Nicholas and the end of the war, the Russians flooded Europe, many travelled to London, willingly go to him, so now, perhaps, precautionary measures are needed so as not to get lost in the visitors. "Your arrival is another matter," he hastened to reassure me, "what you brought is worth a lot. Why didn't you look at the Polar Star?

"We have read it all. Mr. Tkhorzhevsky gave us a book yesterday." - I said "we", "us" and caught his puzzled look: am I really one of the monsters talking about themselves in the manner of specific princes. I am in London with my wife.

"Ladies have a lot to see in London. He was losing interest in me. - Were you on staff or in the ranks?"

"Today - the headquarters, or rather, a fugitive, but in the war, he tried everything. I was in Sevastopol, shot back, built, then built in the north, near St. Petersburg ..."

He was no longer able to seat me in a semi-chair; he resigned himself to the fact that the visitor was just as restless, upright as he was, or maybe he hoped that the conversation would not drag on while standing. He asked about Sevastopol and perked up when he heard that I got to know Totleben and especially Alexander, whose coronation was then expected in Europe.

"There was no decent general in Russia to wage war," said Herzen.

"Some of our generals were three heads superior to the enemy," I objected, "but the most decent general will not save a lost cause."

- Why is it so? Why are Russia's affairs always lost in advance?

It seems that he was giving me an exam, waiting for an answer, so that he could judge not about history, but about me.

- On the contrary, Russia in this century has not yet lost before the Crimea - only victories. However, except for Napoleon, all are Pyrrhic victories. Nicholas lost for the first time; war, and with-it life. But even in the Crimea, how desperately the soldier fought!

Once in a battle, a Russian man fights heroically. But no one considered this war sacred. - And he asked me the same question and in almost the same words as Nadia's father in Warsaw: - And how is Alexander? Will not the beginning of some other time for Russia come with him?

"I was familiar not with the emperor, but with the Tsarevich," I evaded. - He noticed inconsistencies, he sympathized with others in his youth. Will the emperor keep these feelings? Don't know. I confess, I do not rely on the individual."

Herzen's brown eyes burned with the excitement of disagreement, argument, and argument with anticipation of victory, as if he were young and full of faith in the future, and I was outdated.

— Do you have an estate in Russia?

- Manor! I chuckled. - A hut in Novocherkassk, and that is not mine, a weather vane on the roof, and in the cellar - honey and liqueurs.

- So, you are from the Cossacks! Herzen rejoiced. - From this wretched estate? ... And I keep thinking: where from? You don't have a Moscow seal, I can smell Moscow a mile away, but what about Petersburg? He touched your thoughts, thoughts, not feelings, nature did not distort. So, from the perishing Cossacks? Without land, without serfs!

- I would not tolerate serfs, I would set them free.

"Now many people in Russia think so, others do so." Herzen lowered me to the level of commonness, made every day what burst out of me like an enthusiastic faith. - Soon the government will have to comply with this; perhaps a few years will pass, and the serf will become free. After thirty years of the butchery of the Unforgettable, others are ready to rejoice even a little.

- Is it possible to bruise the forehead in gratitude for the most natural, indispensable human rights!

"We are a very grateful people!" exclaimed Herzen slyly, still testing me. "We are so used to being strangled that when we are allowed to breathe for a minute, it already seems to us a great mercy."

We were interrupted. As soon as the conversation began, Herzen's youngest daughter, a lively, swarthy girl, ran into the office, followed by her sister, about ten or eleven years old, and an imperious governess, whom Herzen, it seems, did not dare to argue with; the daughters gathered for a walk and came to kiss their father. The son, Alexander, came in, his father introduced him to me, the shy young man took the book prepared for him and, leaving, looked around in embarrassment at my uniform, now the servant came in, in an undertone, in the same bad French, he discussed provisions with Herzen. The sun shone into the study, and the street was illuminated along, without shadows. I saw Nadia again and thought that everything was for the best; her open nature would have suffered from Herzen's evasiveness and obliqueness.

"On Sunday we have loud battles going on here," Herzen said, releasing the servant. - There are always a lot of Poles, Russians, Italians, French, Germans. We give such concerts that they knock on the wall from the other half of the house.

- Behind the wall - the owners?

"It's not customary to know this here: before ten years of neighbourhood, it's inconvenient to get to know each other. I would be glad to see you and your wife on Sunday as well.

He invited us from the heart, but in the very hint of a possible future meeting, one could also hear the end of this one, the only one for me.

- It's impossible. Tomorrow my wife and I are sailing to America.

— See the New World?

"We are leaving forever, and with this I have come to you. I'm tired of military service under a monarchy."

"You can go back, can't you?" Every shadow of condescension or irony was gone from his gaze. — Have the ships been burned yet?

- In Russia, they don't know that we are forever. But it is so.

"If I wasn't Russian," Herzen said thoughtfully, "I would have left for America long ago. He also got up, but he did not look at me, but saw the same wall and street, and he saw Nadia, and followed her with his eyes. - The kindest Mikhailo Shchepkin kept trying to persuade me and the kindest Mikhailo Shchepkin to hide in America, to consign my name to oblivion, not to destroy friends and myself, and only then, years later, to show up and ride on an apostolic donkey to Moscow University, to the anthem composed by the gendarmerie corps Colonel Lvov!

And we did not favour the Gatchina Orpheus - Lvov, but the mention of him at that very moment, when Nadia was walking by, struck me and closed my mouth; now I could not talk to him about Nadia.

- Outside of Europe, there are only two active regions - America and Russia. America is in motion, there is an excess of forces, the restless republic is reaching further and further, to the West. You can't say anything: what grows is young, - And then he asked with unexpected coldness: - Why are you looking for advice from me?

- I do not know another person among the Russians, whose word would weigh so much for me! Fearing even the slightest shade of flattery,

I uttered these words with gloomy ferocity, and Herzen burst out laughing, as a child or a person whose conscience is forever clear can laugh.

— Well, and among the Europeans? he asked, still laughing.

"I know their books, others I respect them. But this is a science, it needs centuries for decisive changes, and a short time is assigned to an individual. I am your reader, and in some way a victim ...

"To be honest, I aimed at others: more and more at Kleinmichels and at the Unforgettable.

- I have been reading Beltov for a long time, bowed before him, but did not apply to myself. And after the Carpathians re-read and howled. Years go by, uniform gold grows on me like scales on a reptile, but what have I done? What did you do? Then I read From the Other Shore, Kolbasin gave, from the first books that came to St. Petersburg, there are terrible words: "we have no soil at home on which a free man would stand!"

- How are you - beware of books, theories, and do the main act of life according to the book.

- No! I objected hotly. - I am ready to go behind the plough, stand on the porch for alms, but only in the republic; one more time and the monarchy will kill the man in me.

"If I say no, will you listen? He folded his full arms across his chest and stared at me.

"But if you say yes, I will cross the Atlantic with a light heart.

Why didn't you choose Europe? She received many exiles.

- My financial part is too meagre for an independent life in Europe. Europe betrayed the republic, bowed to the monarchy, and I want to see the only republic in our century. - I noticed Nadia hurrying to the omnibus, the shiny croups of the horses and the roomy, moving carriage. Nadia left, and my heart was relieved.

"What do you intend to do in the North American States?"

I will buy a farm, I will sow wheat. I will prove how much a free man can do.

- There will be one more farmer on earth! Isn't it enough, even for an admirer of Beltov?

"This is the beginning: I will strengthen myself and create a commune."

- Book dreams, Mr. Turchaninov! In America there is a spirit of camaraderie, financial association, but neither our Russian artel nor the rural community. There, the individual unites with others only for a certain profitable cause, and outside of it, cruelly and jealously defends his separateness. Boring America! he rounded off the thought.

I was saddened to hear those words. I turned away from the window, I didn't need it now, and stood with my head down.

-I know what's on your mind, Herzen said calmly. - A complete change of views, a betrayal of yesterday's ideal! No and no! I have always believed in the ability of the Russian people, I see from the winter shoots what a harvest can be! In the poor, repressed manifestations of his life, I see an unconscious means to that social ideal, to which European thought has consciously reached. Not returning the past, no! History is not returned; life is rich in fabrics, she does not need old dresses. But if we return to the artel of workers, to the secular gathering, to the Cossacks, he got carried away and pointed at me with his hand, as if I were a living confirmation of this historical opportunity, "having cleared them of Asiaticism, of wild meat," here it is, our vocation.

— Is this possible among despotism, embezzlement, batogs and banners?! When both deed and word are strangled?!

Yesterday it was impossible, tomorrow it will be possible. You are afraid to become sour in fantasies, but the commune in the Fourierist way is the same withering fantasy. A kind of barracks phalanstery (Phalanstery – Noun. a Fourierist cooperative community: a self-contained structure housing such a community.) , salvation for the weary, who pray that Truth, like a nurse, would take them in her arms. And America has tried all this, dear sir, experienced it in Kabet's monasteries, in communist sketes, without drying even a single orphan's tear. You are looking for work in common, why should you not see it in the Russian community?

"She is lifeless, Herr Herzen. Her numbness is complete."

"Maybe it's lethargy?" The tyrant's death gave rise to hopes, much has changed since the middle of last year, even here in London words of approval and participation reach me.

"All the same hopes for a new reign!" I said wearily.

- Nikolai died, and we were definitely drunk. But this joy is already old, now Russia is increasingly occupying and disturbing many strong minds. Alexander can push events, he can delay them, but he cannot cancel history either. God! What cannot be done with this spring growth after the winter of Nicholas, when the blood in the veins thawed and the compressed heart beat more freely! I will repeat to you: if I had not been Russian, I would have left for America long ago. But I am Russian, our word will get stronger even at home, it will be heard not only by impudent ears; isn't it worth the farm? he softened his vehemence and open passion with a joke. His eyes fell on the manuscript, interrupted by my appearance.

— I started writing about Ekaterina Romanovna Dashkova. What a woman! What a full and powerful existence! And this miracle could have arisen on Russian soil: remember the courage of two women, Dashkova and Catherine, when they change the fate of empires: eighteen-year-old Dashkova on horseback, in a Transfiguration uniform and with a sabre in her hands! - He was possessed by enthusiasm, but suddenly something besieged the ardour of Herzen, fettered his quick movements, extinguished his eyes: my gloomy appearance or the slamming gate and the voices that resounded downstairs in the living room, but, most likely, some unbearable thought. - He was broken by the demon of family troubles: family misfortunes are so deeply undermined because they creep up in silence and it is almost impossible to fight them ... They are like a poison that you learn about when a person has already been poisoned.

Before me was a suffering man, a brother in exile, and when silence became no longer possible, and he was still standing at the table, bowing his head and resting heavily on the paper, so that the gray cloth of the sleeves ran over his fingers, I said gratefully:

"I have to go, and you are expected. I took your time, but I am happy that I got to know you before a long journey."

He raised his face, dark from the rush of blood, framed by dark hair, and with the same dark flame of his eyes, approached me, dragged me along with him to the narrow couch and spoke, as to a friend:

- We are all on the road, for a long time ... you can get used to it. - He did not notice that he immediately got up, still holding me in place. - Many are sailing to America, there Gekker is a brave man, and he is not alone. When Europe strangled the revolution, some of its sons found salvation overseas. Well, try it. He looked at me in parting. - Apparently, the Russian people are destined to die for someone else's freedom.

Chapter Six

A propeller ship fully deserved the name of an emigrant ship: however, a rare vessel - under sail or steam - set off from Portsmouth without an emigrant cargo. I have already spoken of the Irish and the Germans: two more young Poles were reclusive in the cabin on our ship - they behaved as if they kept the golden sceptre of the Polish kings with them - there were also French women, a flock of girls sailing towards all the dangers of their despised profession, which and males are engaged successfully, and most of all in the service, crawling from one bureaucratic step to another, until finally the uniform of a court adviser gives them relative freedom and the privilege of selling themselves to one generals and nobles. French women, led by a good-natured fat patroness, took a sea voyage as a respite from worries and labours and tirelessly scurried around the ship, bumping into the scolding of the mothers of families and the feigned contemptuous glances of the fathers of the same families. Only Nadia and I and the skipper, a stern puritan, knew French, and all the tenderness of the Parisians turned to Nadia; after involuntary confusion, she discovered in them good people, with a lively mind and a clear conscience, with all the impurity of the craft.

We didn't see New York right away. Previously, a clean public, American citizens, were taken away from us by a New York boat, and the car, tired to the point of wheezing, dragged us further, to the tip of Manhattan Island - to where the waters of the Hudson and the East River merge, to a gloomy building with the big-name Castle Garden. Once upon a time there was a green meadow lined with windmills; a barely hilly land where sheep moved leisurely during the day, and tired New York workers strolled in the evenings. Then, on the site of the mills, gloomy, with a vast courtyard, a military fort grew up. But no one threatened Manhattan, and one day the city fathers decided to make a music hall here, to revive the old stones with the divine voices of Grisi and Mario; however, the music turned out to be uncomfortable here, the concerts were moved closer to Fifth Avenue, and an emigrant depot was made in Castle Garden. This happened a year before our arrival, but even in one year the huge hall of the Castle Garden, the barracks, corridors and galleries, the walls of the fort and its very air were permeated with the smell of poverty and emigrant knapsacks.

No one was waiting for us here, we didn't need a hall for family visits. I stood alone - Nadia went with the French women to negotiate with emigrant commissioners - I looked at the signs on which crafts were named, inviting immigrants to divide according to work: mechanics, printers, lumberjacks, masons, carpenters, coachmen, servants ... Signs, written in three languages and nailed over the doors, led me along the busy anteroom, where people decided which door to enter; perhaps they knew several crafts and were in no hurry to choose, or perhaps they, like me, were held back by the absence of a craft. My heart was beating anxiously. Here it is, the first lesson given to me by the New World! It is in vain to look for signs with inscriptions here: nobles, princes, gentlemen, colonels, landowners. The buzzing hall, the oak doors polished by thousands of palms, appealed to labour, to the earthly field of man, to his highest destination. Who are you under heaven? What works useful to many can come out of your hands? Can you bake bread, build a wall, chop down a house, forge a horseshoe, weave a rope, sew a suit or butcher a carcass to feed people? What business are you trained in?

The signs above the doors pointed me out to everyone as a person without the right to cross any threshold. And the trouble is not that there were fewer tablets than crafts on earth; be here and rooms for blacksmiths and carters, cable-carriers and furriers, carpenters and diggers, sawyers and locksmiths, hatters and plasterers, masons and shipwrights, lawyers and priests, doctors and merchants, hundreds of other rooms - I don't go to any of them, then I would rightfully enter. Which room to choose; really the penultimate one, with an insulting name - servants ?! From childhood we were taught - this word, dressed up in brocade: sovereign servants, servants of the throne, servants of a merciful god! Vile minds from professorial chairs and printed pages managed to combine the lackey name - servant - with worthy words, and now the servants of science, the servant of progress, the servant of enlightenment, have already flashed in kholuy camisoles and on the backs of noble carriages ... Why should I, the sovereign's servant, not push into an assigned room? But no, I remembered the tablet that Nadia and I needed, about the room where the farmers were waiting. In a difficult hour, only the work of my ancestors could come to my rescue. We will buy some land, seeds, if we hurry, we can sew winter crops, we will buy a horse, we will also make a down payment for an agronomic tool.

I had already opened the door of the farmers' shelter, when a French-woman ran out of the twilight corridor:

- Mister Russian! Save your wife! Quicker! We rushed into the depths of the corridors. To the right, a vestibule with columns and a staircase descending to the exit opened. Here Nadia wrestled with two men: a commission agent of the lowest stature and a sergeant from the guard at Castle Garden.

"In vain you, beauty, scandal!" the agent got angry. – "You're drunk, you see!"

I did not turn out to be a gentleman when I first appeared at the Castle Garden; the sergeant flew off behind a column and fell down the stairs, the commission agent stayed on his feet, but, it seems, with damaged teeth. They came to me; not with fists, - both cordoned off my blow and were in no hurry to measure their strength. In Castle Garden, as in any civilization, power is greater than mere force. And now I was moving along the corridor, escorted by my enemies. Nadia followed us.

"Well, you start life in a foreign land!" the commissioner said. - From prison!

"I am an officer, a colonel, and I am obliged to protect my wife."

They were embarrassed by my English speech and expen-sive London coat.

"You have insulted the government!" the sergeant said.

"The insult to my wife is immeasurable," I protested. - She decided to help French women who do not know English, and you grabbed her. And where - in a country that honours habeas corpus[7]!

My calculation was correct: Nadia, seen separately, with all her pride and insulted look, now seemed to them a woman whom they did not dare to touch.

We stopped in the middle of the corridor.

"Listen," said the commission agent, and put his finger in his mouth under his mustache, "you knocked out my teeth. Getting teeth inserted costs a lot of money in New York, and even more with legal fees.

"Why bother the court with such trifles as your teeth?"

- Do not go to me with knocked out teeth! He didn't pucker his lips; maybe the tooth was there. "It costs at least five dollars, and a dollar is the price of four shillings.

He was afraid that the transfer of currencies would make it difficult for me, and he hinted that he was asking for the British pound. There was a guinea in my overcoat pocket, and I handed it to the agent:

- Here's a guinea. It is equal to twenty-one shillings: give the rest of one shilling to the sergeant, it seems he needs to iron his uniform.

We have left the servants of order; then we have more than once observed how matters of honour were resolved with the help of money.

While we were gone, almost all the immigrants had sorted themselves out by rank, kept close to their signboards, or lined up at the door of a special bureau occupied by unskilled hired workers. The crowd hummed, called to one another, wept through the mouths of children, smoked pipes and cheap cigars, shuffled their soles, hit the stone slabs with matting bales - most did not have the money to take the service of the luggage office.

It was getting dark, fog was rising over the water, it cut off the Castle Garden from the gas fires of New York; in the room with the sign, the farmers lit a lamp. Here an agent with a brick, healthy face was conducting business, which itself spoke in favor of rural life. While waiting in line, I concluded that this agent was special, he represented the interests not of the country, but only of the local state and knew its lands through and through. It turned out that one desire to become a farmer is not enough. What was needed was capital—in cash or in bank papers—or a surety from some worthy citizen and a loan from New York banks, but how could an unspoken seeker get it?

We stood cheerfully in front of the agent; the land of the republic has been reached, now let the servant of the farmer's faith, the red-faced pop Baptist, marry us to this land.

— Russians? he wondered. - Russians!

"You speak as if you were a curiosity, Mr. commission agent. Nadia smiled at him.

— Russians rarely come. - For some reason he stood in front of us, tall, athletic.

"I haven't seen Russians here all summer. Few of whom I remember are originals: a singer who lost his voice, and a prince who lost his fortune.

"That's right, and we're originals," I said.

"So, you need more than just land," he objected reasonably. "And that's a bad sign: the farm demands total loyalty, more complete than marriage, religion, or a banner."

- You do not make an exception for the Republican banner? Nadia asked.

"It is enough for a wife to be faithful at night," he answered jokingly, "a banner in the days of war, and the farmer belongs to the farm day and night, in all seasons, otherwise she throws him away." He pushed up oak chairs for us, they were recently seated in the concert hall by admirers of Jenny Lind, offered me a cigar made of fragrant virgin leaf, scolded me for not converting money into papers, brought it in cash to the Castle Garden nativity scene, advised me not to buy land for the eyes, travel, look after; advised us to move from the fort to the hotel, gave the address of the office where we were supposed to report the next day, and said goodbye, arranging for us to spend the night in a small room at the hall - without doors, like a church chapel in the Manhattan vale of want. From our hiding place we saw part of the crowded hall, they settled down for the night, laid children and old women on benches, on arranged concert chairs, on belongings piled along the walls. Illuminated by scanty wall-lamps, as if from above a tiresome yellow rain of scattered lamp oil, they presented a spectacle beyond Hoggart's fancy; and right there, through poverty and disorder, there was such a wild strength, determination and tenacity, such an ability to endure, which together can both build a civilization and destroy it. I prepared myself for a sleepless night; the suitcases remained in the baggage depot, we had a leather chest with us - I put it on a bench in the corner of the room, there I arranged a bed for Nadia from my coat, and she soon fell asleep.

I stood at the entrance to our aisle and lit a cigarette. The unified hum still hovered over the crowd of immigrants, someone's whisper, quick, elusive speech, the quiet laughter of a woman, as if her child had not taken her breast for a long time and finally took it, the work of suppering jaws, a monotonous, pleading voice - it seemed like a prayer, a strangled quarrel, a quarrel without passion, the noiseless movement of several gray shadows, as if into a hall, for complete resemblance to all mankind, let the insane out of the hospital. The two Poles walked quickly up and down the free floor, hurrying the morning; seeing me, they took me aside and went into a dark corridor. Even on the packet boat they showed me all the measure of contempt; not for this they fled from Poland to hobnob with a Russian master.

A picturesque old man walked by, red-bearded, squat, reminding me of my father with the tilt of his head and the narrowing of his unhealthy, red-lidded eyes. I looked at his twisted back, wide, with a hint of his former strength, at strange, backless shoes that separated from the heel with every step, looked at someone else, and pain over my father left on the Don tortured my heart. With a burning, tormenting feeling, I remembered how I was in a hurry for the last time to Aksai for my parents' blessing. (Aksai Chin, Chinese (Pinyin) Aksay-qin, portion of the Kashmir region, at the northernmost extent of the Indian subcontinent in south-central Asia.)

Mother was lost from a short happiness, she was not up to that to see through the joy of a hard & rocky bottom. Not such a father; he suspected something in my haste, in the fact that I suddenly began to repair everything in the house and on the estate that was worn out and dilapidated, as if I would never find another time. I was just saddling up on the roof, adjusting the ridge and the weather vane, when he appeared below, squinting against the sun and summoning me with such a mournful gesture that I immediately went down to him. He straightened my unbuttoned shirt, pressed his hand to my chest, then slid his hand down, squeezed my fingers and immediately let go.

"Goodbye, Vanya," he said. "Are you going?" "I'm on the road tomorrow, I can't take it anymore."

"I know. The service demands it." He chuckled bitterly, almost acrimoniously.

- The late Pavel Petrovich Turchaninov did not make it in time for his own wedding, or rather, he left the wedding, mounted a horse and went to the Alps, and the wedding was postponed.

He sighed. "Well, at least he died in his own land." He followed the trail, circling and circling, ready to howl, he felt that it was underground, albeit at a fathom depth.

"Wedding when?" He needed a sure point in order to lead his thoughts from it, but I evaded whether there was much joy for him in cohabitation without the sacrament of the Church.

"I do not know yet. I am now to the prince with your blessing." - "So, what, what is the prince?" he got offended for me. "And you are a nobleman, and your rank is equal to him." - "It's not about the prince; he has a daughter - a miracle, she has no couple on earth. I have been waiting for her for the eighth year, but I would have to, and I would have waited my life."

"And how will she receive us?" How she would accept them, Lord, how she would rejoice in their simple disposition, how she would look for my traits in them - but this will not happen, never will. That's why I could not find an answer cheerfully, easily, but with a fake smile squeezed out of myself: "Why ... She is a smart, simple-hearted girl." He looked at me not with reproach, with a plea not to hide, to remember that he is old and has seen a lot in his lifetime, and it is better to meet the hard truth than any lie. What could I answer him, born of him and so far, removed from him? All that remained was what my heart, mind and hands, from which the roofing tool fell out, asked - to hug the old man and press him to my chest. So, I did and heard a sob burst in his chest, and he spoke quickly, stifled, fearing tears: "God be with you, Vanya ... Live, son ... think about God, about us ... Here you are."

The squat old man, the lopsided ghost of the Castle Garden, approached me with each pass through the hall. He was attracted by cigar smoke, and I silently handed him a cigar. I have already noticed how much this gesture facilitates in the relations of strangers.

"Are you a Pole? Young people," he puffed and nodded toward the dark corridor, "are also Poles."

- I am Russian. Came to buy a farm.

- If you have money - buy it. - There was no ingratiation in him: life is lived; it seems he finished it without hope. With money, buy everything.

It seems that this man in cross-legged props, in someone else's frock coat, knew better times. He squinted, puffing out smoke.

- Something I do not remember you on the ship? I noticed.

- I crossed the ocean a long time ago, then only sailboats sailed.

"And still in Castle Garden?"

"Here I earn my bread. It's also nice to see that you're not the only one left in the cold.

- Are you an agent?

- Agent, but on worse terms: I get from the purchased head.

He spoke evil, and I did not stand on ceremony:

"Why don't you dress more decently?"

"You think like a money man, not like a beggar!" He smiled wisely. A beggar would rather believe his poor brother. I do not shove papers to him, but I guard them like a dog of sheep, and tomorrow I will bring them to the owner. Here he is sleeping, satisfied that he has not signed anything yet, and how the mousetrap closes tomorrow is none of my business.

He turned out to be a German, a Catholic, traveled half the world hunting for a man; once to a black African, and now to a European of the same faith.

"And, how are they?" I nodded to the hall, where sleep overcame people. "Will everyone get a job?" Both work and shelter.

- What about freedom?

- Wake them up and offer them a choice - freedom or dollars; then we laugh. Only money gives you freedom.

"But if there is a job, then they didn't make a mistake?"

- All of them are future citizens, but not all of them are full. He saw my bewilderment and, in gratitude for the cigar, continued: "Sometimes the work is difficult, sometimes the journey is so far that you can't even scrape up a ticket." In winter it is difficult: in big cities, especially in New York, so many extra hands gather that if only a couple out of a thousand take up a knife, then even then not a single wallet feels safe. The authorities had to open two gratuitous shelters on the local islands ...

"Why are there extra hands, if everyone has a job to do?"

— In winter, farms and plantations are closed to the worker. One shelter on Black Island - for those who have already become a citizen of the country, there they will not even throw the sick out of the gate, and the one on Ward's Island is hell and hard labour for foreigners. The ardor with which he spoke of the Ward's Island Orphanage had a simple explanation, he himself had to go through this purgatory. - In the morning a cup of bad coffee and a slice of bread, if you drop it, you can't see it on the dirty floor.

The superintendent, John Wayles, the damned bastard, pitted us like dogs, settled us in the halls: separately English, separately Irish, Germans, Swedes, Italians, fed everyone the same shit, and boasted that he fed us differently ... - He puffed on his cigar with special pleasure. - I don't understand one thing: we were all like brothers in this hell, without the barking of a dog: they say, I'm an Englishman, I'm an Irishman, I'm a German!... "So, this is possible!" I exclaimed.

- Only in need, at the very abyss! he said fiercely. - And in satiety, everyone is closer to their own bone. Today I have a good night, I can tell you, you are not my game.

If you buy a farm, other bloodhounds will follow you: banks, creditors, tax agents, envy of neighbors and fever - fever is inevitable in broken land.

What is this, broken earth?

"The earth, untouched from the creation of the world, must be broken with heavy plows, allowed to rot everything that the Lord covered it with before you. On such land, the farmer is shaking with a fever in spring and autumn. The old man looked at me with indulgent pity and with the familiar feeling that relieved him that here was another broken fate. "Yes, and can you hold a heavy plow in your hands?"

- The Russian plow will teach you everything, after it the American settler plow will seem like a feather.

He did not understand my allegory and shook his head: "Few survive. They take the land, leave their last pennies on it and run. He was smart, that scoundrel, and guessed my true trade under the English dress. "It seems to me that you are from the military."

- You guessed it, I'm a military engineer. "There are more hunters in humanity to kill than to sow!"

The old man changed at the last minute of the conversation: he became brusque, something unhealthy came through, tears slid from his red eyelids onto his cheeks and coat, sick, without crying. I held out another cigar, I saw in him a readiness to take, instant greed, but he did not take it, he defended himself with a proud gesture.

"There are a hundred times more sowers than killers," I protested. Otherwise, life would have ended long ago.

Do you think it hasn't stopped? He marveled at my innocence. Do you think this is life? The old man pointed to the hall. "Then call Dante's hell earthly life!" Take the greatest of books - the Bible, and it is a collection of murders, deaths, evil intentions ...

He stepped into the depths of the aisle and stood up, over Nadia. For him, I was young in my mid-40s, but the beard and bald head, it's true, made me look older in his eyes. And on the bench, turning on her back, the girl was sleeping; the old man looked back at me, wanted to understand who I am to her: a father, a husband, or a rich patron? And he understood everything correctly, he was smart and experienced, and my face did not hide anything. The old man dropped the rest of his cigar, intertwined the fingers of his large, unclean hands so that the joints crackled.

- Lord, merciful! he spoke repentantly. Why are we executing them? Why do we plunge into filth and poverty?

He was crying, standing over Nadia. His slanting bent back trembled; I took him by the shoulders, pitying and afraid that he might frighten Nadia. The feeling of the unexpected power of the bony old man's back struck me, like any senseless vegetation of strength in a person. He obeyed, went to the exit, saying through his tears:

"Before you break the ground...life breaks you. It will take away everything dear, everything that was life ... Run! he suddenly exclaimed wildly. - If you have money for a return ticket - run! "

And so he left, leaving me in a state of confusion and anxiety. I guessed his own misfortune, a long-standing misfortune that cannot be overcome, which crushes a person in a way that a gravestone does not crush. I sat down on a chair next to Nadia and did not take my eyes off the calm, sleeping face. Is the old man right and am I taking her to the slaughter? Is it possible that living life, and the flow of its blood, and good shadows near the eyes, talent and thought, and it's very breath - everything will be sacrificed to political speculation, given to rude life, grabbing into hundreds of insensitive hands?

Chapter Seven

Thirty acres of land on the eastern tip of Long Island I bought from the owner of the large Rowland estates. He took dear virgin land, seduced by several acres of forest and a stream. The house built by someone also encouraged me; a prudent person would be alerted by the sight of an empty house, an abandoned hut, with the seal of someone else's ruin, and I rejoiced at the deaf one - with doors, but without windows - a log house, before winter I expected to cut through windows, put a cast-iron stove and lay a second floor. I did not put the rest of my money in the bank, and our neighbors were divided into two parties: some saw me as a poor man, for a money man, before taking his place in the church, would take him in a bank, others were embarrassed by our leather chest, they saw a lot of money. The money really was with me, but scanty, and every day I took a particle from them, they flew around faster than a rusty leaf that autumn from my trees.

Autumn was hired to help us: long, sunny and dry. The sun caught us on our feet, I looked at it in the morning without squinting, and it, the same as over the Russian land, told me: that's all I can do for you, the previous five years of autumn were short and rainy here. The sun was rising to warm our farm and log cabin, one of the two doors, always open, served as a window for us. At twilight we lit a lamp, on a cast-iron stove Nadia boiled beans and coffee, roasted bacon, baked cakes and bread, wheat at first, but a month later tough and hard, with bran, but we were satisfied with a slice of this bread and a mug of barley coffee.

The postman brought the New York newspapers, the regular 'Weekly Tribune' and the Women's Revolution, and tied to wooden arms and legs, we read in the evenings, dreaming of winter, when the earth would give us rest and we would begin to write. This Long Island winter, Nadia wrote 'La Chólera' [8] and two more stories.

The water in the stream turned out to be unhealthy, from a nearby swamp, I began to dig a well, rested against a stone layer, got gunpowder and stopped to break through a stone. We need a lot of water - we have a house, two horses and a cow. I bought a mowing machine, a plow and a harrow, shovels and spades, axes, a saw, seeds and flour, glass and nails - you can't list everything a farmer needs so as not to die of hunger and not to freeze closer to dawn when it cools, red-hot in the evening, cast iron stove.

And in the midst of all these worries, we were happy: Nadia would find me at the plough or in the clearing with an axe in her hands, run through the arable land, get her head under her jacket, to the shirt salted with sweat and breathe loudly, showing how good everything is around: under her own feet the earth, the merciful sky turns blue overhead, late birds sing in it, and ahead is life. And very close - winter; under the snow our field will throw greenery; cattle work up their strength in the barn; seals the stream with frost; through the swamp there will be a direct path for the postman to the country road.

In silence and warmth, we will set about our work: a lengthy memoir about the Crimea, St. Petersburg and Narva was taking shape in me; Before Nadia's eyes stood living images of Russia, and among them the old, Arakcheev villages, an old man in bast shoes and a torn Armenian coat, speaking through eaten teeth about the damned goat tribe of nobles. I was attracted by reflections on the perniciousness of the monarchy, and Nadia's sheets, her suffering word, the stooped shoulders of arable peasants, the grain transparent from dryness, and the parish priest crushed by the anger of the Tsar - all rebelled against the monarchy. We are one soul and one thought, what can be higher than this? Our happiness was complete, knocking down.

I leaned over the rocky, dry-summer earth, deeply rooted, and thought back more than once to the Castle Garden agent's warning. A horse cannot break this land, you need buffaloes, or four horses harnessed to a heavy plow. Two horses and Nadia and I, hanging on the plow, only scratched the ground, pulled up the roots, prepared a shaky bed for the grain. But there is nothing to be done: autumn has given us more than one respite, and winter crops must rise to the snow, take it on their tender feathers. And we threw grain into the ground, having done only half the job, without turning the whole layer.

From time to time Rowland appeared, tall, with an immense, buzzing and wheezing chest, pale and whitish, as if he was dipping his face and neck, piebald mustache and stiff bobbed hair in rye flour or bran. He was on horseback, with a pack of dogs, as suspicious as he was, and remained the master of everything: he chose the road as he was used to, regardless of the new layout, he might not leave the saddle if Nadia did not stand close, as if I were not the owner of the land, and his worker. He is in the saddle, and I have my back to him, - I did not give in to Rowland, throwing words over my shoulder, - let him see that I have no time. A man not evil, unhappy, as they said, in a family, with a sick, tyrannical wife, he was ready to start any dispute, even if lost for him, he could trot behind me for a long time in plowing, so that the horse breathed down my neck, and the dogs poked at the pants. Rowland was teased by rumors about my capital, he wanted to make sure I didn't play a joke with him, I'll take it and pay for the land, but it gets into my head, I'll sell it, and he will look over someone else's estate from the hill until the end of his days. He graduated from West Point, went to the Mexican campaign, was shell-shocked, insulted his commander, and, as a retired captain, took vast lands on Long Island for nothing. Among other incomes, he perfected the income from the sale of plots and the almost inevitable bankruptcy of their temporary owners. He put the matter no worse than the way the agronomist distributes his land - for bread, for fallow, for meadows, but, like a Christian, he regretted his victims and, having achieved their exile, accompanied the departing wagon on horseback and endowed them with good tips for the future device.

— Hello, Turchin! From the shadow on the ground, I saw him putting his hand to his hat. - You are a stubborn person, but I will take a chance and come up with advice: open a bank account, do not keep cash with you. - Rowland dubbed me Turchin, without intent, out of impotence in front of the inconvenient name of Turchaninov.

- They warned me about this back in Castle Garden: Clifton Young.

"Sometimes a man like Yang will suggest a case. She and Yang didn't get along.

"I wouldn't trust our neighbors with even a dollar; they can dig up the whole earth in search of money.

- I'll thank them; horses do not steal, even if thieves break the ground. You don't have money everywhere.

"Under every square foot of earth is golden!"

He did not believe me, and yet he scanned the ground with his tormented eyes, staring blankly, like a man of slow, delayed thought. Once I told him that the money was already in the bank and we were sleeping peacefully, Rowland chuckled, apparently, he knew my every step, had friends who were aware of everything. I stood sideways to him and his horse, cutting into a pine trunk with an axe.

- Nothing, Turchin, you did not cross the threshold of the bank. "Watch out, Rowland, lest you get hurt. – Doesn't look like it. He didn't move an inch. "More likely, you will have to retreat, you are on the leeward side."

Indeed, I jumped back, a tree fell between us, the horse reared up in fear.

"You will exhaust the forest, Turchin," Rowland looked with regret at the branches trembling from falling.

- I need a tree, I have to build a lot. Do not buy timber on the side when there is a lot of your own.

I upset Rowland very much; he jumped to the ground, something thumped in his protruding, asthmatic chest.

- I want to give you advice, Turchin: it's up to you to obey or neglect. I don't know what faith you hold, but you must have your own faith. I nodded: I also believed in something. - A family that no one sees in church will take root here.

"I am Orthodox by birth. Not to the church for me."

"Of course," he agreed after some thought. "You are a gentleman and do not act against honour.

- Honour above all: I'm also a colonel.

He recoiled - am I kidding? In Russia, even the rank of general was not as rare and loud as here the degree of colonel.

"Colonel of the Guards," I confirmed. - A veteran and with a scientific medal for the Academy of the General Staff. Do you want me to show you?

The silver medal is with us, in a leather traveling chest, I had no reason to part with it either, as I parted with my uniform, throwing it into the Atlantic on one of those stormy days.

"Mr. Turchin," said Rowland respectfully, "there is an Orthodox church in New York. Put on your best dress clothes, and the neighbors will be convinced that you have gone to God's house.

"And gutting the farm looking for treasure?"

— I'll take care of the farm.

Frost and snow cut us off from the ground. The house kept warm while the wood burned in the cast-iron stove. Nadia grew thin, her face was weather-beaten, dry lips took on a colour equal to that of her face, an expression of unquenched thirst, impatience appeared in her, a kind of gypsy, flexible desperation with gray eyes and fair hair.

I admired her both in the winter sun, and with a mean wick, and in the jumping glare at the open stove door, I admired greedily and guilt-ily, as if I had taken something that was not mine, had stolen its best stone from a distant land and taken it across the ocean, into the wilder-ness, to the backwoods of Long-Island. While working, Nadia forgot me too; she went into her pages, like a stone into a hole, like a traveler goes into the night, she left all, without hope of returning. Bending over the paper, I processed the thought, I was its driver, its lapidary, I molded it and burned it, and Nadia herself surrendered to the lava flow, where thought and passion, united, reached an impossible heat.

Before Christmas we put on our best clothes and took a stagecoach to New York. Nadia had just started a correspondence with Lucy Stone [9] — She was going to create a Women's Magazine, the tireless American was looking for employees, preached everywhere with the sermon of equal rights for women and the rejection of slavery.

I needed Clifton Young, he promised seed grain at a reasonable price, such that "both in the Sahara Desert and on an unwatered stone it will sprout by itself." We returned empty-handed, having spent the night at the station: Young had left on urgent business, the publication of the magazine was delayed due to lack of funds.

There is silence on the farm, and traces in the fresh December powder: traces, traces, as if a meeting of the surrounding inhabitants had passed here - all traces into the house, and the lock will be unlocked. Gloomily we moved towards the house: we lost good easily, losing faith in people is bitter. By that time, we were getting on well with our neighbors, even with Rowland. After dark they ate their bread with a slice of pork, drank coffee, sometimes without sugar, lunch differed from breakfast only in the amount of bread and meat, and meat was not supposed to be for dinner. What kind of capital is there!

Nadia opened the door and stood up, petrified, and then I looked out over her shoulder. The light from the window fell on a crowd of neighbors who sat at a long, knocked-down tabletop. Nice American women, Irish women and a German widow worked hard and arranged a holiday meal for us on their Christmas.

And we sat down at the table, in the warmth of their friendship, to pies, to the same pork and the same boiled beans; but what a taste these beans had, barley coffee, drunk in the midst of a friendly conversation about business, worries and hopes for a harvest!

The winter dragged on for a long time, gave time to fill out more than one dozen papers; then came the arguing spring, the bright sun, followed by thunderstorms and downpour. Our winter crops did not sprout, and where they made their way, they were not thicker than a hair on the face of a Kalmyk; I began to sow again, only a little plowing the land. And then the worst happened: Nadia fell ill - with the March dampness, she was broken by a fever. Pain, pain all day and doubly at night, chattering teeth in chills, cold feet and hands. I prayed to fate to work a miracle, in a short dream to throw a fever over me, so that her flesh would be cleansed, so that her eyes would not be clouded, and her hair would not split, did not grow dull: I had no reason to live without Nadia, and then we could only die together.

That spring, we became convinced that Rowland was a kind person, or, more simply, a person, until it came to his bank account. From him we had a doctor, medicines from home stocks and a lot of good advice. But when the due date approached, Rowland did not give a reprieve. He already knew that we were poor - Nadia's illness opened everyone's eyes - he took pity on us, but he also took pity on his income. Rowland realized that we could not hold on to the farm and looked for ways to reclaim the land for a better profit. Neighbors advised resorting to the court and auction sale: we could recover up to two-thirds of the money given to Rowland.

We hesitated: how else would the court treat stateless immigrants? And Rowland was cunning, dodging, promising a delay or a decent amount for the repurchase of land, without public bargaining. Before Rowland stood a doomed enemy: I would have defeated him in the saddle, on pistols, in a fist fight, but on a smooth, iridescent, with divorces, bank notes, bills, bills of sale and mortgages, I was an unarmed soldier. Nadia was fading, and I believed in a saving deceit, in every illusion, I was glad to return to her bed with hope, the naked truth was then beyond her strength. I looked at the papers that Rowland carried with him, and through them I saw Nadia, her exhausted eyes under thick eyebrows on an emaciated face, thinning nose over a mouth bared with a fever - did I care about bills!

Thus, we reached the deadline: Rowland—calculatingly playing with my trust, I—hoping to receive a reward for a working well, a barn, a stable, a plowed field, and two acres of uprooted land. And the time came - and Rowland approached me with the threat of the law, and I had to agree to the most disadvantageous and destructive conditions. Rowland paid back the money that we used to start a new life in Philadelphia. Receiving dollars from him, I also sat in the saddle: as leaders of two armies, we towered on a hill, from a height there was a view both of the log house where Nadia came to her senses, and of Row-land's ivy-covered house.

"Where are you going with your horses, Turchin?" He already looked at my horse as if it were his own property. The city doesn't need them, and I'd give them a reasonable price.

"We'll ride off in saddles, Rowland; let not a single soul think that they walk away from you like beggars. We are already rich in your bounties ..."

"And you gave a cow to a German woman for a pittance!" Rowland's resentment broke out.

- I am a hunter, I hunted saigas (noun: The saiga antelope, is a criti-cally endangered antelope which during antiquity inhabited a vast area of the Eurasian steppe spanning the foothills of the Carpathian Moun-tains), a wolf, it happened, and a wild boar. I know a lot of snares, traps, traps and holes, but you, Rowland, have invented an excellent tackle.

He pretended not to understand, but meanwhile, I gestured around this tackle - the soft folds of the earth, the still rather dry pattern of the oak forest, the useless stream among the bushes and young grass. The horses were drawn to each other with their muzzles, our physiogno-mies did not express this interest.

"It's hard for me to handle all the land," he complained. — We'll have to look for a buyer, and it's not easy. Everyone wants to go to the West, to free lands... And you could join the army; I'm looking at you - you would get along with junior officers, and with soldiers, perhaps. - He experienced his exile from the barracks paradise through the years. "It's not like in other countries: in our army, strangers are tolerated. In Russia, I think, in a different way?

"Our soldier is Russian," I reassured him. - And the generals - every third - is German.

- I looked at the map, Turchin, there Prussia and other German lands are no larger than Kansas, where are there so many Germans, wherever you look?

"And if a German buyer comes along?"

- Not!

"Will he suddenly offer big money?"

- No, Turchin! If a German settles, it is forever!

Two weeks later, having bypassed the neighbours farewell, Nadia and I went to New York. There it was possible to profitably sell the horses and go by rail to neighboring Philadelphia.

A blush had already touched Nadia's cheeks; she was waiting for a new life with childlike curiosity.

Nadine Lvova Turchin
wife of Union General John Basil Turchin

Chapter Eight

I was returning to the end of dirty Pearl Street, where the painted spokes of carriages seldom flashed; here wagons and covered wagons with sacks of grain, flour and bran creaked slowly in a rustic way. I went to the post office on foot, without extra cents for the omnibus, and in the same manner returned home, from the main streets, where New Yorkers, excited by the approach of Christmas, kneaded wet snow with their soles.

In the morning in the Sun, I did not find myself in the column of persons addressed to the poste restante letters, and yet I went on my way; I was waiting for a letter from Nadia from Philadelphia, she wrote to the post office, not to the Nizhinsky printing house.

A letter was found for me, and this time, a local one, from Sergei Alexandrovich Saburov, I met him back in London. Saburov was handsome, smart, brilliantly educated and elusive as a person; he could just as well be a jaded diplomat, adventurer, gambler, or European-dominated landowner. In London, he ended up in the same boarding house as we, ardently, without wasting a minute, asked for friendship and on the third day, testing my generosity, not needing, as it seemed to me, asked for a loan of money. I gave, a little by his standards, but tangible for us. Saburov was older than me, he served in the Caucasus, there, in the mountains, he knew Nadia's father. He got into service for challenging a senior officer to a duel and a successful shot, got a fake passport, fled to Austria, from there to Prussia, to Paris and London, thousands of miles from his father's rich estate near Kursk, Saburov roamed farther and farther, sometimes littering with money falling into dire need. He then begged for permission to return to his homeland, then, having been refused, he accepted French citizenship out of annoyance, changed it to English, was hired from under the vaults of Westminster or from the taverns of Paris in the soldiers of the Indian or Algerian army, never, however, reaching his destination.

Then in London I did not know his life, I gave money for one evening, and he put the banknotes into his vest pocket so casually that I involuntarily looked at his feet, had he dropped the money? I never saw him again in London.

And so, imagine, in Philadelphia, in the spring of 1858, when our affairs were so bad that even ten dollars seemed like a fortune, Saburov called me from the springs, put me in a fast carriage, announced that he had converted to Catholicism, acquired the friendship of famous priests and was fussing about the pulpit in one of the main churches of the New World. He was smartly dressed, explained that he was in Philadelphia on other business, and when he returned to New York, he would assume a look decent for Catholicism. On the same day he returned the debt to me and asked 'pardon that then', in London, someone else's secret and the honour of a woman in the middle of the night dragged him to Normandy, across the strait. Unexpected money allowed us to move to New York.

I stood in the piercing December wind, with Saburov's notice that I would have him tonight, and looked at the gloomy building of Middle Dutch Church, equipped as a post office, at the now unnecessary bell tower, from which Franklin once surprised the townspeople with electrical experiments. I would have waited for the evening mail, a letter from Nadia, but circumstances hurried me to the press. Two other workers, the engraver Balashov and James Bell, were finishing the work of the outgoing year, Mr. Nizhinsky left for Boston with his wife the day before, I was left alone and could complete the work I had begun in secret. Bell will go to his family for the whole Christmas, Balashov will return to his closet not earlier than he has sold everything in the tavern, to the last penny, and our servant and cook, janitor and messenger, with the big name Napoleon, will stand with me at the printing press, we will print the last half page of the book, and I will immediately scatter the lead letters of Nizhinsky over the cash registers.

That's what my plan was. I chose Nadia's short story "La Chólera". Nadia's childhood memory was expressed in it: in 1831, as a girl, she traveled with an illustrious relative to the Arakcheev paradise, to military settlements, and there, with the start of the Polish campaign, cholera spread, and a bloody riot broke out. The memory retained the spectacle of the wild crowd, the cries of the tortured, the distorted faces of the officers who closed themselves in the house. You know French and you can understand what difficulties I have encountered, not having in the English alphabet all those dashes, squiggles and acceptances, without which a different Latin letter will say nothing to the French eye. I could translate "La Chólera" into Russian, but where can you find a reader for a Russian book? And English, although I wrote political plays in it, has not yet opened up to me from the side that allows me to move in belles-lettres. I worked for a long time as an engraver, almost chewing lead with my teeth, extracting French lines and accents, now I had to lay down the last two pages of the story in lead. My short play "On the Republic" was typed as well. The bloody events on Novgorod soil showed the perniciousness of the monarchy, I spoke about the advantages of the republic and its American vices. We did not expect money from the book, we longed for deeds, a push, the beginning of something that would change our life.

I found Napoleon on the porch. The Negro was sitting in front of the locked door, in indestructible shoes on his bare feet and in Balashov's tattered overcoat.

"Why are you here under the snow?" I was surprised.

- I locked them up. He held up a key in a swarthy pink palm. - Mr. Balashov gave me a coat, and Mr. Bell - a mug of wine, and now we are playing.

Light filtered through the shutters, and the printing press for vignettes and lithographs thumped softly.

"What are you playing, Napoleon?"

- I watch them, Mr. Turchin, and get a dollar for it, and they prepare Christmas presents. Mr. Balashov wants to buy a new coat and give his to me...

He fled from Alabama, from the shores of Tennessee, and more than a year ago he reached Philadelphia, where our friends got him a document in the name of Napoleon - it delighted him and stuck to him, as if he had been baptized by Napoleon. Having settled in New York, we signed Napoleon up on Pearl Street: now not only every dollar, but even small cents, he saved up for the common cause of black fugitives. His only weakness was his constant plans for marriage. In this Babylon, he sought out Negro widows, and the more curly-haired offspring hung around them, the more persistently he strove for marriage. Usually he came not to me, but to Nadia, asking her to see the widow and take part in it. "Mister Bell," Napoleon complained to us in his intricate English, "advise me not for marry dis lady, 'cause she hab seben chil'en. What to use? Mr. Bell can't love for me. I mus'lub for myself, and I love she!"[10]

"But they won't deceive you?" — I went up to the porch.

- No, Mr. Turchin! They didn't lock themselves away from me, I locked them.

He innocently believed in the game. "They print a picture, I saw..." Leaning towards my ear, Napoleon whispered: "There is Bethlehem, the birth of Jesus and a star in the sky, and below, Mr. Turchin, our song: "Jesus make de blind to see, Jesus make de cripple wok Jesus make deaf to hear. Wok in, kind Jesus!"[11]

Let me in, I'll help them.

"Mr. Turchin open it yourself, here is the key," and he sat down with his back to the door, believing that in this way he was fulfilling the conditions of the game.

The lock clicked, but the door did not open - it was locked from the inside. I knocked on the window frame.

What do you want, Napoleon? Bell called from outside the door.

- It's Mr. Turchin back.

They fiddled around in the printing room, rearranging the lamp, the light inside alternately dimmed, then pushed harder through the shutter, and finally they let me in. Balashov rinsed his hands over the basin, Bell, as always, left without washing off the paint, letting us know that he has his own house, that he is a master and despises the rusty basin, sticky remnant and Nizhinsky's dirty towel. Thin Bell stood in the door of the printing house, blocking me from Balashov and a brand-new bag, standing on the type-setting cash desk.

"May Napoleon come in too, Bell?" I asked.

"You all care about the negro as if he were your own son.

"I used to be an officer," I joked, "that's why I honour Napoleon."

"This is sitting in Russians tightly: respect for the Black Sea," said Bell. - Balashov is the same.

Balashov is kind by nature and lived as in a dream, no one knows why.

"And I'll tell you the reason," Bell looked for quarrels. - All because they do not live with you; from unaccustomed to their smell.

- Oh, my friend, - I objected, - and I don't smell his separate smell.

- And mine?

- Yours is bad when you get drunk.

— Mister Turchin! Napoleon pleaded, fearing that Bell would forget about the promised dollar. - The day after tomorrow is Christmas, before the holiday you can drink.

Bell, clowning around, bowed to Napoleon, who scanned the tables and printing presses, looking for a lithographed Bethlehem.

"You're looking for your dollar in the wrong place, black!" Bell took two fifty-cent notes from his purse and handed them to Napoleon. - Here are the ones I chose: clean, like the conscience of a newborn Jesus.

- What are you, Mr. Bell! Napoleon was scared. - Is it possible to say so!

Balashov fished out a coin from his vest pocket:

"I promised you a silver dollar, Napoleon. He glared at Bell. No, keep the papers.

I started dialing, listening to their conversation.

What would you do if you were to become rich? Bell asked.

If I had two hundred dollars?

- Not two hundred. One hundred thousand dollars!

The Negro laughed, but there was nothing to be done, he had to answer.

"I'd give them to the same place as those two."

- To whom?

"To a respectable widow, Mr. Bell," Napoleon said enigmatically. - She has a lot of children. Even a hundred thousand dollars is not enough to buy new shoes for everyone.

"So, if I die, you will take care of my widow too?" - Bell's cock-like profile deceived with militancy, the engraver was a spineless fellow.

"White ladies disdain blacks."

- Oh, you're a freak! Bell laughed. "Sometimes I wake up in the middle of the night and think: I'll get up in the morning, but there is no America! He spoke from the passage, stroking his unshaven, sharp chin. - The houses are in place, and the Hudson does not flow back, and the sky is as it was, but America is gone! I mumble, but no one understands me: Irish, Germans, Poles, Negroes, Chinese are all around. Is it really not enough for a man, no matter how proud he is, to have a mother tongue?

He aimed at me, but I dodged.

"The horse runs properly in the harness," Bell continued, "does not break the shafts. Yesterday's wild buffalo will get used to his stall, but a person is clever at everything, rushing around the world.

Why do you equate a man and a dumb creature! Balashov was offended.

— The Lord equalizes us; Jesus was not born in lace, in a manger, on a pile of hay.

"Goodbye," Balashov bowed to me. - Nadezhda Sergeevna will return, bow low to her. Home to Christmas time, oh, how good! In an inside-out sheepskin, in a mask, to knock on the neighbor's houses ... Good!

"Excellent, Balashov," I smiled too, imagining the Christmas season of Novocherkassk. But man is not a bird. You can't fly across the Atlantic. Don't wait for you?

He had a woman who devastated him, there were taverns and taverns.

- You better not wait. He retreated into the twilight of the vestibule, his voice was so sad that I looked up from the set. - And I will return alive and well - rejoice.

They left, and Napoleon locked the door.

- When you and Mr. Balashov speak your own language, Mr. Turchin, it seems to me that you are both good priests ...

I sent him upstairs for the time being, Napoleon's steps creaked on the steep stairs and died away in the closet. I was left alone, next to the lamp, hearing its buzz, and when, deceived by the silence and the rustle of lead letters, a cricket sang in the corner of the printing press, the feeling of home became aching.

Even in my idle time, I knew how to instantly settle in any passing housing: an adobe hut on the Don during the summer military games, and a chopped Carpathian hut, and a sufficient house abandoned by someone, and the rooms of a Sevastopol assistant doctor smelling of medicines, with a wall pierced by a French one stuck in an oven, core. Nadia did not wean me from this, on the contrary, wherever we close the door behind us, we are at home. This is what happened in Nizhin-sky's press. He traded in advertisements and vignettes; Serviceable, partly unused typefaces gathered dust in the typesetting cash desks, he had a meager income, but he also reduced his expenses to nothing-ness, paying employees not two dollars, but one, but it happened even half a dollar for a long day. There are five rooms at the top of the print-ing press: two are locked, Nizhinsky went up there only to start the loud striking clock, the rest were occupied by Balashov, Napoleon, and Nadia and I. Nadia's father would have called me to the barrier, seeing what kind of life I had doomed his daughter, the Lord musician would have anathematized me from the lectern, and the sun shone through our window more generously than through the two-height openings of the St. Petersburg mansion. Nadia was in Philadelphia - she and a dozen other young women were being awarded diplomas from the Philadelphia private medical school - and I was typing Nadia's story, intending to give her a public life. The secret recruitment had been going on for a long time and was now coming to an end; the Russian emperor appeared before the crowd of yesterday's rebels, and in a few lines the whole action of "La Chólera" reached the highest historical meaning and tension. I read these two pages aloud, lost in excitement, from the uncomfortable lump that was rising to my throat. Before me the crowd came to life and the emperor-deceiver, young, arrogant, frightened, outwardly merciful, but plotting meanness, having already blessed the executioners at the execution, to drive the peasants through the mortal line of sticks. Only once, having seen the living Nicholas in the Kazan Cathedral, Nadia managed to depict him in such a way that through rudeness and cruelty, a nervous, cowardly nature, although not yet broken from the inside, shone through. Ominous was the appearance of the pictorial figure of the tsar next to Count Orlov and His Grace Timofey, vicar of Novgorod, in front of a crowd of military settlers and arable peasants: a dull voice, when, looking at the report filed by the surviving colonel, and finding there eight officers shown "on a business trip", but in fact of those killed, he said: "This is in the far one!", He said with the promise of execution and vengeance; and a hint, just a hint of the Senate Square and another standing of the crowned sovereign in front of another crowd.

Full of hypocrisy and threat was his refusal to accept bread and salt from the hands of the knelt villagers, his quick passage to the exercise house, and then to the church, where the village priest, not prepared for the service, having no singers with him, was silent until the emperor did not yell at him: "Serve!", and turned sharply from the church to the square, to the guilty people. And the quick, calculated lie of the king that the Lord sent them cholera for disobedience and rebellion; and the complaint that he himself lost his brother from the same disease, as if this death lay on them; and swearing from the mouth; and the order to shoot on the spot at the first disobedience; and a sensitive, vile ear, fear of the crowd, pathetic words, in addition, spoiled by a piece of bitten bread:

"Well, I ate your bread and salt, of course, I can forgive you, but how God will forgive you!" - and an order for bloody fun, for the massacre of hundreds, and a quick, cowardly departure for Petersburg, under the pretext that the empress felt the approach of childbirth; the whole picture, raised to biblical grandeur by the silence of the people, through the eyes of the villagers, already suspecting ramrods, batogs and death under banners; and a pathetic dumb question in his eyes:

"What, brothers? "Come on, is that the sovereign?" Didn't the per-egrinet come to us from cholera poisoners? and a dreary female voice, that the rye is all dried up, the grains are already glowing, and the hayfield without mowers; and, finally, the words of an old man who, in the midst of the crowd, answers the cantonist, frightened by the riot and cholera:

"What is there to say! for fools, poison and cholera; but we need their noble goat tribe to be gone..." The story contained refinement, soft colours of the Novgorod region, spaciousness, cleanliness of the ripened fields and all the filth of estates. If this story alone had come out from Nadia's pen, even then I would have known that I had found a woman worthy of standing next to the famous names of Europe. I believed that her fame was near, but her life was lived, and who knows whether future people will read her papers?

I finished typing, called Napoleon, and made an impression of the last two pages, when I heard the carriage at the porch and the impa-tient knock of the knob on the door. Saburov!

He was in everything Broadway, from Stuart, and slipped a coin to Napoleon for opening the door to our lair for him. He came with gifts, brought delicious breads, spicy crackers, peaches, ground coffee, sausages, French wine, and for Nadine, a dyed ostrich feather that had come into fashion then. Saburov apologized that he allowed himself to make a gift to a noble lady, but this day is special, before Christmas, and for him it is doubly special, he is leaving, and we will never see each other again.

The master was in every feature of Saburov - in beautiful, veiled black eyes, in a mocking grin, in overhanging eyebrows, the colour and silkiness of a sea beaver, with well-groomed sideburns, where not a single hair dared to get out of line; in clothes that knew no accidents, flaws or mistakes against the best fashion. A gentleman, but of a special kind, consumed by anxiety; something strangled in itself from a young age and suffocating every day and every night, when conscience, this merciless nocturnal animal, tears the liver with its claws. He was thirsty - and he drank without offering me, he knew that in vain; he

109

wanted to eat and drummed his fingers on the table, urging the dazed Napoleon; wanted to sit comfortably and asked for an armchair - it was an expensive stone and required a frame.

- You live badly. Saburov waved his hand, pointing out the printing press, its dirty walls, the shutters carved by generations of engravers, and me in old boots and a dubious suit. "Both you and Mrs. Turchina deserve better."

"We were looking for this, Mr. Saburov.

- What were you looking for? he protested. — Poverty?

- Live by your own work.

But why look for work for a beggarly wage?

"We are full," I said shortly, but he was silent, looking at me reproachfully.

"Nadia will return from Philadelphia with a doctoral diploma, she will also find a paid business for herself."

- And for this you left Russia! For the common life! He was at his best, and I was inspired with great calm by the pages scattered in typographical solitaire, prints with a sheen of wet paint.

"For civilian life," I said. — "In the republic".

- Have you tried it on the tooth? Isn't it false, your republic?

Sick but true.

"But maybe her illnesses are such that I would prefer a wise monarch?"

"That's where we'll part ways," I objected calmly.

— What, however, nonsense! Saburov exclaimed. - After all, I chose the States, and I do not often regret it.

- You seem to be a citizen of the republic?

He waved his hand wildly: what to be, that cannot be avoided.

- And I'm just getting ready, just going to the goal.

— What do you believe, Ivan Vasilyevich? To a printing press? In a book? After all, in Russia it was possible to pour sweat on such a scoundrel! Nizhinsky is not worth your thought alone, a tuft of hair!

Thank you, although the price is not high. I am now working for myself, not a single quarterly officer will allow this.

I handed Saburov a print.

- Nice! Nice, he said as he read. "It's only a pity – after him, to the grave, he should have read such a thing during his lifetime ... Ah, scoundrel, I, they say, forgive, but how will God forgive ?! And the old man is not bad: you have to come up with a goat tribe! It's you and me - the goat tribe! For some reason he was happy. Napoleon brought supper, roast beef appeared on the table and the pride of our cook - an apple pie. I began to wash my hands: Napoleon poured me out of a jug, and Saburov was consumed with impatience. — By whom is it written? he strained his memory. — Not new feathers from the Russian third estate? They will never dream of such a French language.

"You're right." - I sat down at the table with Saburov.

— Really Herzen? Saburov asked. — You're not?

"Nadia Lvova! Imagine, my wife, yes, yes, Nadezhda Lvova-Turchina. It gave me pleasure to repeat it."

After the enthusiasm, Saburov began to cool down, he noticed that this was a stone on a decayed coffin, and not a dagger in the heart of a tyrant. He did not listen to objections that literature is not occupied with personalities, but with ideas, he complained that in Russia, if they kill a tyrant, then quietly, and thought is suppressed and impossible, and service is not valued.

I also showed my already printed pages with thoughts about the republic; Saburov chuckled, "is it possible to call a North American republic a form of popular government, and politics a complex science?" It is a quackery based on the superstitions of the crowd.

"No one will forbid you this fun," he returned the prints to me. "Is it worth keeping censors, spending money, if no one will hear you even without censors. Turn blue from the cry - they will not hear. I know this feeling, when in solitude you add up angry words, and it seems that you hear them already on the lips of mankind, and he does not care about you, it's good if the sister-in-law reads. All this - into oblivion, unrealizable in utopia, but in life you need to do business. You think I'm an ignoramus, a playboy, right? Not well-read in literature?"

I was silent. He did not need my words: they were all figures of speech, rhetoric.

"I was reading! And old-fashioned Adam Smith, and Fourier, and Saint-Simon, and any Russian, forbidden; he read, thought, was executed, and the more he thought, the longer he lived, the more clearly, he saw that books have their own path, and life has its own. And there is no such holy thought that people would not turn against other people, like a sword that knows no mercy either in the hand of the monarch or in the hands of the Jacobins. From sublime thoughts weave hemp for loops to the gallows ... Here's my life, without hiding: in the Caucasus I hit the beast, the executioner, who spared neither the highlanders, nor their wives, nor children, nor the Russian soldier, - I hit when it was impossible not to hit , otherwise - a bullet in the forehead. We fought, I hoped to die, but I wounded him and was thrown to the soldiers. Well, I did not break, I fled, through Tiflis and Turkey, to the Austrian possessions, I was educated, I dreamed of the modest fate of a tutor. And so, the place was found, I am on the estate of Count S. - I don't want to name it, his widow is still alive, and the memory of me, I hope, lives in her ardent son.

I read to a teenage Rousseau, Voltaire, the latest German philosophers, and one day I am convinced that even in the home nest everything is based on a lie: the count inclines his stepdaughter to cohabitation, and from me he planned to make a machine for the love affairs of his retired wife. And I run again, run to Prussia, they grab me, they open that my passport is fake, and they decide to send me to Russia, on the basis of a cartel signed between Prussia and Russia. The Prussian king is ready, like a worm, to crush a man who has done him no harm! I had money, I bribed the chief of police: it means that this force is rotten and dead, and even a high cartel, held together by two powers, can be bought out for a handful of gold! I go to the conspirators, look for them, offer my life if needed, and I am severely beaten for a prank, for an oversight that even the priest will excuse.

I accept Catholicism - remember our meeting in Philadelphia - I dream of a church pulpit, and they want to make me a spy for the Jesuit order ..".

"Excluding the Caucasus, you yourself were looking for trouble," I remarked when Saburov got up and went between the machines. "As a player."

- What did you find the game in?

"You go to the conspirators without sharing their faith. You were afraid to serve the order, but you were not afraid to change your faith, like a dress. You were looking for a service, did not agree on the price - and immediately Latin out of your head. You spend too much energy on living a carefree life.

- Damn me if you're wrong!" - he sighed with relief, - They said too much, in Russian, but true. I have a business for you, this time for you, just a business for you. You know, Ivan Vasilievich, that the republic is successfully expanding to the west. Beyond Texas, beyond the Red River, untrodden lands all the way to the ocean. But before civilization comes a bullet and a bayonet. You know the terrible fate of the Indians, and all because selfish people serve there, the dregs of humanity. And if a noble officer came there, why shouldn't the republic make friends with kind-hearted tribes? I was a lieutenant, you are a colonel, under your command there was a corps, here you will have a region the size of Little Russia! You are the law, you are the court, you are justice, you are the sovereign prince, and instead of destruction, you bring peace and reason.

"You are looking for shoulder straps, but you are talking about the world; send a missionary."

- And if the tribes do not submit? Saburov was amazed. If they are the first to use their weapons? "So, it's murder. For a lot of money, many must be exterminated, otherwise they will not pay."

"And if the tribes call you?" Saburov asked.

- Their business is closer to me. But do you have powers from them too?

" No! But sometimes I thought, I dreamed: if only I could come to them with rapid-fire weapons, with cannons, because they, what good, would also push the republic of merchants. They have an eye, speed, dexterity - no worse than the Circassians."

- And they argued that he was not a player! You can't do a mean deed in a noble way. He wanted to speak, but now I stopped him with a wave of my hand. "I lived for thirty-six years, and two-thirds of them studied war, commanded guns that did not always shoot at state targets. This is over, and if you recommend me to anyone as a colonel, I will say that this is a lie.

Something interested Saburov on the floor, he bent down and picked up a fifty-cent bill.   - You have money lying under your feet, where can I tempt you with a salary. - His face suddenly changed, his eyes widened, he smoothed the banknote in his palm, put it in his pocket and ordered Napoleon to go upstairs.

We were left alone. Saburov looked around at the printing presses and typesetting cash registers as if seeing everything for the first time.

"I didn't think, I didn't think ..." he muttered. — Ah, nice! What kind of artists! What risky heads! He took out fifty paper cents and showed them to me in the palm of his hand. - Have you seen it? Well, what about now? Saburov covered the banknote with his palm, turned his folded hands over and opened them again. - Focus?

A rectangle was blindly white on Saburov's hand. A counterfeit banknote, a piece of paper processed on one side.

"Test print..." he said. "I keep wondering where Nizhinsky got his money from; a crappy establishment, he would die like a church rat, but he buys houses, a stable from him, a woman ...

I reached for the banknote, Saburov let me hold it, but he didn't release the whole piece of paper, he valued it. A counterfeit banknote, criminal, shameless, like a living, naked creature, it attracted the eye and caused awkwardness, confusion, as if a murder had been committed, and here is the corpse. And Saburov scoured the printing press, looked into the slots of the type-setting cash desks, rushed to the closet.

- The plate is definitely here! Tell me, is this printed from copper or stone? Do not look like a beast, I know that they are not guilty, they will not take you into such a case! But what is it printed from?

- From a copper plate. The stone is rough.

- I thought so. - Saburov's gray checkered trousers are soiled - he was crawling on his knees - the braid that trimmed the skirts of his coat was also soiled. - The record is more difficult to find, the record is thin, but there is, it certainly is; such an amazing resemblance, only a madman would destroy the record.

- I live here, all the work is going on before my eyes ... - I was perplexed. - Nizhinsky in Boston, Bell and Balashov printed the Nativity of Christ in Bethlehem. Well, Napoleon...

Here Saburov also waved his hand; black and will be killed for less.

- In this venture, Nizhinsky: like this, little by little, without risk, in a festive mess, when even a neat German puts a banknote into the cashier without looking. Think fifty cents!

- You are already accusing, but what, like a banknote by chance here?

- A flying bird! Saburov laughed. - Martin? Or a snowman?! - Saburov put on his coat, prepared to leave, but remembered the purpose of his visit. - How do you decide, Ivan Vasilyevich? I'm talking about the West; think I'll wait.

"There's nothing to think about," I replied sharply. "I'm not fit for either the Cortes or the appanage princes. Farewell and be merciful to the Indians.

And imagine this actor, a buffoon at the Theatre of his own destiny, answered seriously and sublimely: "I promise you this! I swear!

We locked ourselves up again. Now I was disturbed by the anxiety that every day in Pearl Street, behind the wooden shutters and the oak doors on the hook, trouble and a speedy judgment might come. To get into the pocket of the treasury, to rob shopkeepers, innkeepers, keepers of brothels, to encroach on the main shrine - the dollar, putting its caricature into circulation, this meant arousing unremitting fury. Who am I to local justice? A fugitive, a bankrupt farmer, a precocious engineer, after a short course at Philadelphia College? Nizhinsky is a citizen, and Balashov is a citizen, who in two years will be called to the ballot box, only Nadia and I are disenfranchised immigrants, we can be returned to Castle Garden.

Napoleon laid sheets, by morning we could finish printing and bind part of the print run, without being upset by the rough, flawed seal. There are books that should be born in sackcloth, books - foundlings and conspirators, rejected from birth, not waiting for brocade or morocco. It seemed to me that the form of our book inadvertently, by itself, entered into a mysterious harmony with its style and thought - it could have been born like this or not been born at all.

I did not know what Saburov would do, whether he would rush to Boston, blackmail Nizhinsky, demand his rightful noble share, whether he would disturb James Bell tonight or find great profit in cooperation with the police. I would like to leave, but where? Philadelphia College gave me little, but there was also an innovation, the only thing that opened up hope for engineering work - calculations of railway bridges, embankments and track laying. The republic quickly led roads, cut into forests, erected embankments, laid sleepers, sewed them together with rails, built steam locomotives, and what she herself did not have time to carry from Europe.

Suddenly I remembered that Napoleon had two new banknotes, as clean as the conscience of a newborn Jesus! Is that what Bell said?

— Napoleon! I shouted. "Show me the money Mr. Bell gave you!" I reached into his pocket. - Where did you put them?

He rolled his eyes in dismay, pointing up at the ceiling.

- Quicker! — I rushed to the stairs.

Together we reached the closet, Napoleon took out money from under the pillow, I began to look at it at the window. The moon was breaking through the clouds, but there was not enough light, and I rushed down.

Banknotes are like banknotes: I looked at them this way and that, I looked at the light, I just didn't smell it; I have never seen the best fifty-cent notes, if they are false, then millions of other banknotes are also false. I was ready to embrace Napoleon: if Bell and Balashov had printed fake ones, they would have paid off the Negro with fake ones.

"It's all right, Napoleon," I said.

- If you need this money, Mr. Turchin, take it. He suspected that the papers had a secret meaning for me.

- Leave them to yourself. I pushed the banknotes into his hesitant hand, but suddenly changed my mind. — No, I'll give you a dollar in return.

It was a dead night. It would be necessary to let the last half-leaf get weathered and dry, so that the paint would not be smeared on the august monarch, and on the crowd of military settlers, and on the old man who screwed up a word about a goat noble tribe; I should have, but I couldn't, time was rushing. The work progressed, the mood improved, we were like two stubborn nocturnal animals: one is light, with blue eyes, the other is black, more suitable for the night. And I could not fail to notice how proportionately a large black creature, how beautiful his movements are, how much good power is poured into his members.

" Why did you choose such a name - Napoleon?" I asked.

— It's beautiful, Mr. Turchin. There was such a saint - Napoleon.

Saying the same name, we thought about different things. What was his faith?

"God is one," said the negro after a moment's thought. There are many saints, but one God.

- Many people think that they have their own god, and he is higher than others.

- But it's a sin! Napoleon was upset. We are not pagans, we are Christians.

"It is a sin to force others to believe in your god.

"If everyone believes in Jesus, sin will disappear.

"Those who chased you with dogs are also praying to Christ. Your former master, isn't he a Christian? Christian, but wanted to kill you.

Napoleon thought about it too:

He sins because he does not believe in the soon coming of Jesus. We Negroes know that our children will see him, that's how I see you, Mr. Turchin.

And those of whom you are talking, think that he will come in a thousand years, but for now you can sin. I feel sorry for them, Mr. Turchin.

He took pity on me too; Nadia and I did not make a prayer, nor a cross, like Balashov before meals, we did not go to church and prayer houses. The Negro feared that it was God who had taken the Long Island farm from us. Nothing stood in the eyes of Napoleon as high as the earth. Life was on soft ground under bare feet; a man is born on the earth, he goes to the earth.

"Ah, Mr. Turchin," he sighed, hearing about the bankruptcy on Rowland's land, "why didn't you call Napoleon? He would plow your land and throw grain into it."

"I didn't know you," I consoled his innocent conscience. "Good people hear each other, just as fish hear each other in still water." "Are you sure that the fish can hear?" "They are God's creatures, why would Jesus punish them?! First of all, there was earth, and only after many years did God allow us to live on it. And if he banishes people for their sins, the earth will remain."

By lunchtime, the books had been sewn together, cut by hand, and stacked on the table. A letter arrived at the post office from Nadia, ahead of her arrival by one day - the next day she was returning to New York. I bought rich bread and sugar, having spent one of the banknotes; My nightly fears were in vain. Another night has passed. Driving up in a cab to Pearl Street, I laughingly told Nadia about Saburov prowling around the printing press and showed her the second banknote.

We drove up to the house, and the police were already in charge. They were waiting for us: two policemen on the porch, two more with the commissioner among the machine tools and typesetting cash desks. "La Chólera" was thrown to the floor, dirty boots trod on it. Handcuffed, beaten, Napoleon met us with a sorrowful look; he didn't understand what had happened. The counterfeit banknote was found in my coat pocket, not in my purse, not together with other, holy dollars, but separately, like a thieves' shell, as they keep poison separate from bread.

We were taken away: Napoleon in a covered police cart, we were allowed to ride in the same cab that brought us from the station; I had to make room and give a place to the policeman.

Bell and Balashov were already languishing in the police station - Nizhinsky, who was captured in Boston with a supply of counterfeit banknotes, was, as it turned out, released on bail and was looking for a lawyer. Hops descended from Balashov. When arrested, he resisted, and his appearance was terrible: his clothes were torn, his face was bruised, one eye did not show at all from the tumor. I coolly waited for the accusation and worried more about Napoleon; Until now, the fake papers of Napoleon have not been opened because no one has asked about them. Now the truth will be beaten out of him, the light hand irons will be replaced with shackles, and his throat will be squeezed with an iron collar. More than a year has passed since the Supreme Court of the Republic delivered its verdict in the case of the fugitive Negro Dred Scott, declaring to the whole country that the Negro, the slave, the slave, is the property of the master as much as cotton bales, buffaloes, a sack of maize, a musket, a cart wheel or a kitchen utensil. And the property is subject to return; if this rule is neglected, chaos will enter the republic, the poor will take away from the rich, the foundation of justice will collapse. The new law, like a sword, stood over the whole republic; hosts of lodging-houses, tavern-keepers, knights of dens and backyards, shopkeepers and solicitors, thousands of people from the stinking, festering city pits of the North, rummaged around with unsatisfied eyes; the hunter risked nothing, and the reward was high.

Bell also tried to get out on bail, but this was not the first time he had dealt with justice, and you can't pay off the police with fake tickets. Bell often demanded a commissar to the cell, insisted on a small deposit - he was possessed by fear, in his eyes there was flight, the leaving road, the steps of the wagons, the flash of horse legs - he breathed as if he was already running. When Bell was convinced that he could not get out on bail, he again asked the policeman, saying that he had an important statement to make, and was taken away from Balashov and me. Soon they led us too, the commissioner announced that James Bell confessed to complicity: according to him, I engraved and printed the money, but I forced, by blackmail and threats, complicity with Balashov and him, and at the last minute before Nizhinsky's departure for Boston, I put it in suitcase owner of a pack of counterfeit banknotes. The scoundrel agreed on everything with Nizhinsky; with the help of an expensive lawyer, they will sanctify any lie. Such a turn gave hope to Balashov, Bell extended his hand to him too.

It became quiet in the commissar's office, and in this silence Nadia's surprised voice was heard:

"How can you bear it, Balashov? Why should an innocent person suffer... God is with you, but it's terrible..."

The commissar mistook Nadia's Russian speech for a conspiracy of thieves, shouted at her, but to the Russian ear in the office Nadia's voice still beat and died away. Balashov raised his surviving eye to her.

"I will reveal the truth," he said.

Balashov was indomitable in drunkenness, but no less in repentance. When it entered his heart, he could even give the judges the evidence that directly led him to the chopping block. Nadia had long ago become his household judge, the red corner of his reckless existence, Russia in the form of a woman settled next to him by the will of providence. Now this woman was looking for his help.

Bell stood his ground, even when the police, at the behest of Balashov, found in the printing house, in a cache, plates of three banknotes of different denominations cut by Bell, and in Bell's woodshed a bag with counterfeit money. It was impossible to stop Balashov: in a purifying impulse, he also killed Napoleon, declaring him a fugitive.

When the commissar let Nadia and me go free, without bothering to apologize, we saw the unfortunate Napoleon through the glass. In shackles and an iron collar, he was kicked to the cart and thrown on dirty boards. The light has gone out for us, we have not even experienced the joy of freedom.

Pearl Street was run by Nijinska, a mournful, taciturn hypocrite, who had never paid a visit to this house before. Nijinska burned the papers, the last copies of La Chólera burned out in the oven, the crimson pages writhing in the flames, disintegrating into ashes. Even Nadia could not feel the loss as fully as I did: I gave life to her pages, they already existed, lived, appealed to people, they were no longer a sketch, not an attempt, but a daring, rough-looking organism of a book, and suddenly they turned into dust.

We packed up our belongings and left Nizhinsky's establishment; the hostess was raking the ashes out of the stove and did not honour us with a farewell. We had money left over for bread and for tickets to Mattoon Station in Illinois; my colleague at Philadelphia College passed through Nadia a letter of recommendation to the engineering department of the Illinois Central Railroad. We hadn't even taken ten steps along the Christmas, sunny Pearl Street, when a magnificent cab overtook us and Saburov's loud greeting rolled out of it:

— Where are you going to, gentlemen? Welcome to me! Please visit me!

He must have been guarding, looking out like a hunter for game, and, noticing us on the porch, poked the driver in the shoulder: "Catch up!" He is wearing the uniform of a lieutenant, the blue cloth made his figure look younger, the army coat is thrown at the back, carelessly, with a claim to the romantic Caucasus.

- Ivan Vasilyevich! he called. – Agree, and the colonel's uniform is yours. He pushed his overcoat wider. - These scoundrels offended me, did not give me the next rank:

I asked the captain, if not the major.

"You should have demanded a general right away," said Nadia. "Humility is not the price here.

Saburov laughed benevolently:

Let's go, Turchaninov! I guarantee you wealth!

"Get away! - I hit the nearest horse with my palm so that passers-by turned around. So I hoped that I would not see Saburov again."

Chapter Nine

From a letter from N. Vladimirov to his father
dated January 17, 1901.

"I hasten to write about an important event, until the details are gone from my memory. Pay attention to the new address: it's on the envelope. The widow Fergus found me a cheap apartment with a table a hundred paces from the bookshop. We see each other often; her daughter Virginia willingly fills the position of a messenger between our houses. Yesterday she came running in the middle of the day and mysteriously demanded me to see Mrs. Fergus. In a bookstore, I was introduced to three Yankees: one of them was Senator Foraker of Illinois, another was former Senator Mason, and the third was an agricultural machine manufacturer, Mr. Johnston, small, with a cast silver head and a strong smell of wine. All three are friends of the general. Senators served as officers in his brigade, Johnston as a simple volunteer.

"My friend! Widow Fergus told me. "We won the war!" It turns out that they have been fighting for the general's pension for many years. Sometimes the Congress retreated under the onslaught of veterans, but the War Department intervened, and everything went into a quagmire. In October 1898, friends forced the general himself to write to Congress about granting him a disability pension. A year and a half later, Congress passed a private resolution paying the general fifty dollars a month, finally equating him with the sweepers of the Chicago streets. But even these dollars were lost in the passages and passages of the War Department. And then the last fortress on the Potomac fell: the century passed the milestone, we entered a new year, 1901, and bread came into the hands of Turchin. How long can the general chew his stale bread!

"How I would like to go with you to the old man!" exclaimed Johnston, but Mason cooled him off: "It's impossible, George!" ! "The general won't take handouts!" "You know what to do, Horace," said Foraker embarrassingly. "I have everything prepared, but it's terrible, Senator Foraker, yes, and Mr. Mason, and you, Johnston, I want you to remember that this is godless ..." - "Take a sin on your soul, Horace!" - "What do you to my soul! What do you care that George Fergus is waiting for his wife in heaven, and she will go to hell?!" Horace Fergus put a decanter of cherry brandy on the table, and we raised our glasses.

Virginia came with us. She rushed to the general, hugged him, and he patted her on the cheek, not noticing us. Then he sat the women on the bed, me on a chair, and suddenly compulsion entered the room. In Turchin, there was a strained expectation of something, an old man's unwillingness for news and annoyance that the pen is cooling down at a torn line; "I see, dear general, you are unhappy! - the widow walked right through. - If decency allowed, you would have pushed us out ... And I brought an angel to you, Virgie ... "

" Where did you get that from? Turchin, taken aback, muttered. "On the contrary... such a joy, such an honour..." "Write, write!" the dragonfly sought its goal, it had to knead its stubbornness in its fingers. - You cannot wait - and write, and we will sit, and admire you..." "How can you! Turchin refused, tilting his heavy head to one side. - Is it so decent, Nikolai Mikhailovich; guests to me, and I write? "He sought support from me against his own temptation to listen to the widow."

Look what happened to the general when Virgie broke the news to him! He got up, leaning his hand on the table; at that moment we were for him accomplices of his enemies and detractors. "Mr. Turchin," the widow began, "I understand your feelings, but this payment is from the people, not from the rulers..." — "The people trusted bad people too much; and if the people are a slave, they have no money of their own." "But you asked for a pension. You wrote to Congress..." "This is my last mistake, Horace. I was forced by gentlemen senators! Mason and Foraker! Foraker and Mason! He called out their names, as if calling them to a court of honour. "They believed in the nobility of the enemy even during the war..." "Silence! The widow dared to interrupt the old man. — You yourself presented them for production. Don't go too far, Mr. Turchin!" "Thank you, Horace," he said after a moment's thought. You can't be unfair even in old age. Especially in old age," he amended. Yes, I asked for a pension. But in the petition, I warned that I would not wait more than six months! Years have passed. I am free!"

Sighing, the widow put a piece of paper on the table, telling Turchin that he had signed this letter of renunciation of the pension; signed, as I later found out, a simple power of attorney addressed to Mrs. Fergus.

Virginia and her mother left, the general left me. He easily dismissed the idea of state alms from his mind, and his story about his acquaintance with Lincoln convinced me that Turchin could not have done otherwise. I will write down this story and send it to you, but in the meantime, I am sending an excerpt from the general's pamphlet "On the Republic" in the edition of 1865. Right from the station, I hurried to the Fergus's, asked me to show me the paper signed by the general and demanded that I not be deceived ... The good widow has a very practical mind, she did not understand me for a long time, but Virginia understood. Youth saw meaning in such concepts as dignity and honour. The widow had to yield; she promised that she would not give a move to the paper unless she received Turchin's consent. And this is impossible."

THE FIRST INTER-CHAPTER IS ABOUT THE REPUBLIC [12]

A republic embodying the idea of self-government can have only a representative form of government and no other; as the government of the people, it may justly be considered the best government possible. However, the theories formed by great people are destined to be tested in practice, applied practically by ordinary people.

And the most ideal principles, when they are followed formally, when they are only mechanically hardened, turn into mediocre measures for correction and improvement. The broad path to freedom and emancipation, measured and marked on the map by the founders of the republic, branches into a thousand paths of narrow and selfish interests by those who follow the founders; the people who elect their leaders believe in them, follow them blindly for a long time, and only then notice that they are lost and turn back. But what's done is done, you can't fix a mistake, and the devil, happy, laughs.

The Declaration of Independence demanded the granting of human rights to Americans, and a republic was born on the basis of the drafted constitution. Separated from the Old World by the wide expanses of the Atlantic, it did not have strong neighbors with their age-old monarchical prejudices, so that no one prevented it from developing as it wanted, no one threatened to invade its borders in order to strangle the newborn people. And the monarchies, after their surprise at the appearance of this miracle passed, seeing that this republic was a weak and harmless creature, soon forgot about its existence, and the United States of America was left to its own devices. The policy of strict neutrality adopted by the government was deliberately designed to isolate the federation from other powers so that it could quietly build up its strength. For three generations born under free institutions, the country was inhabited by free people who knew about the monarchies only by hearsay; The country's population has grown from two and a half million to thirty million people, and our people have vast territory and inexhaustible wealth at their disposal.

But the isolation, which was so conducive to the formation of a free people and the development of the vast material resources of the country, also favored the emergence of such a vice as political demagoguery. The people who made up our constitution belonged to the English civilization of the last century; they were statesmen and great men; but the following generations, having created magnificent technology, almost did not put forward other equally great statesmen. And those few who appeared, not having the slightest interest in foreign affairs or diplomacy, turned all their energy to domestic political affairs, to problems that were not yet clearly, but brewing in the country. They connected these problems and created political parties, each of which had its own views on methods and solutions. Crowds of supporters joined the opposing camps, and an internal war began between them - a war between politicians.

Politics became a complex science that had to be diligently and tirelessly studied. Politicians became great masters of "pulling the strings" and weaving intrigues, and business or working people hardly understood anything in this art. Political science taught: in order to achieve some important advantages for a given party, all means are good; freedom of speech and freedom of the press, which great-grandfathers intended for lofty and noble purposes, great-grandchildren began to use for dirty purposes. There were no boundaries for the rampant demagogy; people were forced to vote this way and not otherwise by all honest and dishonest means, not disdaining either bribery or fraud. Shameless politicians began to subdue the obscure and unenlightened masses of immigrants to their control in order to use them for their own party purposes; the feeling of patriotism gave way to devotion to one party or another, political opponents, imbued with increasing hatred for each other, became real enemies, ready to fight for their interests not for life, but for death. In public places, at meetings, direct attacks on freedom of speech were committed; ugly skirmishes between representatives of the great people began to occur more and more often in the halls of the Congress; demagogy grew into betrayal, and betrayal became openly arrogant and was committed openly everywhere - at private gatherings and in the halls of Congress; and while the politicians bickered and fought among themselves, the people rushed headlong down the river of life, without a rudder and without sails, met and overcame the thresholds, and no one tried to stop them, to warn them of the abyss that opened before them ...

Abraham Lincoln campaign button for the US Presidential election.

Chapter Ten

Mr. Turchin, they have come to you!" I was awakened by the voice of the messenger Thomas.

Three people were waiting on the porch, with the air of conspirators after a sleepless night. One glance at the editor of the Mattoon newspaper, Teddy Dawson, at the post office director, and Mr. Hansom, the station master who ran the branch of the young Republican Party in Mattoon, I understood why my only employee, the son of the mistress of the house, half of which was bought out by the company, woke me so hastily. Illinois Central for the office: one sight of three famous citizens of the city made the young man tremble.

"We need you John! Damn necessary!" the editor announced, shaking my hand with damned vigor.

The other two were silent; their silence called to me more than Dawson's words.

I buttoned up my jacket: September was mild, but for now the night wind from the Great Lakes and the plains of Canada cooled us. The wind carried the iron shriek of engines to the porch, the smell of working smoke, oils, pine logs from neighboring Iowa and Missouri, paints McCormick generously laid on new threshers and reapers.

— Mattoon needs you today, Mr. Turchin," the director of the post office also confirmed.

"Oh, Teddy," I whispered, afraid to wake Nadia. "How I dreamed that the whole city would need me!"

— Today Mattoon will become a historical arena. You are the first one we came to!

I humbly bowed my head and said:

"But I didn't sleep, Dawson. I was dragged home at night by a freight train, riding on the Missouri pines.

Teddy and I were friends: I cut vignettes and caricatures for him for a pittance, the Mattoon Courier is still poor and young. Teddy did not yet earn money, but lived on his own, he also printed Nadia, her notes on the women's movement, signed N. T.

— No resignations, soldier! exclaimed Teddy. "Freedom does not take resignations.

The stationmaster, a short man with big greenish eyes that sparkled, looked at me without respect. We served the same company; Hansom commanded dozens of employees, from machinists and conductors to sweepers, paid me a salary as well, and could not in any way persuade me to join the Republican Party.

"Freedom! Yesterday I visited its ruins, - I cooled the ardor of editor Dawson. "Do you know where your friend Edwin Ramage is now, Teddy?"

— Behind the editor's office.

"In the hospital, Dawson. Unconscious, in the hospital. The printing presses were smashed, the typefaces were trampled into the ground, the printing house was burning down when I arrived in the city...

Ramage graduated from Harvard the same year as Dawson, returned to Missouri and began printing the Herald of Freedom in a town across the Mississippi, fifty miles from Illinois. Ramage printed leaflets, speeches by the Illinois lawyer Lincoln, he agreed to publish my play on the ideas of freedom and personal independence; I hurried to him with a list ready for typing, and ended up at a funeral feast, in the ashes of an unheated conflagration. I told the early guests how hordes of Missourians had gathered outside the newsroom; foot towns-people and horse-drawn guardsmen, galloping almost from Arkansas, from the banks of the White River, how they kicked Ramage with their boots, how, having been mad, they smashed a bookshop, burned papers and books.

The Mattoonians bared their heads.

During the night word came to Mattoon that the next day—and tomorrow is already here—both presidential candidates, Stephen A. Douglas and Abraham Lincoln, would be coming to town. Their duel would take place in Mattoon Grove, and the Republicans wanted to decorate the stage with a picture in favor of Lincoln. So, I was needed, my brush, my ability to convey serious objects in a caustic way.

"The passions are irritated," I remarked. "I hope I don't suffer the same fate as Ramage.

"You are all maneuvering, Turchin," Hansom pricked me. "You yourself will have to vote this time, you are now a citizen of the States.

"If anyone's tacking, it's Abie," I protested. "I listened to Lincoln and studied his Political Debates with Douglas.

- What did you find bad there, mister engineer?

"He's too much of a lawyer, lawyer's hooks just get in your eyes.

- Enough for me! Fed up with! Hansom backed away angrily, but Dawson grabbed him by the cloak.

"Don't you know Reverend Turchin," he intervened. "If Jesus himself had appeared in Mattoon, he would have found flaws in him too.

"Lincoln's face is well known to me," I gave up, "but how do I draw Douglas?

- He's short! exclaimed the messenger Thomas. "Mr. Douglas is even smaller than Mr. Hansom.

"I hope," Hansom said in an icy voice, "that we do not stoop to playing on the candidates' physical defects."

- Draw it from the back of the head! it dawned on Teddy.

More than others, Thomas discovered quick wit: only a republic could make teenagers mature politicians.

Stephen Douglas is a drunk, he must have a red nose. He smokes big cigars...

"Abie smokes cigars too," Hansom said.

"But when a big cigar is smoked by a small person, it's ridiculous, Mr. Hansom. And Douglas always rides in his own car, which our road gave him.

"Don't draw a wagon, either," Hansom grumbled.

- You can draw a toy, and let Douglas drag him along on a string! found a young man.

- And on the car, I will paint four letters - I. C. R. R. [13], - I lit up.

- Mistakes are possible in politics, Turchin. Douglas was presented with a wagon when our party was just born. However, - said Hansom, you will not understand this, you were brought up by despotism.

Once again Dawson stopped us, separating us with an outstretched hand, like a striped barrier of Illinois Central. He said that it would be nice to draw Lincoln with a Bible in one hand and a woodcutter's ax in the other, so that everyone would remember that he came to them from the wilderness, appeared to the voice of truth and God.

Teasing Hansom, I said your Abie, but I could have said ours. I traveled a lot across the state, to Chicago and Springfield, to stations that didn't exist yet; only piles of logs, yellow bricks and tents. I met Lincoln at the headquarters of Illinois Central and on the way - Lincoln rushed through the Illinois lands: either in a wagon with nosy businessmen of his headquarters, then in a carriage, then in a wagon harnessed by a quadruple, into a quarry, as if he had escaped from the wilds of the forest and hastily changed the dress of a lumberjack on an alpaca coat and waistcoat.

I am attracted by people who have reached everything with their mind, bypassing academic colleges, who have not sinned even with one book read out of politeness. They will read a hundred books in their lifetime, but precisely those where the wisdom of the world is. Everything in them is one, everything melted like in a crucible, and there is no need that you are ugly, thin and your legs are crooked, and your cheeks are sunken, that your dress hangs on you like a scarecrow, and thin brushes are torn from the cuffs and your head is disheveled, like my messenger Thomas; you spoke - and the crowd is yours, and no one knows what everyone gets from you, even their native grammar. He was reputed to be a pagan, but he knew the Bible as well as any churchman of the New World, Abie's long fingers blindly searched for any projectile on its pages and launched, as from a sling, at the enemy. He hated slavery, but publicly did not encroach even on the vile law on fugitive slaves. He was a child of young America, and although he had the nickname old Abie, although he was over fifty and the years had worked hard on his face, working roughly, sometimes with a chisel, and sometimes with an ax, he seemed to me young and passionate.

Now, having nailed a wooden, papered shield to an old elm tree, I drew Lincoln.

The sun was shining on my back. Working with a carpenter's pencil and brush, I was convinced of how great Lincoln's fame was and how passions were divided. "Mr. Turchin! called the owner of the Mattoon hotel. "You are painting a holy seer, and Abie is a descendant of a gorilla!" Draw a tail too! From the crowd I was advised to give Lincoln an umbrella in his hands, extend half a foot of the top hat, which I reached when climbing onto a chair, draw glasses, let them see that Abie is a learned man, add hair to his bushy eyebrows, paint a watch chain with gold trinkets and put on Abby is wearing French gloves. I drew the Egyptian pyramids in the background, with Abie striding toward them, peering into the distance from under his arm. I sketched Douglas from the back, and from a slight turn of the head in the direction of Abraham, one could conclude what malice consumed the little giant, as the newspapers called him. I moved away from the shield to look at it with the eyes of the future crowd in the grove, and many of my viewers called me to retreat even further, to the fence, advised me to put on a hat so that the sun would not burn my bald head, invited me to dine and volunteered to run to a restaurant for roast beef or roast beef with beans. I was part of this crowd, a Mattoonite, a citizen of the States, who yesterday was not needed by anyone, except for Saburov, today is in great demand. In New York and Philadelphia, I was the observer of the country, here I joined the life. The Mattoonians believed that their city would outdo Chicago and St. Louis: who would throw a stone at them for their holy faith?! In New York, neither the mayor nor the chief bankers of the city could account for the daily addition of wealth and sins; in Mattoon, messenger Thomas could turn in a full report by the evening, without making a mistake of a hundred dollars. Like two arteries, the railroads throbbed as they intersected in Mattoon—Illinois Central and Alton R.R. room, with purses and pistols under the pillow, from post offices and business offices, from hastily built houses and mansions in which every stone breathed security. The moneyed and active Northwest found itself in Mattoon, at the crossroads of the railroads, in constant mobility and working muscle contraction. Russians here - not a soul; in Mattoon and did not understand: why are we Russians? "Most of the Mattoonians are unable to point our land on the map.

Our English, with St. Petersburg evenness, managed to get better, pick up Yankee jargon, the smell of beans and corned beef, and we gradually became Mattoonians. Wonderful, unmeasured, but mattoonsy. I drew and drew, drew and drew, as if I had just been born. I succeeded in portraits both from memory, and from daguerreotypes, and with the most cursory posing: better than others, I was given faces captured on the fly, in a tavern, in a coffee shop, by a pile of tarred sleepers; but I gave too much to learn how to sell. And can it be learned? The longer I live, the more I am convinced that this talent is innate.

I fed on drawings; performed them quickly and without errors. The Mattoonites came to me for various needs: to draw a plan of possession for the execution of a bill of sale, to sketch a drawing of a future house in stone or from logs, a wooden bridge over a ravine or a small water-driven mill. I was provided from need, I had time for writing and for the violin; the owner of the hotel and restaurant once admitted that he wanted to hire me to entertain the public, yes, my music was painfully sad. "The soul is exhausting," he admitted. "I drink from your music..." "Well," I replied, "it's not bad for you; the public will drink more, and the income will rise." He shook his head: "I am a meek person, and a brawler, what good, will smash the tables. If only you were merry..." "It's impossible," Nadia explained to him. - The violin was made by the Italian Gasparo da Salo two hundred years ago, when his wife died. So, he made a violin on which you can't play fun, it will immediately fall apart ... "He looked sideways at the demonic product of the Italian, I played a dance after him, and he kept looking around to see if the violin and our house would survive. The movements of the last war took away from me more than ten Mattoon portraits of Nadia - my treasure, priceless sheets, in which I, an amateur, seem to have risen to art, expressed Russian nature in the movement of time, in a turning point, in the emergence of a new character. On my sheets, she was Nadine Turchin, and at the same time the alluring mystery of Russia, and the image of motherhood, but with longing and bitterness, because motherhood did not leave her, but remained in her like flour; and in the look, in the turn of the head, there was already the woman of the New World, ready for action, ready to vote on a par with the man.

Nadia healed the Mattoonites; women went to her, called her to children, and to the peoples, and everything that the Mattoon people brought in payment for her work, they took away with them. It seemed to her that the residents were doing her a favor, trusting her, giving her practice, choosing her over the old doctor, the easy-going Yankee. He was not angry with Nadia: in the active Mattoon, there was enough work for everyone, except for the undertakers.

Summer has come, and with it the day that changed our lives. Nadia received letters from New York and Chicago; Lucy Stone's friends wrote, invited her to publish, sent an issue of "Revolution" with Nadia's play about the possibilities of women in medicine. And, to top it off, George came to Mattoon. B. McClellan, chief engineer and vice president of Illinois Central.

That day, I brought Hansom the plans for a tower to supply water to the locomotives. Hansom decided to brag to McClellan about the drawing sheets and the people he had contrived to find in Mattoon. I was invited into a room, where a short, graceful man paced about, his hands hidden behind his back, under the folds of an expensive frock coat. His face seemed familiar to me, the blond beard embracing a feminine chin, intelligent and anxious eyes, an expression of polite interest in them, and meanwhile also of suffering boredom. He became interested in the drawing of the tower, and Hansom took my other sheets from the closet, showing off the hefty work of the Russian. At the word "Russian" McClellan looked at me with a long, searching look and held out his hand to me:

"Couldn't I have seen you before?" In Chicago, for example?

"No," I replied. "I have a perfect memory for faces. Maybe in Paris or London?

— I was in Crimea during the war. If you were in the military...

I didn't let him finish.

— Captain McClellan!

Here it is, the strength of native tabernacles, handfuls of her father's land, her sky, the voice of her bird! What would it be like for my heart to meet my own mother now, or at least the bottom artilleryman, with whom we were dragging a cannon, if the sight of a strange captain, whom chance brought to Sevastopol, made my heart pound so against the ribs! We sat down on chairs as equals - however, memories brought us back to times when we were not equal: I am a colonel of the Guards, he is a captain of cavalry from the North American States, at the head of a military mission, a quiet, courteous man, as if embarrassed that behind his graceful figure is a huge, young country. McClellan climbed the length and breadth of the trenches of the allied troops near Sevastopol, tasted the Russian bread - he was a free bird over a foreign battlefield, without a suffering soul and without true sympathy.

I did not like McClellan in his uniform—everything in this officer was weighed, and weighed not all at once, but long ago and forever, as if life were so negligibly poor that one could foresee everything in advance. But as vice director of Illinois Central, he was simple and friendly and gave the impression of complete decency.

"I remember that you were preoccupied with something in the Crimea," he showed the strength of his memory. – "Somehow very sharp."

We were losing the war, it doesn't happen every year.

- Yes! he nodded sympathetically. — Sad, but also heroic days. Everything is in front of your eyes.

I preferred to laugh it off: then I was still a tumbleweed, a disenfranchised immigrant at the Mattoon crossroads.

- I did not know if the bride would wait for me, would she marry another?

I was about to take my leave so that the former colonel of the guards would not seek protection from the retired captain, but, apparently, this man was kinder and simpler than me. McClellan offered me a job as an engineer and began seeing me occasionally. He honestly served the company, but his ambition was flattered by descriptions of European battles. McClellan sent me as a gift a military memoir published in Stuttgart in German, some recent American pamphlets, and then asked my opinion on the manuscript, which he was preparing for publication in Philadelphia. It was his subsequently known volume of 'The Army of Europe'. The book was published in 1861 and served McClellan a bitter service - the White House decided: 'if McClellan knows other people's famous armies so well and criticizes them, maybe he can handle ours, the federal one.'

When I returned the manuscript to him, I remarked:

"You often admire fortresses simply as majestic bulks, without considering their strategic importance or tactical uselessness."

- Is it? McClellan was worried.

"I will take our Russian fortress Novogeorgievskaya. This is in Poland, at the confluence of the Vistula and Narew.

Ah, Maudlin! Modlin is an exemplary fortress, of which there are few in Russia.

"Fortunately, not much," I said. – "It was built by Napoleon against Russia, it is extremely beneficial for anyone who would go to war against Russia."

— A very unexpected thought, Mr. Turchin.

"The bridgehead on the left bank of the Vistula is open from the flank, and the bridge across the Vistula too. The Russian army cannot hold out even for a day against powerful enemy batteries; and if the bridgehead falls, then the value of the fortress will be reduced to zero." He remained silent with the inflexibility of a pedant.

"How long were you in the fortress?" I asked. - More than a day?

The beautiful, slightly languid eyes of McClellan seem to have seen the wide Vistula, which takes in the Narew, heavy walls, and all this is pleasant to him, as a traveler who has been where his compatriots have never set foot.

"So, you have a strong stomach, Mr. McClellan!" If you had to go out of need in Modlin, you would have noticed that this fortress, or rather a huge fortified camp for many thousands of soldiers, did not have a sewerage system, although it stands on two rivers. A Russian soldier would have died of disease in Modlin if he could have stayed in it for a long time.

I offended his delicacy, he restrained himself. I broke his toy - he thanked me but did not think to correct the manuscript. I broke a lot of his toys - he endured. He was offended only once, when I said that in his book the soul of an engineer, historian and statistician, but no warm heart.

"In your opinion," he said, "the heart is only in being with Russia?!" "The heart of a Republican is with the people." "The people are dark," McClellan waved him off. "You see how much hostility is around, how close we have come to rebellion. John Brown wanted freedom for slaves, but he took the life of free people. "He gave his life too!" "Is that an excuse? Would sacrificing one's own life justify killing someone else's? If a conflict starts and we encourage the slave, give him a musket in his hands, he will destroy the South, without analyzing the right and the wrong.

I will say to McClellan's credit that my criticism did not quarrel us: in the summer of 1860 he started talking about my appointment as a topographical engineer in Chicago. We decided to change. Life in Chicago promised more interest, the hope flashed that Nadia would find a field for her studies in literature. Now I was returning my last debt to Mattoon - in front of his citizens, from oblivion, from the white depths of the paper, I brought to life the images of two candidates for the presidency. Lincoln looked victoriously from under his arm at the distant pyramids, Douglas hurried after him. The choice of Lincoln threatened war, I understood this as well as others; Douglas' choice aggravated slavery, took the essence out of the very concept of a republic, for a republic with the addition of a slaveholding is mockery and disgrace.

Chapter Eleven

Douglas was the first to arrive in Mattoon. A locomotive a few miles from Mattoon trumpeted triumphantly, calling his supporters. Members of Douglas's staff and himself, with a top hat raised in his hand, descended into the crowd to the sounds of a brass band. Douglas limped as he reminded voters of his service in the Mexican War; he had the right features and looked like a noble person, so that even when I saw him, I would prefer to draw Douglas from the back.

Soon Lincoln's staff also arrived. They jumped onto Mattoon land like corsairs on the deck of a boarded brig. They were too similar to the people of Douglas's headquarters, so similar that if I didn't know one of them by sight - the emigrant Gustav Kerner - I would have decided that they were all recruited by one arshin *('Arshin' is a Russian unit of length equal to 28 inches.)*

It was getting dark. The Mattoonians were so close together that it would have been easy for a band of thieves to rob the city, but it seems that the god of America kept her children during the hours of political rites. Two brass bands rattled, intersecting, the bells of the Mattoon church rang, the locomotives toiled: two railway companies deafened each other with iron throats. Pyrotechnicians lit up the sky above the grove, firecrackers fired, in the flame of torches. My drawing, nailed to an oak tree and for the time being covered with a cloth, turned white. Teddy Dawson was languishing, worried about Lincoln's lateness, and there came a moment when the crowd murmured, but then the sky brightened with lights, and a wagon pulled by six well-fed white horses burst into the grove. A man with an umbrella in his hand, in a tall top hat, which he had to hold with his hand, jumped out of the van, settling on crooked legs.

Lincoln and Douglas stepped onto the platform, and Thomas pulled the canvas from the shield. The crowd saw two Lincolns and two Douglas's at once: the light of the torches fluttered, and the drawing came to life in their flame. Lincoln hung his umbrella on an oak branch and, taking off his top hat, was looking for something in it, as if he were rummaging through a traveling knapsack, and, not finding it, he raised his hand and spoke, stretching his long neck out of his snow-white collar.

- You can't retell everything, and there is no need; The Yankees picked up every word of Abie, any of his jokes, whoever needs it, will find it. Wherever Abraham came with a speech, Douglas appeared there, and the performance began, and just give Lincoln a sharper injection. So then in Mattoon, Douglas pointed to the drawing and shouted:

- It would be nice for Mr. Lincoln not to draw pyramids, but to draw a counter: he was a grocer and sold whiskey!

"It is true, friends," Abraham said, anticipating a brawl. "The difference between me and Douglas is small: I stood on one side of the bar, selling whiskey, and he whipped it on the other side.

Douglas, next to Abraham, was a stern, restrictive, and fearsome god. With great skill, he showed how easy it is to lose what has already been acquired, acquired, put into a warehouse or bank account, but showed little ways to accumulate new wealth. He called on God more often than Lincoln, and this is a strong helper in the States - where there is sin, there is repentance - but not as strong as profit, a thirst for movement to the West, a passion for the circulation of money. And old Abie, screwing up his eye under the cornice of his shaggy eyebrow, dropping his frock coat into someone's hands, Abie, with the rough face of a woodcutter, not softened by years of advocacy, called the Mattoonians to activity.

He promised them movement and hope, advocated the construction of a railway to the Pacific Ocean, promised to give government money for the construction - part, but most of it - along with the railway, new lands would fall under the plow, they would be given away for free, for the mere labour of ploughing and building houses. He fascinated Illinois farmers with talks about the steam plow, about the benefits of fertilizer, deep plowing, and seed selection; and, looking at him, people thought that Abraham was a born farmer and his stoop, and his crooked legs, and his tanned, wrinkled face - from long years of walking behind a plow. He was indignant that many positions in the country, from ministers to tax agents and solicitors, were in unworthy hands, that they were sold and became an instrument of bribery; and this, of course, meant that with the election of Abraham, the posts would be vacant and put up for a new auction. Lincoln raised the importance of the free man, saying that he could not remain a wage labourer forever. Labour, he proclaimed, stands above capital, because labour goes ahead of capital, and if there were no labour, there would be no capital. He spoke to everyone, everyone heard him together with the crowd, but also separately, everyone had the feeling that he understood Abraham better than his neighbor, that he wanted to exchange a word with Abie, and if he weren't here, the lumberjack deputy could and pass Mattoon.

Nadia listened to Lincoln with her head thrown back, her eyes fixed on the movable Adam's apple, menacing eyebrows over kind eyes and large, mocking mouth. There was something in her look from surprise at the meeting with the antipode and from the fact that something close was hidden behind the complete dissimilarity of the appearance. I still remember the moment when I noticed how Nadia was listening to Abraham: a feeling of envy, stupid orphan hood, a buffoonish desire to remove Abie and speak from the stage himself - that's what rose in me. When Douglas spoke and Abraham stood with his long arms clasped on his chest, Nadia looked not at the speaker, but at him, how he twisted his neck in a wide collar, how he wiped his face with a large handkerchief, how he screwed up his eyes and smoothed his hair with his fingers, how he looked at his watch, pulling them on a chain from a vest pocket.

In my stupid jealous bitterness, I didn't immediately hear Douglas talking about the slaves—until now the candidates had been firing with small guns, now it was going to explode—Douglas's voice took on a snarl. He talked about Negroes, but every other stranger, even with a white skin, born of an Irish or Scottish woman, had to tremble and remember that America does not belong to him completely, that this land has true owners. Stephen Douglas flattered the local Yankee, insisted on his pre-emptive right to own land, wealth and the future of America. We are free people, he flattered them, we are equal to each other before God and before the law, but a slave is not equal to us, property cannot be equal to his master. He boasted about how many Germans and Irish people America had given shelter to, and then, with a heavy plow, he plowed a furrow of inequality between the true children of America and the newcomers.

Lincoln raised his hand and said with a grin:

- Once I asked an Irishman why he was not born in America, if he were a real American, he wouldn't even have to spend money on a steamer. Do you know what he answered me? Abraham paused. - "I swear on my honour, that's what I wanted," said the poor fellow, "but my mother did otherwise!"

The anecdote slightly sprinkled the soil, but it was a serious matter, an angry plowman was walking behind another plow. He announced to the crowd that the Republicans were going to civil war, a slave uprising that would end in unprecedented blood: the noble gentlemen of the South would not yield to a band of traitors, and fratricide would become inevitable. Like a clever matador, he rushed around Abraham, stabbed banderillos into him, hoping that a careless word would come out of the mouth of the republican idol, and cautious people would fall in favor of Douglas. "What do you want? he demanded. "You hide the truth even from your own wives, you want chaos, ungodly freedom, the ruin of the planters!"

Douglas waited, and the crowd waited, and Abie seemed to be trying to get the crowd's alarm and apprehension. "If I knew what to do," he finally began with hidden sadness at the imperfection of the world, "I would come to you as the messenger of the Lord and resolve your doubts. But I don't know much, though more than the venerable Stephen Douglas, and that's so little. I am opposed to slavery coming into the free states; every healthy person is against infection. Douglas's friends say: Mr. Abie, you put a dirty boar, your property, in a wagon and take it to a fair in a free state, even to the White House lawn. Why should we not bring in our property, our slave? He turned things to his own disadvantage and seemed to stumble to his feet. "It is logical, absolutely logical, if there were no difference between pigs and blacks!" If the Negro was not a person for whom we are responsible - before God and conscience. What do we do? Free blacks?

"He stirred up the crowd, a sly one, a soul-catcher and a hater of slavery, he took a detour. - Equalize blacks with us? My personal feelings do not allow it. - He plowed the field and came to a dangerous line, beyond which he could be left alone. "The noble Henry Clay dreamed of returning Africa to her black children, and we dreamed about it together with him. They were taken away by force, their grandfathers and fathers were taken away, we would return the children, and they would bring the fruits of religion, civilization to Africa ...

- And the scars from the batogs! (*NB. a BATOG is a rod or stick about the thickness of a man's finger traditionally used for corporal punishment in Russia.*)

Nadia's voice was heard.

"That's right, lady," Lincoln said, "and bloody welts on their backs. But freedom heals wounds, and blacks would be free in Africa. What do we lack to give a gift to Africa? Nothing! Some two billion dollars, the slave is property, and the government must buy him ...

"If he were president," Douglas shouted, "he would use your money to ransom the slaves!"

"It is better to spend them on blacks," Lincoln continued, "than to the war, which Douglas scares you." But the South will not sell the slaves, just as we will not sell our houses and our working cattle. In the cities of the South, volumes are being printed proving what a wonderful thing slavery is, but we have never met a joker who would ask to be a slave. The curse of society is that some try to shift the burden of work from their shoulders to the shoulders of another. But if a certain Smith can prove that he has the right to enslave Jones, then why can't Jones one day prove that he can bridle Smith?!

The cold wind forced Abraham to put on his frock coat, he did it with an awkward arch, not caring that there was a crowd around. "They object to me that Smith is white, and Jones is black. Beware of this people! You can become a slave to the first person you meet who has a lighter skin than yours, especially you, Stephen Douglas, with a good Indian dressing face. The frankest supporters of slavery talk about the law of profit - it is profitable for me to make another a slave, and I do it. Beware of this too, the day will come when another will prove his strength and his advantage, and you will get slavery as your inheritance ...

Above the grove the sky darkened with rare stars, the oak forest rustled with leaves, people stepped closer to the stage, Douglas again frightened the crowd with the spectacle of wild, free slaves. Poverty and chaos will become the lot of the country. In the South you will not see a black slave with an outstretched hand as a beggar; in New York alone you will meet more beggars in a day than in the South in a hundred years.

"Douglas is right, we still have too much poverty," Lincoln agreed, "we are only willing to take on the burden of administration because we hope to do everything necessary for the people, provide for the sick and minors, build roads and bridges, make everyone respect the laws. But Douglas was imprudent to speak of slavery here, and I will answer. - He seemed to be weighed down by this necessity, but he did not find another way out. "While we think that slavery is evil, we can still afford to leave it in the position in which we received it from our grandfathers. But can we free people let it come here, to other states and territories, and overwhelm us? To this we say: no! No and no! If we do not protect freedom, it will take a little time to turn into a thing and the poor whites, each of you, as soon as you stumble or run out of money or credit. So, let us be filled with faith that the righteous do what is right, let us stand to the end and fulfill our duty!

He did not drive the Mattoonites into a narrow corner, like Douglas, the decision was postponed, and what else is so pleasant for a person busy with day labours as a delay in serious and blood-threatening matters! He was with old man Brown in denouncing slavery but admitted that the execution of the old man was accomplished according to the law. He led the crowd by the hand along the twilight corridor, everyone had the support of his hand and voice, and at the same time he felt free to choose; the path was long, and Abraham did not hide it, and kept pace first in front of passing doors on the right and left, and opened those from which light came, and propped up dangerous doors with a log and hammered them with nails.

Mattoon gave his heart to Lincoln. In the morning he had to travel in the direction opposite to Springfield, and he decided to spend the night in Mattoon, but the owner of the hotel, an opponent of the Republican Party, did not have a separate room for Lincoln. Invitations poured in from Gustav Kerner, Teddy Dawson, Hansom, and the director of the post office.

- Friends! Abraham raised his hands. "Does anyone have a bed so long that I can lie down?"

And then Mr Thomas emerged from behind Hansom:

- Mr. Lincoln, you could stay at our house: we have a huge sofa in the office, we will be very happy, my mother and I, and the engineer, and his wife. This is Mr. Turchin. He drew you on a poster. He is Russian.

Thomas laid out everything with one spirit, filling us with a price.

"We'll be glad to see you at our place," said Nadia, too.

Next to the eminent Republicans of Mattoon stood their wives, mistresses of rich houses, and a newly-made American woman jumped forward, a doctor's assistant, ready to take birth from anyone, not excluding black women. The ladies were dumbfounded, and I probably looked stupidest from surprise and annoyance at Nadia. But Lincoln held out his hand, guessing that Nadia's hand was already striving for him to get acquainted, shook hands with me and moved after Thomas.

"You have been kind to me," Lincoln told me. — Newspaper artists brutally deal with my physiognomy.

- Because, it is true that they are real artists, and I am self-made.

"It's convenient for me to draw, nature took care of it.

143

He waited for the usual objections, and I kept quiet: I saw how far nature had gone in creating a living caricature.

- However, if a person does not give a profit to a pencil, is it good? Lincoln chuckled.

We walked with the crowd, occupying the pavement and broad grassy verges. On the right, along the road, marked by a dark line of trees, several wagons passed, riders skipped with torches in their hands.

- Do you belong to the Republican Party? Lincoln asked.

He is with us, with us! said Teddy Dawson ahead of me. "There is no one in Mattoon who hates slavery more than this stubborn Russian couple.

"I'm a Republican, Mr. Lincoln, but without a party," I resisted Dawson's flattery. "Messenger Thomas and I have our own party, one for two. - Changing his foot, the boy was attached to my step. - I would oblige by the constitution any president who sits down in his chair to leave the party. - Lincoln was silent, but his face, attentive eyes asked and inquired. "To serve the people, the country, and not the narrow interests of one of the parties."

— And you think so? Lincoln asked Nadia.

"If the benefit of the White House party and the benefit of the people diverge, it is fair to favour the people."

The people from Abraham's headquarters said goodbye to us at the entrance to the office, the Republican activists took them home. Teddy Dawson did not stay with us for long, in the morning the residents of the town should receive the number of the Mattoon Courier with a report on the rally. Dawson put me to work too; at dinner I cut out a drawing for the newspaper, repeating the caricature. When we sat down at the table - with knives and forks in our hands, and I also with an engraving engraver - Abraham already knew a lot about us, that we were from the nobility, that I fought as a colonel, and came here for justice; that we were in dire need, but the case helped - a meeting with engineer McClellan. Dawson managed to whisper all this in the minutes when we were away on household chores.

While Dawson sat with us, all we could do was eat and listen; exhausted by someone else's eloquence in the grove, the editor spoke incessantly, promising Abraham a quick victory. He ate greedily, quickly, as if not fully aware of what he had eaten, and interrupted Dawson only once.

"I told you, Dawson," he said, wiping his big mouth with a napkin, "only events make presidents. Not a clique, not the businessmen of the electoral headquarters, only events. Nobody knows ahead of time what will happen. Remember last October? Sunday afternoon, America calmly goes to sleep, and on Monday each of us will know that the country is no longer the same, one impudent old man changed it all.

"Why did you admit to the crowd that his judges were right?!" It still burned Nadia.

He did not answer right away, he waited for more of her words, he waited and wanted, as they sometimes want, the sound of someone else's voice, just a voice, even without the meaning of words.

"Because John Brown fought evil by breaking the law."

- God will justify him! Dawson said. Without getting up, he put his hand to his chest and bowed to Nadia.

"Brown believed that he had not sinned against God," Abraham continued thoughtfully. "I don't know, I don't know ... Mary paid ten years ago for a place in the church, but he," Abraham raised his head to the ceiling, "doesn't favor us and takes away something dear. Mary is my wife," he explained. Sadness was in his voice. Not a complaint, not a request for sympathy, but precisely the sadness that visits strong people.

"I remember a lot of what Brown managed to say, and I'm not the only one," Abraham continued. "But over time, his words about blood came forward. Remember?"

One Dawson nodded, but without conviction.

"Here they are," Abraham turned his face to Nadia, atoning for her recent lawyer's evasiveness: "I, John Brown, now know for sure that the crimes of this sinful country can only be atoned for by blood. Now I see that I hoped in vain that this could be achieved with little bloodshed ... "Yes, gentlemen from the Order of the Iron Collars hope that a can-

145

nonball is stronger than a ballot ball. - He turned to me: - After all, the weapon has the last word?

- Never! - I was waiting for a question. "Generals are flattered to think so; one out of stupidity, the other out of the impossibility of otherwise justifying the mountains of corpses. — Dawson was no longer with us, pushing the plate, I engraved a drawing for the newspaper and felt the eyes of Abraham on me. "You can win the war and immediately lose everything in the first manifesto, in the first treacherous bill, in the first meeting of Parliament."

Something I did not give to Lincoln; he needed time to convince himself that it was conviction and not stubbornness that spoke in me.

"A weapon changes a lot," he said after a pause. "Especially in the hands of bad guys.

"Mr. Lincoln," Nadia intervened, "you hate slavery, I conclude this from many things. Why do you recognize the boundary beyond which the crime of slavery becomes legal?

"We received this border, this wall from our ancestors and we want to dismantle it stone by stone; it collapsed suddenly, it will crush too many.

"So, slavery forever?!" Nadia grieved. – "The only republic on earth, and in it - a shame, a recognized crime, which cannot be found anywhere else."

Isn't the peasant a slave in Russia? I know the privilege of Russia, the privilege of open despotism, without democratic masquerades. But isn't the Russian peasant a slave?

"A slave and a martyr, but only a higher grade," I intervened. "It's as if he's not full of cattle, at least his marriage is considered sacred by the church. The time comes, and the peasant is entrusted with weapons, allowed to load cannons, to die for his masters. There are laws in his favor, forgotten, expelled from the courts, but there are.

"And yet you are here," Abraham said. - We came to the States, live below your title and ranks.

"We left these trinkets across the ocean. On Long Island, my wife and I would have traded the colonel and the princess for a couple of strong working oxen, but there were no hunters.

"Your life is ahead of you: maybe the colonel will still be useful.

- This is over for good! I said hotly. "Only honest bread. With these hands, a hump, a plow, a drawing pencil, a violin ... - I pointed to the open case and the brown soundboard of the instrument. - Let it be alms, but not murder!

I thought that, left by his comrades in a strange house, he would soon find an excuse to go to the office, where a bed and a candle were waiting for him, but Abraham showed more and more interest and participation in us; he sat freely in his chair, now folding his arms over his chest, now lowering them along his body, so that his fingers almost reached the floor.

"People run to us out of need, out of hopelessness," he said, as if offering a possible explanation for our appearance here. Sometimes they run from the law.

- From lawlessness! Nadia said.

- Most often - from poverty. He looked at us with narrowed sly eyes. "But there are people who deliberately ask themselves the most difficult lessons: I would not be surprised if I knew that you were running from wealth.

"We didn't have big money or big sins, Mr. Lincoln. Nadia felt at ease with this lawyer from Springfield. "The Russian police are not looking for us, but we had trouble with the New York police. "And she told him about our crash on Pearl Street.

Abraham handed her a business card and said jokingly:

- Take care of it! If there is a need, I will come to your aid. I can win cases.

"But if you become president, you will shut down your law firm.

- This will not happen, Mrs. Turchin, and by the next election I will become an old man. My name is already now ... you know what? "It seemed to me that he just wanted to hear his common name from her lips.

"Old Abby!"

Against her, he was an old man, clumsy, with a wild head. Life broke his face with furrows, as I failed to break the soil of Rowland with plows. The change of mood on this physiognomy reached the sharpness of a grimace: everything was large, definitely, expressed to the end. In years, he was quite suitable for Nadia's father, but they were not a daughter and a father, but precocious friends, so busy with each other that it seems that if I disappear imperceptibly, like a frightened brownie, they will not be missed. I soon became convinced that this was not so, but in those moments, I suffered, and how beautiful was this suffering, the only one of all - saving suffering, and how poor is the person who did not experience it.

- I think you have been called that for a long time ...

— You knew it! he said seriously. "Somebody told you!"

Nadia shook her head.

- You guessed it yourself?

She did not have time to answer: the window glass shattered with a clang, the stone flashed and hit the wall. And then the second. I grabbed the lamp and lowered it down. Nadia was already hugging my head, closing me from the window. Abraham sat motionless, leaning back slightly.

"They can shoot," I said. - Move to the corner.

"You don't know the country well," Abraham replied. - Whoever decides to shoot - shoots immediately.

Outside the window, screams were heard, the barking of the master's dog, fuss and blows. I hurried out into the yard, and soon Thomas and I were back, pushing a tall guy with an idiotic smile on his freckled face in front of us.

"My hat is in there," he told Thomas, nodding toward the window. - My hat fell there, look for it, you brute ...

It's amazing how he managed to beat Thomas. Yet joy danced a jig across the messenger's bruised face.

- Mother fell asleep, and I ... walked ... - he said. — I was standing under the window. I wasn't eavesdropping, I was thinking about my own... and here he is – bang!..

"You did him well," Abraham said, although the guy was standing unharmed, only with a torn collar. "Bring the gentleman a hat, lest he burst into tears."

— I have a new hat! the guy called after Thomas. - They gave it to me today. Don't hesitate to change.

Abraham got up, frightening him with his huge size; the lamp was still on the floor, our shadows stretched out, broken on the ceiling.

Who gave it to you, mate? Abraham asked.

"Mr. Benjamin Trope," the boy boasted.

Trope's name didn't tell us anything. I replaced the lamp.

Why didn't he teach you to throw stones? I wouldn't pay a cent for this kind of work.

The guy smelled a catch, but the idiotic grin did not leave his tight, bluish face.

- Nobody gave me stones, I picked them up from the ground myself. Aha! Let's! He snatched the brown silk-banded hat from Thomas's hands and put it on. The face was hidden under the drooping brim.

"Come on, kid, take it off," Abraham ordered. "There's a lady here, and you're wearing a hat."

The night visitor said rudely:

- Just think, white mud! .. Take off your hat in front of them!

Thomas tore off his hat, threw it on the floor and began to trample.

"Let me shoot him, Mr. Turchin," he pleaded. — I'll take my father's gun.

- Let him live. Abraham drew Thomas to him by the shoulders. "If you kill him, they will kill you too, and it will be very painful for us. We'd rather take him to court.

"Damn it, you'll sue me!" Idiots are not judged. He looked askance at the hat but was afraid to bend down.

Who told you to throw a stone at our window? I asked.

- I have a disease: as soon as I see where the light is after midnight, immediately - bam! ...

"What if the pastor had a light on?" Abraham asked.

"What, I don't know the pastor's house?!"

Something bothered Abraham, something that had to do with the guy himself, with his unhealthy flesh. Nadia and I looked at each other, we were so struck by the expression of suffering on Lincoln's face. He picked up his hat, dusted it off, and handed it to the boy.

Take it and leave this honest house.

Abraham turned away, went to the table, and we all turned our backs to the door. Thomas's dog could be heard throwing himself at the night visitor, but soon the yard was quiet too. Only streams of night air reached us, slightly swaying the fire in the lamp, reminding us that the window was broken.

"I would like my sons to have a friend like you," Abraham said to Thomas, watching Nadia wipe the blood from his lip with lint, carefully lift his swollen eyelid. "You see, Mr. Turchin, how much trouble I have caused you; if you were going to vote for me, now beware. So far, stones, but they can also from guns.

"I'm not afraid of fire, Mr. Lincoln, I'm afraid of another. We agreed to move to Chicago, didn't Editor Dawson tell you? Higher position, more money, it seems to be good. And what if later even higher and very high, and there is so much money that even the wallet is small for them, they need a bank account. What then remains of my concept of democracy? Did you think about it when you agreed to seek the presidency?

- I think about this all my life,

And yet they agreed.

— My concept of democracy is very simple: I don't want to be a slave, but I don't want to be a master either.

... An hour later, when I was returning from Dawson, having driven an engraved plate to the set, I saw a light in the office window. Stretched out so that even our long sofa was not enough for him, Lincoln was reading a book in the light: his head was uncomfortable, his neck was at an angle, his chin rested on his chest, his left arm was extended along the body, in his right book. And in the weak light - a dark gap in the cheek, a dark eye socket, a shaggy shadow of hair, the face of a commoner, and at the same time a missionary martyr.

BOOK TWO

Union Cavalry Officer.

Chapter Twelve

We rode through crowded Chicago to two unruly regiments: Gov. Richard Yates in an open carriage, accompanied by a scowling army captain, and behind, dusting and picking up the crumbs of street vivats, I and Adjutant General of the State Allen K. Fuller jogged on short horses. The captain now and then took off his cap, stroked his hair with a not quite firm hand, holding his palm on the overgrown back of his head.

When we came across a gang of volunteers or everyone's favorites, Chicago Zouaves, when crowds of townspeople approached the carriage closer, the captain straightened his shoulders and put on his cap.

At the governor's house I did not see the captain: there was a taste of undress, the depth of longing eyes and a feeling of concern - but everything is fleeting, incorrect, like the impression of carelessness in clothes; not a challenge, but precisely neglect. At the governor's gate, I was not up to the captain: my patrons, the co-editors of the Chicago Daily Tribune Joseph Medill and Charles X. Ray, did not let me go a step, fearing that I would destroy their cunning plans by word or deed. In their dreams they saw me wearing the uniform of a Union army general, but Governor Yeats cooled their friendship and agreed to entrust me with a regiment of infantry. He deliberately seated the shabby captain in the carriage, on leather cushions, and watched how I, a Russian colonel, put my foot in the stirrup, how I waved it up and what a hard-Indian saddle I would have. Medill and Ray started to push their carriage behind the governor to take care of me at Camp Long, but Yeats soon turned them back. It was a matter of removing two regimental commanders, and the governor did not want witnesses.

So, we were taken to the shelves: they were taken to the bride and to the quick bivouac matchmaking. There are two of us, and two regiments - the 19th and 21st - what to worry about, just a shelf per living soul, but Medill and Ray advised to stick to the nineteenth. Five companies of the regiment represented the counties of Cook, Cass and Stark, the cities of Malin and Galina, five were formed in Chicago, which is why the regiment was also called Chicago. It was not German, like the famous regiments of Hecker and Koerner, Irish, like the regiments of J. A. Milligan, Scottish, like regiments numbered 12 and 65, miners, like the 45th Illinois, farmers, in the manner of the 34th under a big name The Rock River Rifles, a teacher's room like Bloomington's 33rd, was not all Methodist believers like the Preachers' Regiment's 73rd, Chicago mixed languages, beliefs, and crafts. The pride of the regiment were the companies, sometimes called the Chicago Mountain Guards, sometimes the Fire Zouaves of Ellsworth: they had already faced off against the rebels, captured the city of Cairo at the confluence of the Mississippi and Ohio, and guarded the Big Muddy Bridge.

We followed the governor, now at a trot, when the carriage rolled without restraint, now turning up the muzzles of the horses; I feared that my grinning little horse would take hold of the uncut back of the captain's head.

"It's a pity the captain was late," said Fuller, "we should have assigned to each his regiment in advance, so that it would not be decided by chance.

- Who is this officer? I inquired.

— Captain Grant. His name won't tell you anything.

"In other armies, a good officer remains a captain by his age.

"The Republic is reluctant to produce generals!" Fuller retorted arrogantly. She doesn't care about uniforms.

- The war will change this too: in the war, generals become more expensive.

- Rather, they are getting cheaper. Shakespeare noted that the price of nails and virgins falls in war: I would add generals to them. Do you think Grant is old?

"He is at an awkward age for a captain.

"But he's younger than you," Fuller smiled, apologizing for the upset he brings me. - For three months.

The North, like the South, hastily folded companies, at the head of regiments and divisions, it happened, there were people who first tried on a military uniform - a colonel's, and even a general's - by right of a wallet and a bank account, with which such a patriot paid for three months in advance several hundreds of volunteers. Some of the career officers, even those with the best reputation, having received Union shoulder straps, suddenly found themselves in the South, among the servants of the rebellious Richmond. Fuller signed appointments, sent papers to Washington, not knowing whether the seeds of treason were ripening in the officer's soul. What did he know about me? That I left the monarchy for a republic? Alas, Fuller himself did not endure the republic, he received it at birth, like a Chechen his mountains, and a Bedouin his desert. For him, I am an apostate, and whoever cheated once can change again. He also knew that I was a newly-minted citizen of the States, an engineer and an engraver, traveling for weeks from one railway station to another, arguing everything with conductors and machinists, with sawyers and rail layers, squeezing political plays in the Medill and Ray newspaper, encroaching on orders and reputations. Illinois. Now Fuller must entrust me with the lives of eight hundred of his fellow Midwesterners, and with them the Federation flag.

"I'd like to put you in the Twenty-First Regiment," Fuller admitted.

Why at twenty-one? I asked nonchalantly. Is he taller or shorter than others?

"The regiment is difficult, prone to indignation: a strong man is needed there," he flattered me.

I thanked him with a nod, albeit a short one.

"I was told about you as a person who is determined, but who knows how to get along with people.

The praise of the handsome Fuller was not to my liking: I am not hired as a senior clerk.

"Any regiment will suffer with me," I snapped. "That's something you should keep in mind when you start getting complaints.

From the soldiers?

More like officers. And most likely from my superiors. What's in the twenty-first?

— The Colonel is a bitter drunkard. And you, I know, are a teetotaler.

He seemed to know everything about me.

"That's right, and the captain doesn't drink," I quipped. "He looks so sick.

It was Fuller's turn to chuckle too, and not just, but with the superiority of his youth, career, and caste over a vain world filled with mere mortals, drinkers and non-drinkers, badly combed, in worn-out shoes and greasy hats.

"He promised to quit," Fuller said. And on the very first day...

What about nineteen?

I knew that the regiment demanded a more experienced, elderly commander, the current, twenty-two-year-old captain, as if humiliating them with extreme youth.

- There's something else. The regiment is a young man," Fuller did not lie. — Volunteers would like to keep him as a deputy, but the commander is required higher. Yes, they demand, they demand," he lamented.

We stopped at the Chicago arena, where the loudest volunteer depot in the Midwest had opened; the governor walked through the crowd into the arena towards the rumble, the commands, the roll call, and Fuller, dismounting, followed him. We were left alone with the captain: he looked at the soldiers, the old cannons on which, before parting, the couples, the red trousers of the firefighters who stood under the gun, the hitching post closely furnished with horses, looked from under swollen eyelids, without surprise, a little boring, then a glance, the same, with a hint of boredom, rose to me. He held out his hand to me, and I shook it, firm but impassive, like his eyes.

— Grant!

— Turchin!

Do you know how we do it?

Apparently, he was thinking with displeasure about the upcoming market in Camp Long, but I did not immediately understand what he was talking about.

— Election of the senior officer of the regiment. Grant chewed his parched lips, took a deep breath, as if the crowd at the arena was taking his breath away. "Looks like a cattle auction. And instead of an imported bull - an officer, just don't look in the mouth.

"It's in my blood. I caught his eye, heavy and mocking to my cheerfulness. - In my homeland, even the chief ataman was elected by free votes: the Cossacks will converge in a circle and decide who will be the military ataman, who will be the clerk.

Why did you leave your Eden?

- There are no more former freemen.

"Hoping to find her at Camp Long?" He looked with suspicion, as if he were a gullible, if not self-serving landsknecht (Noun, Landsknechts, German mercenary pikeman of the late 15th and early 16th centuries. At the height of their success, the Landsknechte **ranked among the most-effective foot soldiers in the world.).**

"I have seen the old armies in abundance, I have finished with them: only the new army could call me.

"Well, are they all so bad—the old armies?" - He did not believe my everyday tone, without the loudness that would have suited this occasion.

- They were well organized, but, with rare exceptions, criminal.

"Ah, yes!" he conceded languidly, shielding himself with a skeptical grimace both from me and from the noisy spectacle of the arena, the corners of his mouth, hidden by hair, drooped in disgust. - You are also a full volunteer, no one is forcing you to serve.

"Conviction compels me," I said unkindly. "The victory of the slave owners will abolish the republic.

What are you waiting for, Colonel? Desperate heads are already at war.

- Waited! Waiting for you all to hurry up. I expected that it would become like a war, and the matter would do without me.

- And how is it not possible? he revived.

"I don't like the way the North fights: that's the whole answer.

He laughed, almost inaudibly, inwardly.

— What a coincidence, Turchin! I don't like it either, and it also seemed that things would not work without me. You're playing? he asked with a quick gesture, as if dealing cards.

- No.

"Today we will be players," he invited. And fatalists. Let everything go as it goes.

Do you agree?

I nodded.

In the camp, standing on the parade ground, I understood why Grant was restless on the road. At first, the 19th was called up at the veche, but the volunteers on the 21st crowded right there, stretching out in bands from company tents, from wagons with long drawbars thrown to the ground, from fires. The officers stepped forward, but the soldiers did not keep a low profile either; yesterday's diggers and carpenters, red-breasted firefighters, Chicago artisans were tearing their throats. Captain Bashrod B. Howard, having reported to the governor, retreated behind the other officers of the regiment, who numbered about forty; tall, thin at the waist, with a boyish blush on his cheek and the eyes of a newborn antelope, Howard was defiantly young.

And now we were placed in the middle of the dusty parade ground, two old men - next to Howard we seemed old men - and the choice had to fall on the one who looked older, the only way to mitigate the offense of the captain, against whom no one had a grudge. In my thoughts, I retreated from the 19th, but the bargaining lasted, both of us had to pass all the tests of the volunteer circle. Fuller spoke about us: he praised Grant, his Mexican experience and military talent. Then it was my turn, and the glances of the volunteers turned to me, a gentleman in a frock coat, with a large forehead and a bald head over blue Irish eyes. They recognized me as an Irishman - after all, every third in the companies is an Irishman - and how could they not recognize the consanguineous stubbornness and temper, albeit in a stranger. And then the jaws fell off: Russian!

How does he deal with us? - loudly surprised in the ranks.

"He knows English as well as we do," came the officer's thick voice.

The daring men whistled in surprise and disbelief.

He can't even swear!

A red-haired fellow, with an unshy face ready for a quarrel, made his way forward from the ranks, and an officer who knew me warned him:

"Easy, Barney!"

"Yes, Captain Raffen! shouted the volunteer.

— Barney O'Mullen! repeated the captain threateningly. - Keep up appearances.

"Sir, we don't choose a wife. You can kick your wife out if you're not a Catholic, and the colonel will push us around. - And he started to me: - Are you Russian?

— Natural, from Russia.

"Mr. Turchin is a citizen of the United States," Fuller said.

- Your country, then, is similar to England and Turkey? Barney pretended to be a simpleton.

- Probably not at all. "I didn't know where he was going.

"Why, England is oppressing the Irish, Turkey is oppressing the Kandyans, and you are oppressing the Poles!" His logic worked on the crowd. - I only love these three nationalities; they are equally unhappy.

"I know the Poles, but here I also recognized the Irish; These peoples are strong, they deserve not pity, but respect. The serious, unflattering tone offended them.

"Now you left Ireland—you or your father?"

Father is buried at home!

- So - you: why do you like the local order better?

"And you can't compare: the United States is here!"

Why didn't you make them in your own country?

"I couldn't make my chicken coop in Ireland, let alone the free states! If only I were a colonel!

Laughter ran through the ranks, mild and equally at me and at Barney.

- And if you knew that Ireland had declared independence and was preparing to defend its freedom with weapons?

- From the same place and would have flown back! The Irishman waved his hands, showing how he would do it.

- So, we will get together; not for the sake of money, but for the sake of freedom, we are here.

"I need the money too, Colonel," Barney said as he got back into line. - I have a talent - to remain penniless everywhere: at least elect me for president.

I told them what I wanted to say, and only under whose command they would go, mine or Grant - this I did not know, and, to be honest, I did not care about it. Volunteers, I accidentally won over. The senior officers are another matter: I noticed Grant's evasive, devastating contempt, Fuller's displeasure, Howard's narrowed eyes with mockery - it seems that they suspected me of a persistent seeker of a place.

The captain held himself in front of the line, like a quick-witted artel leader, took up a new cigar, languishing the volunteers in silence, gradually examining company after company, not very pleased with the inspection. Now I fully understood his invitation to be gamblers and fatalists so that everything would go by itself. I retreated from the agreement - without intent, wanting to fence myself off from the persecutors of the Poles, Irish and Candians. Grant doesn't need it, he's a Yankee, breathes his own air and holds his gun.

Things were going along according to Fuller's calculations, no matter how the independence of tone Captain Grant removed from the regiment - and I did not find in myself annoyance or envy; I was only suddenly cooled by the thought that even in the army of the North my life would not be easy. Mattoon became a prep class for Nadine and me; Chicago completed the change: we now said our country, our republic, our order; now not only strangers, but I often called Nadia - Nadine. She led the house, wrote a lot and well, practiced in one of the city's clinics, her fame in Chicago grew. Editors Medill and Ray fell in love with our house - and they were not alone - and here Nadia's personality, her intelligence and hospitality came forward.

Meanwhile, our candidate Lincoln moved to the White House. Sensitive people, wives of indomitable Republicans, those who praised Abie's victory and expected profitable government jobs from him, loudly rejoiced that children had finally moved into the White House: President Pierce, when he lived there, had no children yet, and Buchanan was unmarried. The ladies saw in this the finger of the Lord and the promise of eternal peace to America, and in the meantime the war was bursting through both windows and doors. The North remained too calm in the consciousness of its strength, the South heated the horse. The South fell away from the Union, dragging along the frontier states, the South arbitrarily took away the Allied forts and arsenals, tempting officers with the image of chivalry against the merchants of the North, the South did the job, the North messed up, inflamed with speeches. I saw the military advantages of the South; his rider, born in the saddle, his forests and hills, suitable for partisan attacks, his desperate determination to protect his black property. Yet the South yielded to the North: it grew cotton and sold raw materials, while the North poured iron and steel, made cars and laid railroads. The South also built roads but laid northern rails on top of the sleepers; When building dwellings, the South drove northern nails with northern hammers, dressed his children in northern dresses and shoes, and if he nailed his dead into coffins of Arkansas pine, he lowered them into the graves with ropes woven in the North. But the North hesitated and shied away, its military commanders seemed to be afraid to enter the road of war with its horrors and blood. Now there was only one way for me to stand up for the republic - by taking up arms ...

The marriage was delayed. Fuller ventured to hasten the regiment, seeing that the volunteers had positioned themselves near Grant. And they, barely sensing the compulsion, changed - from the discordant mouths there was a breath of obstinate coldness, one of the junior officers said to Fuller:

- Why do you keep Grant as a captain? In other armies, apparently, officers are better appreciated!

"We are a high-ranking captain," Fuller said calmly. "Not in Illinois alone, in many states captains are at the head of regiments.

They shouted in response:

- Howard is a captain too!

"Our captains have already measured themselves against the rebels!"

The arrogance, the arrogance of a soldier who had already exchanged shots with the enemy, in front of one who had just received a handful of paper cartridges for camp firing, flooded out.

"What the devil have you not found one of your captains suitable for?" I intervened rudely, an officer who had rightly departed from them.

- Look what good fellows are standing!

They did not move, did not look at me, ten company commanders, ten captains. Skylarks sang in the sky, over elm clumps, over tents and camp equipment, scattered in a fair amount of disorder, behind our backs, in the kitchen, someone was chopping wood with a grunt that is the same on all continents, the officers were silent, not honouring me with an answer.

This is when Governor Yeats intervened:

- And really, guys, I don't understand you. Started a change, and now - what do you want?

Captain Ruffin stepped forward again.

"Mr. Yates," Ruffen said, without undue reverence, "none of us would agree to accept a regiment, none of us would want to be subordinate to any of the company.

To a connoisseur, his words said a lot: it means that everything has already been discussed and thought over more than once.

"You obeyed, damn it, Bashrod Howard!" Yeats exclaimed.

"We appreciate Captain Howard and would like to keep him as Regimental Deputy.

From the group of officers came the captain with a nervous, tired face. Before that, I singled him out among others by the tenacious, penetrating perseverance of his eyes and the general expression of determination.

"I am Captain James Guthrie of Cook County. I am a redneck, Captain Grant, and I want to ask you bluntly: will you leave Bushrod Howard as your second in command?

Grant looked around with cold disdain, as if seeing the camp for the first time, the new wagons that had not yet been on the road, the horses unharnessed in the meadow, the volunteers lined up, the officers opening their uniforms in the June sun.

"I'm tired of talking, Fuller," he said ungraciously. "Even the cattle auction has an auctioneer's hammer. Once! Twice! Thrice! He cut the air with his hand, in this movement the rage that possessed him was expressed. - And we are all trampling around, chatting! ...

James Guthrie of Cook County shook his head defiantly.

"And you, Colonel Turchin, will you agree to keep Captain Howard as deputy regimental commander?"

"I won't let you push me around!" I couldn't resist either. - Do you need a regimental? Well, they brought the goods - look. Still look, and how you decide - then blame yourself.

We left them, went to the officers of the 21st, to find out where their bitter regiment was, but no one knew this. From the 19th there were loud voices arguing, now Grant's name, then my name came to my ears, and soon three captains, already assembled, pulled up, - Alexander Ruffin, James Guthrie and James Hayden - announced to the governor that companies chose a Russian colonel.

- Congratulations on your first win! Yeats shook my hand.

"Did I break the deal, Captain Grant?" I asked, dissatisfied, not rejoicing at the preference given to me.

- At first, they distorted; then got better. "He has already lost his heart. Do you know why they chose you? Firstly, colonel, five minutes to a general, and generals are loved not only by generals. And one more thing: you arrived in the saddle, I am in the carriage; do you think they didn't notice? A soldier sees everything, a soldier is an amazing organism, whoever does not understand this, it is better for him not to meddle in the army.

Captain Ulysses S. Grant took over the 21st Regiment.

Chapter Thirteen

Soon we split up: Grant stayed at Camp Long, and we moved our tents to the outskirts of Chicago, to Camp Douglas, closer to the bustle of the world. I was convinced that my regiment was most molded of Irish dough, but these Irish are already half Yankees, Union patriots, their historical sense of revenge would have been sated by the defeat of the Richmond rebels - having defeated the rebellion, they would have avenged the desecration of the ancient land of Erin. Next to the Irish in the regiment are the native Yankees, whose great-grandfathers were born on the American mainland; the Germans, who are plentiful in the Midwest; a handful of Frenchmen, a few Poles, a Serb rider; There was also a Spaniard, but not for long. When needed, I could speak with each of my volunteers in the language of his homeland, excluding the Serb and the Spaniard - this opened many hearts to me. Only Bashrod Howard did not come to meet me, he asked for a company, where the Pole Tadeusz Drum was the senior lieutenant, and John R. Maddison, a wit, a fighter and a servant of Bacchus, was a lieutenant. I was expecting trouble with this company and I was mistaken: they served jealously and often came forward. I always felt Howard's outstretched hand, a gesture of excommunication.

Nadine's appearance at Camp Douglas was received with condescending derision by the regiment, only Howard and his officers deliberately ignored her; in this, the army charter was not with me, but with them. But when Fuller rode up to the camp to reason with me, I knew that it was not Howard's men who told him about Nadine, but the regimental chaplain Augustus X. Conant. The Adjutant General of the State inspected the regiment, praised it, marveled at the rapid change, and then, when we were alone in the tent, had the inevitable conversation with me. "The charter does not allow the maintenance of a woman in the troops, with the exception of nomadic girls," he said. "Probably you didn't know that, Turchin." - "What kind of regimental, not knowing the charter!" I replied. "If you're having trouble as a new person in Chicago, I'll have my officers find an apartment for her." "My trouble is elsewhere, Fuller; give me good guns, even smoothbore ones, if you don't have Mignet or Colt carbines, give them bullets, I can hardly provide guard soldiers. "Soon the regiment will march." "And she will perform with us. She is a brave person, and also a paramedic." "Did you give her a position?" Fuller was concerned. "Mrs. Turchina is rich enough, she is a Russian princess, and is ready to defend the republic without monetary compensation." Fuller had heard of Nadine's nobility, but had no idea it was so high. "We are not defending the republic, Turchin, but the Union. - He went around. "The rebels do not encroach on the Republic." "Here we will part ways.

I see a conspiracy against the Republic." "Richmond wants secession from the Union, not the monarchy. Neither General Lee nor Jefferson Davis is going to be crowned." "If the rebels gain the upper hand, the country will change, money and slavery will call for a tyrant, a despotism will arise, the most sophisticated and vile." He looked at me with regret, looking for traces of long-standing crashes that did not belong to America, the scars of past defeats that made me look so gloomy. "You will change your mind, Turchin. The American military Theatre is not Sevastopol, not sitting in a trench, but throws and marches. "We got a saddle for the lady; here, look, - I pointed to the corner of the tent, - a light, shaped saddle, through and very comfortable ... "We ended there, believing that we outwitted each other: Nadine, Fuller thought, would soon change the saddle for a place in the Chicago car, I calculated otherwise - battles would start, and there would be no hunters to come to us with trifles.

Word of me reached Mattoon, and Teddy Dawson and Thomas showed up at Camp Douglas. They came with the intention of staying in the regiment. Dawson's paper had to be shut down, Mattoon Courier readers poured into volunteer depots by the dozen, Teddy agreed to volunteer too, but I turned him down. Long Dawson tossed about the tent, ran outside and, forgetting where he came from, hit the tent canvas in the most unexpected places. I told Teddy that he deserved a better fate, promised to call him as soon as I got the fonts and the machine. I took Thomas. He wanted to stay with us, as in Mattoon, and we missed him, he was our Mattoon boy, although he had grown up a lot. But I did not indulge either him or myself; out of servanthood, even voluntary, one cannot make vocations, it is not a field for a brave person. "Do you want Zouaves? I asked, cutting off his path to batman or orderly. "Of course, they have the widest trousers."

Clever, he understood everything, in his thoughts he said goodbye to our tent, but he found himself, did not agree with me: "I wanted to become an artilleryman." - "I have an infantry regiment, but there will be guns: then we'll see." "I shoot a gun, Colonel," Thomas said proudly, "and I think I can aim the gun pretty well. If possible, I would like to have a repeating Spencer carbine." "I myself dream of such a thing."

Thomas remained in the regiment, with the right to choose which company he liked, and already on the second day the adjutant of the regiment, Chauncey Miller, informed me that Thomas asked to be assigned to Howard's company. I didn't object: if at seventeen I had chosen one of the ten companies, who knows, maybe I would have chosen Captain Howard's company, courage and tenderness, the gait of the proud and the bitterness of the disgraced, the sharpness of judgments and the deliberate dispatching were so bizarrely combined in its commander company commander. "I liked Captain Howard," Thomas said when I asked if he had found any acquaintances in the company. "And this Pole Tadeusz, too." And no more words: as soon as he touched this company, he borrowed from them a silent secrecy.

A week later, we were in formation on the streets of Chicago: the regiment received an order to depart for the front. With a handful of officers, holding back the pace of the horses, Nadine and I moved ahead of the regiment, which was on foot. Street after street, down Franklin, down Webster, Washington, Jefferson, down the narrow Black Hawk, named after the recalcitrant Indian leader, to the music of Chicago bands, step by step, past the excited crowd, to the fireworks - as if the city was welcoming the winners, and not seeing off another of the unfired regiments.

The blue army colour of the North had not yet gained the upper hand over other colours, my infantrymen were walking - some in caps, some in bright hats, in tunics and caftans, in frock coats, in gray trousers and crimson red, with a soft satchel behind their backs, with assorted muskets and carbines, but all tied together with brand new belts obtained from the Philadelphia gimp factory. The mules, drawn by six, pulled heavy wagons and several cannons, given to us before being sent to the front. The regiment stretched out, the Chicago companies, feeling the special love of the townspeople, sang, or rather even shouted, to the tune of the song "The girl I left behind me" [14], their own, folded under Cairo and on the Big Muddy Bridge: "One, two , three, four, five, six, seven - Tiger!

The North did not yet know its future: the July sun, even at sunset, warmed the earth; barges, boats, sailboats, and steamships plied the Michigan plain; the captains in saddles and the lieutenants in front of the companies were arguing with bearing and composure, and next to the regiment commander, through the city of sixty thousand, a woman rode. Nadine had waited two wars for me, now the military campaign of the Republic allowed her to prove the equality of woman: who could stop her? Adjutant General Allen K. Fuller? Governor Yates? Army charter, folded according to the customs of foreign armies? Consciousness of mortal danger? Had Nadine perished, I don't know whether I would have had the strength to live on, but I didn't even look at her to stop her from moving to the regiment; Samuel Blake, the regimental physician, found her an assistant in the most difficult and dirty work. The bandages are not yet bloody with wounds, but camp life with its exhausting science gives the doctor an opportunity to intervene at every step. Legs covered in blood, back and shoulders worn out, cut up by the straps of a clumsy satchel, fever, especially common among guys from the lower Ohio, checking provisions, boilers, linen - Nadine worked on a par with me, and it also happened that the dawn was won back , and the flag is lowered, I am ready to turn off the lamp in the tent, but it is still not there. She spoke without difficulty in several languages in the garrulous soldier Babylon, behaved with dignity and affectionate simplicity, which will baffle both the reckless wit and even the man of low morals. I stopped worrying about Nadine; she entered under the moral protection of the regiment.

Nadine did not see herself through the eyes of strangers, this was her strength and weakness. What will Atlantic passengers gossip about her friendship with an artel of French women, how will the poorest Mattoonians, accustomed to the idea that everything true is given for money, accept her free medical treatment? - what a misfortune, is it worth thinking about this, if activity is possible. With that she came to the Long camp, with that she pulled off the sweaty stocking from the worn-out dirty soldier's leg, put her hand on the fever-sticky forehead of the pockmarked volunteer from under Cairo, with that she sat down in the saddle, in rough boots and a long cloth skirt, to ride through crowded with Chicago people. She leaned with her right hand on the pommel of the saddle, in her left she held the whip and hat so low that the ribbons almost touched the ground, her head was slightly thrown back, as if her heavy hair styled high were tilting back, in her gray eyes there was happy surprise, and her lips turned pale, they live they take air, so clearly and so greedily, as one can take water; the blood drained from his face to shade his tenderness. Meanwhile, she is all strong and not alien to the saddle, orchestras, she is like the wife of a craftsman or a woodcutter: broad in the shoulders, broad in the hips, a woman, a woman in everything.

At the arena, the state administrators, headed by the governor, were waiting for the regiment. Richard Yates invited me to stop and look at the marching companies with him. He bowed to Nadine, and she dismounted, not knowing to go further. Behind them, horses whinnied, at the hitching posts and harnessed to carriages; the wives of famous Chicagoans remained on their seats - from the pillows, as from a stage, everything is visible, you can call out to a friend whom a patriotic impulse pushed into volunteers.

There were companies. Approaching the arena, the volunteers equaled the pace and yelled songs at the top of their voices. The Zouaves, even before me, were distinguished by good formation - they assisted me in training other companies, I alone could not have corrected things in three weeks. Trumpets blew, the state parade band passed at a quick goose step, wedged between the companies; in the roar of the soldiers' throats, I already distinguished individual voices: especially one - high, angelic, at its sounds the very thought of war flew away. This was sung by George Johnston, the perfect boy. The riders froze on horses, harnessed by six to cannons and wagons. Light Philadelphia belts gave a certain uniformity to the variegated soldiers in boots and leggings, in high boots, under the very groin, or in sharp-nosed, almost closed trousers with stripes. The fiery Zouaves, on foot, tinkled their spurs, which they had arbitrarily appropriated—I found these spurs in Camp Long and did not take away their toys from the Zouaves; The South imagines itself to be a knight's nest, without a spur it will not measure even a step, a northerner in his eyes is a huckster, a lout, let the Zouava's spur ring, reminding the southerner that the North is not alien to romance.

It was a good moment: the street resounded, forming a square near the arena, houses and lonely elms did not prevent the evening sun from mixing red gold into the swarthy faces of the volunteers, into the mirror-polished copper of the pipes, from spilling it on the canvas of the wagons, on the grains of the horses and the trunks of the guns. Fate gave me not a regular soldier, but a volunteer, a volunteer, in whose knapsack you often find an envelope and writing paper, or even a book, than a flask of wine or a deck of cards, people, soldiers who mixed beliefs and languages. We lived by faith in the brotherhood of man— the hour was approaching when our faith entered into a great trial.

Slowing down - there was a traffic jam ahead - Bashrod Howard's company approached the arena. The captain was in the saddle, his horse, as if in a vaulting arena, raised its thin legs in white stockings, and next to him, close to the rider, walked a woman with her son on her shoulder. The first impression is that next to the captain is a slave, a slave who fell in love with her master, and therefore free at least in feelings, in passion, in devotion to the one who kept motionless in the saddle, did not turn her head to us, annoyed at the delay of the column. Resinous, straight, like an Indian, her hair fell to her shoulders, a swarthy face with a wet mouth and a coarse, sensual nose turned to the captain and to her son, seated on her shoulder. Tight fuller bodice, unbuttoned at the top; as if her breasts were tight in him, a dark leather belt, not feminine, wide, without trinkets, intercepting the waist, loose folds of a skirt without hoops; it seemed that to complete the picture, a woman should walk barefoot on the pavement, as befits a slave. But her legs were in expensive buckled shoes, the dress was made of the best material, the pectoral cross on a black ribbon sparkled with stones, the green stone also played in her hair, on a black velvet, almost indistinguishable in the resinous strands - this stone would have been enough for that. to feed my regiment for a week.

To Howard's annoyance, the column stopped, and he had to rein in his horse, turn a gentle face to the governor, and take off his hat. The woman also turned, letting us see her son, tiny Captain Howard. The resemblance was so complete, as if the sacrament of birth had taken place apart from this swarthy beauty, or rather, she was only a devoted receptacle of a new life, not daring to add a single stroke or colour of herself. The similarity of the faces of father and son was striking; before us were two boys, only one of them was enlarged in size and put on a saddle.

Nadine was shocked by a sudden vision, a perfect image of motherhood, a swarthy Illinois Madonna and Child next to her warrior father. An artistic eye appreciated the completeness of this group, the mixture of severe and tender colours, and the heart, a poor heart wounded by childlessness, passionately responded to someone else's happiness. I knew only too well those moments of mortal anguish, when the world was darkened, and the spirit seemed to be broken - Nadine's slumped shoulders, her impossibly held breath.

"Good to see you, Howard! Richard Yates said. "Hopefully we'll hear from you soon.

Howard bowed his head silently.

"I introduced you, Howard," Fuller informed him. "Hello, Elizabeth, I think your husband will return from Missouri as a major, if not a colonel.

"I would make him a general even now," she answered cheerfully, with her free hand she raised the cross and kissed him. - God! Spare him and the soldiers, be merciful!

The column started moving.

- Nice couple, aren't they? Governor, Nadine said, unaware of her dismay. "She's from Arkansas, near Batesville. Famous family, famous and rich. All are rebels, a whole company would be made up of them. Howard is from Tennessee, and his older brother is also with the rebels. And they," he nodded after the Howards, "for the Union.

"Hurry up with its production," said someone from Yeats' entourage. "No matter how much resentment drives him to Richmond.

"The captain deserves a promotion, but not for the sake of bribery," I interjected. "Howard won't change out of resentment; on the contrary, she appeals to his pride.

"Don't vouch, Colonel," the man retorted gloomily. - We promote officers, entrust them with soldiers and weapons, and in the morning, we are short of either one or the other.

Nadine followed the Howards with a devastated look; Thomas, in short, almost under the floor intercepted by a belt with a bandolier, uniform, tried in vain to attract her attention. I smiled at him, narrowed my eyes in a friendly way, Thomas answered with a nod, but still waited for Nadine's look. She went to the hitching post, led her horse forward on the reins.

Again, Nadine and I rode at the head of the column to Michigan, and from there, after speeches, a picnic, and fireworks to the station, to two trains brought by Illinois Central for my people. I didn't disturb Nadine's silence; these tormenting hours were then invincible for her, like bouts of yellow fever or an eclipse of the mind. What was she thinking? What images tormented her mind? The gray slitted door of the sakli taken from the highlander, behind which her mother was dying of smallpox? A pile of sheets of her La Chólera, blazing in flames, blackened and mute? The vanity of her life? Unborn children? I never knew that.

Without her, I was standing at the staff car, in the crowd of my Illinois friends and unfamiliar admirers: Nadine climbed into the car. I involuntarily remembered our departure from St. Petersburg, and then from Portsmouth, where only Tkhorzhevsky saw us off. Were we to open ourselves up to our friends on a lifelong departure, if we called everyone together with our relatives, and there would be a fair crowd in St. Petersburg: the Russian heart knows how to respond to friendship. Here we were still not quite our own, the Russian meta was still on us, but meanwhile my comrades surrounded me, and they saw me off not only to one duty, like Fuller, or a food commissar, or a governor with a governor's wife, but also out of true friendship; having lived for four decades, you distinguish it well in the bustle of the day.

Members of the 19th Illinois Volunteer Infantry regiment, 1861.

Chapter Fourteen

Excerpt from a letter from N. Vladimirov
to his father in St. Petersburg:

"... I mentioned in one of the letters George Johnston, a young volunteer in the past, and now a wealthy businessman. His recklessness led us to disaster: Johnston went to the general with a bunch of gifts, with a box of expensive cigars, with chess, which Johnston is obsessed with - he decided that the extreme poverty of the general had passed and now it was possible to poke his nose at him with gifts.

The general is cunning, and Johnston is simple-hearted: the old man drove him into a corner and extorted him about cheating with his pension.

The plans for reconciliation converged on me. Virginia volunteered to accompany me and wait at the gate. For two days in a row I sought a meeting with the general but was not accepted.

The general's story approached the war, and I noticed how hard he parted with papers, how his hand trembled, how a sad look followed the sheets and letters into my doctor's bag. He made me his executor.

I do not know whether I was a good executor of his plan; but, apparently, he needed a Russian and Russia for his papers, and I knew how to listen - people of reason listen better than sensitive natures. It can be seen that nothing can erase the homeland even from strong hearts, even among citizens who honestly served their second homeland, a young and forgetful people. The immensity of the separating versts does not extinguish the melancholy but makes it longer.

And everything was broken; and the general is bad, and the denouement may come at any hour. Offended by friends, with rushes of blood that paint his face and forehead unhealthily, he may die without finishing his confession. I asked Horace Fergus if Nadine was alive or was she all alone? The widow looked at Virgie, and I heard one answer: "Alive." Alive! Could it be that he had driven her away from himself, vomited up the difficult character of the general? There were moments when I was indignant at Turchin.

And suddenly a short letter from him - addressed to Horace Fergus but meant for me. There is a note in the envelope, an invitation to come as soon as possible.

I found a changed old man: his step became heavier, the movements of his swollen hands slowed down, his face was touched by a pale yellowness, as happens with living tissue after a bruise, when the dark colour disappears, and life has not yet rushed to this place. Again, I turned to the ear and collected archival tribute, plucking up patience when the moment came to reconcile the general, if not with Congress or senators, then with Horace Fergus. Of all the papers handed over to me this time - and among them are letters to commanding generals and even cash receipts from the owner of the steamer that transported Turchin's soldiers across the Mississippi - of all these papers, I am sending you only Turchin's letter to General Hurlbut. It was written a few days after arriving in Missouri and shows how far-sighted Turchin was.

Artillery travelling to Camp Turchina, 1861.

INTER-CHAPTER SECOND

**Headquarters of the 19th Illinois Volunteers
in Missouri. Camp Turchina, Palmyra
July 17, 1861**
To Brigadier General S. A. Hurlbut

Sir,

*From the information I received yesterday from the Federalists, I definitely
conclude that 18 miles northwest of Palmyra, near a place called Marshall's
Mills, on the Philadelphia road, on the Fabius River, there is a camp, and
possibly several camps of secessionists. Some people say that there are about
1600 people there, mostly horsemen. I have heard that there is another camp
at Warren, 10 miles west of Palmyra; I sent a reconnaissance party yesterday, which approached this place and received information that there were
three companies stationed there, but they had departed for Shelbyville. There
is a third camp between Marshall's Mills and Warren, but I couldn't figure
out exactly where. Now, as I write to you, it is possible that all these camps
have already been moved to Shelbyville or somewhere else. Many citizens of
Palmyra are constantly in our camp, and some of them come and go every day,
delivering provisions and reporting news ... In general, with very few exceptions, the citizens of Palmyra are secessionists to one degree or another. My
opinion on the war in northeastern Missouri is as follows. In addition to the
troops guarding the railroad, here it is necessary to have a reserve consisting of
one or two regiments, as well as parts of the cavalry and artillery, which, being
independent of the stationary troops, would constantly move into the position
of the enemy to deliver surprise blows to him. With this mobile reserve, there
would always be an officer, a representative of the command of all our forces,
who himself would receive all the information on the spot and act quickly and
decisively. If, however, we occupy only the railroad, without being able to attack
these camps, then the secessionists, having increased their forces, can easily
cut the railroad line, isolate our dispersed companies and destroy them piece
by piece.*

In my opinion, this reserve should include a cavalry regiment, an infantry regiment and a battery of several cannons. The topography of the country is very favorable for infantry operations, but, as the enemy force consists for the most part of cavalry, they, even when we attack them by surprise, can easily break away from us and retreat, although if we ourselves had cavalry, we would be better could beat them, cut off their retreat and destroy all these semi-organized bands. In my opinion, it would also be better for us to have lighter and more mobile artillery.

I myself am not in a position to obtain a good map of the state of Missouri, and I have already respectfully asked you to send me such a map, on a large scale, immediately, but have not yet received any. And we need the best possible maps in order to always have them in front of us and put on them all the details about the enemy that we get from the information and intelligence reported to us. The enemy knows these places well, but we, having no maps, are like people wandering blindfolded over rugged terrain.

Delays in getting us weapons, blankets, uniforms and other provisions are discouraging us, especially now that the regiment is so dispersed.

We had some riots on the first day we arrived in Palmyra, but I stopped them, and I will send you a special report on the results of my investigation.

Sincerely

Your humble servant

Colonel J. B. Turchin,

commander of the 19th Illinois Volunteers in Missouri.

Brigadier General Hurlbut with General Macallum.

Chapter Fifteen

The Spaniard stood in front of me and the company commander, James Guthrie, his white-toothed mouth bared in a smile.

"Captain, I stayed in the regiment for an extra two weeks in addition to the legal three months: I am entitled to money for these weeks, so be it, there is no time to wait.

The regiment loaded into Quincy at the river landing on the banks of the Mississippi, beyond its expanse the war began, the plains and groves of the state of Missouri lay, where the rebellious gangs of Harris and Green galloped.

The Spaniard is wearing worn, oversized boots, a uniform made of moth-eaten remnants of the Mexican war; his unarmedness, quick, irreverent speech, with Romanesque passages in the middle of English, already in themselves seemed to have removed him from the company lists.

"Money is empty," Guthrie said grimly. - The military treasury will pay everyone to the last cent.

- Whoever has a lot of money, that is empty: he will drop a dollar and not be enough. The Spaniard winked at the volunteers. - And at night the widow sewed up a tear in my pocket so that the cents would not fall out.

"Aren't you a coward, boy?"

The Spaniard drew himself up, frowned, showing how much strength and determination in his frail, in appearance, body.

- He's a brave guy!

"Easy, captain!"

- The fourth month without a gun - everyone will get bored! ...

Even the volunteer who had an old musket but had used up a handful of paper cartridges issued to him, was now as good as unarmed.

"Without a weapon, you'll be a coward!"

"We were not recruited as the target!"

"Where are our guns, Colonel?!"

- All of them, as many as there are, were sent to Washington!

Have you heard, Miguel? Run to Washington for a gun!

This was not the first-time companies had lost men; they were given a three-month period, set by politicians for the training of a volunteer, for the entire war and victorious fanfare. Somehow, I missed the sentry too - in the middle of the night, the volunteer figured out that the day before he had completed his term of service, left the post and calmly put linen, tobacco and biscuits in a soldier's bag. But then the conversation went public; it could have consequences, because of the Mississippi the green land of Missouri, a free state, was looking at us, bloodied with enmity and passions to the very border with Iowa - it is better to cross the Mississippi, leaving fifty unsteady volunteers on this coast, than to risk the success of the battle.

"You seem to me a man of determination," I said to the Spaniard. "And if you decide to leave, go away, go away, don't start a quarrel," I urged him. "A soldier fights badly under the stick, and Captain Guthrie and I not only have guns, and we don't have sticks in stock.

- In the volunteer depot, I signed a contract for three months ...

— And with God, with God, soldier! - I scolded him cheerfully: it would be better if he had a set on edge, a taste of rejection and hidden resentment; that was my Jesuit plan. "Here's my hand and hurry up." It's time for us to get on the steamer, no matter how you get on the deck of the "Jenny Dennis" and end up in battle.

The Spaniard moved away, but it seems to me that when the flat-bottomed ""Jenny Dennis"" rolled away from the landing stage and splashed with her plates on the water, moving away from Quincy, his figure loomed among the bales and boxes for a long time. We sailed down the Father Rivers to Hannibal, Missouri, from where the railroad track ran to St. Joseph on the Kansas border, dividing the state of Missouri into two unequal parts; less than a third of its land to the north, to the border with Iowa, most of it south of the road, with the capital of the state Jefferson City, with Lexington, Springfield and the town I remember, where I happened to mourn the destroyed printing house of the Herald of Freedom and its maimed editor Ramage. The "Jenny Dennis" loaded five companies, the rest followed us on an open boat. It soon got dark, the songs died down by night, the Mississippi resounded in breadth, on the port side, on the Illinois, lights flickered, but the Missouri was dark and deaf. It seemed that this free coast of a free state also took the side of the enemy, the black paint of slavery, threatened us, warned us not to land, not to tempt fate.

I climbed into the helmsman; E. L. Shible, captain and owner of the little "Jenny Dennis", was on watch. We were silent for a long time; I didn't even hear Shible breathing; only the chugging of a ship's engine, the slapping of wheel plates, the quiet murmur of jets, the darkness into which we were sinking deeper and deeper. And when I was about to doze off to the unaccustomed music, Shible spoke, dropping the words into a fair beard falling on his chest:

- After he was hanged, the war came up to us; every death will be counted in heaven, and America will pay in full for the damned murder in Virginia. Once upon a time, peoples knew how to honour the prophets, and we?! How many days would we remember Christ if he appeared among us?

"This war, Mr. Shible, will change many things in the world.

- This war! he said with deadly contempt. "It won't change a damn thing and it will never end; centuries will pass, and the white cattle, forgetting God, will still consider his ass more beautiful than the physiognomy of a negro. That's how a man works, Colonel.

My name is John Turchin.

"And at least you were called Moses!" No one will follow you, and the Lord will not give you new tablets in your hands!

- A shelf will suffice; They would give everyone a Winchester.

"We are all modest, and modest ones don't win wars," the captain said angrily. - You are happy with eight hundred idlers, and he opened his heart to all of America, to millions of slaves; guns were needed - he took over the weapons factory and arsenal at Harper 's Ferry ...

- That's what you're talking about! I bow to John Brown.

"It would be better if you didn't bow, colonel," he turned my words around, "but galloped that morning to the gallows, cut the rope. Your Abie will not live long enough to wait for peace and peace in America," Shible said his prophetic words, and that night I considered them irritated chatter. "Some people call Brown a righteous man ... I don't know: this word smells of lamp oil, but he was a man, he had a simple mind, no catch, and all from directness and truth, there is nowhere to turn away from such a mind. You don't understand this.

"Because I'm a foreigner?"

- Our house is large, unfinished, even the doors are not hung, everyone will come in to us, who wants to.

- Are you sad about it?

"We are still a people without memory - you will end up in Missouri, you will find Paris, and Milan, and New London, and Mexico City, and even Troy, and so on everywhere," someone appeared from three to nine lands, put up a house and a latrine - so Troy is ready, and even Athens or Petersburg. The whole world of God is in our pocket.

He was worried and angry, fed up, apparently, with the Missouri war - marking time, confusion and losses to the rebels, while I, although tired, was seized with a secret joy. Where would she come from, in the middle of the black Mississippi, with old muskets and summer soldiers that made up half of my companies? But there was, believe me, there was joy, and hiding it from Captain Shible, I felt it even more sharply. The boards of the "Jenny Dennis" were gently jarred by the machine, reminding me of sailing across the Atlantic, the darkness of the ocean, the deck of the packet boat, and talking softly to Nadine so as not to disturb the sleeping Irish. And now I had Irish and Yankees with me, Germans and Poles; five peaceful American years closed in one day and flew off into the past - I moved from the ocean packet boat to the Shible river boat, came to the States to taste another, right-wing war, to prove that my republic is not an idol of faith, not a Belt's fruitless dream, but flesh and blood.

"I'll tell you, Colonel, this is what I want to tell you," Shible spoke again, "when in battle, in a difficult moment, the flag of the Federation looms in front of you and you sigh: help has come, they say, don't rejoice ahead of time. Point your guns, Harris' gangs and southern generals are ready for dishonour; they keep our flag in reserve.

- A dishonourable deed admits vile means; and we will fight," I said with faith.

- You still dream about the right fight; hell, everything went wrong here, as you rarely see on the Mississippi: whirlpools, rips, whirlpools, go figure it out.

Thomas slipped into the helmsman—Howard's company was floating on a pontoon, but Thomas was with us—Nadine had asked the company commander before loading. Thomas came for me: the July dawn was approaching, and with it the piers of Hannibal. Not finding the lights of the boat on the river, Thomas became alarmed:

"They have an old machine, no matter how the boat gets stuck.

- There are islands, the river goes in two channels, so their lights are closed. Did you miss the company?

"Captain Howard is having a hard time already.

"Has anything happened, Thomas?" We stood at the cabin door. - You do not have a company, but a secret lodge.

"We have a friendly company," he defined in his own way. "Maybe it would be ignoble for me to tell you about the captain?"

"Look, I'm not pulling your tongue," I answered, taking hold of the door handle.

"Someone slipped a newspaper to Captain Howard in Quincy. From St. Joseph or Lexington; it is printed about another Howard, about his brother. The rebellious Howard with a gang of horsemen is operating between the Missouri River and the railroad. Now, behind every hillock, Captain Howard's own brother can be waiting: here whoever kills whoever, and mothers - tears. And it's not easy to kill, Mr. Turchin.

- Not just, Thomas, keep you fate from this! - Before me appeared the image of a Pole, sprawled on a bench in a Carpathian castle. "What did Howard say about mother's tears?"

- Nothing. The captain, as he read the newspaper, bit his tongue; not a word to anyone during the day. He and Madame only nodded when she asked to let me go to the "Jenny Dennis".

— Madame? — I was surprised by the new word.

- That's Mrs. Turchin's name. Tadeusz came up with it, and now all the companies have begun: madam.

- Tadeusz wanted to play a joke on her, but it turned out well.

"Tadeusz is a noble man," Thomas defended him; he definitely did not concede to me the company and the honour of its officers.

I lost Thomas. Where is the reason? In the youth of his new comrades, which gives a sense of equality despite the inequality of rank? That I rejected him, did not make him a batman and a servant? Suspecting that I acted cruelly in agreeing to take the regiment from Howard? Thomas's footsteps have died away on deck, and I can still hear the word madam he uttered, respectful and gentle. Madam! Thomas, a young man educated by the primitive and rude Mattoon, could not even understand how much Tadeusz put into this word when he first put it on a sharp tongue, twisted it in mocking jaws and pushed it into the light of day.

Here is a hint about the princess, and recognition of her independent character, her nobility, her rights not recognized by anyone, and next to her is a doubt about these rights, a clownish bow to the self-proclaimed queen of the regiment and a suspicion that it is not a lordly whim to bother about bloody heels and unclean underwear. soldier.

— Madam! I said as I entered the cabin. - I'm listening to you, madam!

- Do you already know? she replied with a laugh. - Boys!

— Madame! This is higher than a regimental commander.

"And it doesn't look like a candy girl at all?"

— Madame! I balked at the comparison.

Did Thomas tell you about Howard?

I nodded; with Nadine's question, with her worried look, my worries returned to me.

- Civil strife, a dispute between your own people - this is not a trip to a foreign country. It's about the structure of society; in such a war even, brothers are at enmity. But in this war, one side is certainly right: not every war can be said like that.

The regiment did not linger at Hannibal; the boat had barely arrived before we moved northwest to Palmyra, and from there, after a week of fighting, to Emerson and Missouri Philadelphia, on the wretched post road that connected Palmyra and Shelbyville. Reality appeared to us in blacker colours than those of Captain Shible. We saw a state torn apart by passions, a lost state, lost in the changes of military happiness, in mutual threats, in fear and uncertainty. Moving along forest clearings, with traces of recent wheels and cavalry hooves, you did not know what awaited you behind the trees, even if a faithful scout promised friends or a farmer willing to sell food to the Federalists. The Missourians were divided into opponents and supporters of slavery, the farm threatened the farm, the settlement - the settlement, the horseman - the horseman, the threat sounded not only distant shots, but also the creak of wagons, the shout of the driver, the crunch of branches or the rustle of footsteps.

I was immediately convinced that there was nothing to think about the actions of the whole regiment, here the one who knew how to conduct partisan battles was in time - deaf, invisible to the world, hidden by forests, without final victories and fixed territories, with meager trophies, and even without them. Gangs of rebels happened to get their weapons, wagons and ammunition at the expense of allied companies taken by surprise, but if they took away food from a farmer or merchant, then screams of curses and accusations of robbery rushed into the Missouri sky - what should a robber do if not take someone else's! But as soon as a northern soldier was seduced by a handful of someone else's maize, taken - out of need, out of hunger - a pot of ripening potato tubers, and then screams of protest shook the air, newspaper pages and the sensitive hearts of the northern generals. To make peace with the enemy, to take them under your protection, to show a friendly, corporate attitude towards them - that was the imaginary nobility of these generals. They were ready to bring to court-martial or demote a brave officer who refused a slave owner a sentry to guard his potato field, they were ready to drive out the commander only because he had taken away from the rebel a bull for hungry soldiers. And any officer who dares to take away his slaves from the rebel, to give them reins, a kitchen shell or a shovel into their hands - God forbid even to think about guns! - such an officer would be considered an enemy of the nation.

It was a bitter time for us. The whistle of bullets and the roar of exploding shells will get on the nerves of anyone who is not used to war, and many of our generals in the hot summer of the first war year looked to see if there was a bush somewhere to hide. Well, after all, Frederick the Great somehow hid under the bridge, and Henry IV of France cursed and shamed his sinful body shell for cowardly behavior in the first battles!

So, the first battles of some of our generals looked like a daub by a novice artist. They believed that if an army occupied a city or county, then they could be considered conquered once and for all, like Mexican lands or countries taken from Indian tribes; they forgot that war is a game of chess, and it only takes one wrong move for an army that has won a victory today to be driven far back tomorrow. Missouri gave him hourly lessons. Shoddy uniforms were sewn, the manufacture of which is the worst rag, turned into rags in a month of camp life; boots of rotten leather fell apart in one march; clumsy knapsacks cut the shoulders with straps, crushed the chest, bloodied the back; cartridges with sintered powder ("Sintering is the process of coalescing a powdered material into a solid or porous mass by means of heating without liquefaction"):or half full of gunpowder, so that a soldier for forty shots had to carry eighty cartridges with him; shells with tubes embedded so badly that they exploded at half the distance, killing our shooters; cannons that exploded from a powder explosion, crippling gun servants and soldiers; saddles that disfigured the backs of cavalry horses; undershirts that barely reached the waist - all this our generals endured, put up with everything. The Sevastopol soldier, betrayed by Petersburg, was still standing before my eyes — even in my new place it was not difficult for me to understand why the soldiers were stripped and given over to the defenseless fate of the soldiers not of the monarchy, but of the republic.

I took up the reports. During breaks in fighting and skirmishes with the rebels, I pestered my Brigadier General Hurlbut and General Pope, who commanded in this Missouri area, with the demand for weapons, horses, company wagons, cannons, cavalry, topographic maps, blankets, uniforms, which are especially important when the regiment is dispersed, when and the company is divided into small detachments and only the form will allow you to recognize your own.

Civil War arsenal, 1862.

Chapter Sixteen

Brigadier General Pope appeared at Emerson suddenly, with two staff officers. He had just been with General Hurlbut, had read two of my reports, and must have found that my reproaches did not apply to Hurlbut, but to him, the commander of the Union regiments in this part of Missouri. But he appeared imperturbable, he appeared bravely; after all, without courage you will not set off with two officers on a journey through the country, where the rebels can open up at every step. As the center of the camp, we chose the farm of a farmer devoted to the Union - a slippery man, zealous in slander, in an attempt to deal with neighbors with the hands of my volunteers; he hung around behind the officers, slandering, rolling his pink, rabbit eyes in his brown, smallpox-pitted face.

Pope praised my two companies - Howard and Raffen, who met him along the way, advised to keep the regiment closer, in view of the danger of a blow from Shelbyville, where up to two thousand mounted rebels converged.

The camp was as it had been before Pope came along: I didn't stop shooting practice in the meadow behind the farm and the bayonet lessons of James Guthrie's guys, I didn't blow the muster and I didn't line up the volunteers. The July sun was hot - uniforms were wide open, chickens were walking around, clucking and throwing up their wings at each training salvo; volunteers with attached bayonets moved against the backdrop of oak forests and a frightened herd of cows pressed to the edge; two Negroes were zealous at the kitchen; the wounded soldier was sitting by the hayloft, gazing at his superiors in a rustic way.

"I remember you complaining to Hurlbut about the lack of amm-unition?" Pope remarked at the frequent volleys.

- That's why I asked, because I thought about teaching.

"Your soldiers have no shortage of live targets.

- This target on a horse, in order to correctly take it, samples are needed. I pointed to where the shots were fired. - A rifle is dangerous only in capable hands; an inept soldier sends bullets at random, at random!

My marksmanship lesson made the officers look at Pope, expecting him to put me in my place.

"You are right, Colonel. With his gloved hand he took my arm, and in this cramped position we walked a few steps. - We have a few officers who fought. Prefabricated soldiers, prefabricated officer corps; the old, glorified armies of Europe were homogeneous.

He was open with me and I responded in kind:

- I am glad that my regiment is like this - from the forest and from the pine, as they say in my homeland.

"But uniformity gives strength. "The purpose of the war gives strength when a volunteer, whoever he is by blood, knows that he is fighting for the republic."

He took his hand away, we pulled away from each other.

"It's all good for the volunteer depots," Pope said disappointedly. But money is more important there. Few will rise to the goal. He grinned gloriously and pointed to the cooks in tattered blouses, through which a black body could be seen. "I hope they are not in the regiment?"

I didn't have time to reply that I had taken seven Negroes to the kitchen and as riders: their owner, the owner of the lands between Warren and Woodland, fought against us, burned his own farm - which happened very rarely - and locked the black property in the barn, and galloped off to Harris - We managed to save seven from the fire. Before I could reply to Pope, a gig with a tan handsome man in shafts rolled up to us, stopped a few paces from us, and a thin, handsome old man, without a mustache, but with a gray beard around his tanned, wrinkled face, came down to the ground.

"Your soldiers have taken my horses, General. The old man took off his hat with dignity.

"That's who the commander is here," the general pointed to me.

The old man had a calm, unflattering look and an equally unser-vile figure. He saw the owner of the farm appear behind me and responded with a contemptuous movement of his nostrils, as if he had smelled a stench.

"Your people have taken four horses from me.

- You won't choke! shouted the owner of the farm, and Pope turned to the voice of hatred. It's Scrips! the farmer announced, as if the name alone said it all. "Fitzgerald Scrips, his sons at Price, in mutiny..."

Pope looked at the farmer for so long, until he moved away, with a gesture of reproach and indignation.

"Are you accusing my people of robbery?" I asked.

"Determine for yourself and return the horses to me."

He held out a piece of paper. It was a receipt that the horses were taken for military purposes; I recognized the hand of Stephen Hill, first lieutenant of Captain Stewart's company, recruited from County Stark. I gave the receipt to Pope.

- What do you want? This is written by my officer.

"I want horses," said the old man. - With your piece of paper, I will not raise a chill and I will not bring my harvest to the market.

"But do you have any more horses?"

"I won't let anyone count my property: neither this ferret, nor you, Herr Colonel.

"They'll give you horses," said John Pope.

- If they do not fall from the bullets of your friends! I added.

In this case, I will demand money.

He did not speak in haste, knowing that the general was on his side, and along with the general, the law.

"Horses taken for reconnaissance for the Union army, against the rebels. — The scraped face of the old man became disgusting to me.

"I demand the protection of my property from looting.

"Are you seeking protection from Union officers?" Then tell me: do you recognize yourself as a citizen of the American States?

"God knows, I'm a Missourian, and Missouri is still in America!"

Do you recognize yourself as a citizen loyal to the White House and President Lincoln?

"I am loyal to our constitution.

He was a master of dodging, but already I noticed his weakness. Scrips was losing his temper, it's hard for old people to hide it, a little blood rushed, and his eyes weren't the same anymore, and his white hair fell sharply.

"Where is your honour if you evade an answer that your sons would not find it difficult to answer?!

I felt Scrips' seething hatred and the general's displeasure; what could he do? Even the kitchen Negroes stopped wielding knives and throwing logs on the fire - everyone listened to us.

- Why are you silent! - I did not let the old man enter into even breathing. - Tell us, are you a citizen of the United States?

"Not exactly, Colonel," he began sullenly, clasping his trembling fingers with his right hand with his left and about to launch into space. "I, you see...

Already I truly despised him and turned to the old man rudely:

- Well, then, go, turn to your consul, stranger, let him take care of your business. And I turned my back on him.

There was a creaking of springs, a whistle of a whip, a snort of a buckwheat, and the gig rushed away.

"You will see me off, Colonel," said John Pope.

This man knew how to hold on; they reported about his tough, desperate disposition, but he knew how to behave like few. He showed displeasure, perhaps only one thing: Pope did not say goodbye to Nadine. Could I have a heart for this on Pope? Hundreds of officers, equal to me and above me in rank, parted with their wives for years, and with bad luck forever, - Nadine is with me. They lived by the rules - I broke the rules. We rode past the skirmishers in the training circuit, past the soldiers of James Guthrie, smashing straw enemies with bayonets, then the heat turned into a coolness, even the horses breathed deeply the air of the shady oak forest.

"Has the regiment been in contact with the rebels for a long time?" - He himself knew the answer: Pope not only had a regiment; a company was on the account.

- Eight days. On the fourteenth of July we arrived at Hannibal; the first two days there were small skirmishes, on the eighteenth I attacked the camp near Palmyra with four companies. Without cavalry and artillery, my regiment was scattered throughout the enemy area. I wrote to Hurlbut about this.

Pope reined in his horse, fearing that I would be too far away from the camp. He took a map of Marion County from the officer, studied it as if he were checking the road, then folded the map and handed it to me.

- You asked for a map, if you please, here it is. You write a lot to Hurlbut," Pope chided me. Does he have time to read?

It was a challenge: outwardly playful, but a challenge.

I only write when absolutely necessary.

And then Pope went to the scolding:

"Today you turned Fitzgerald Scrips away from the Union. He never hesitated in choosing the flag. I made him respect the Union. And hate.

"And he came with what he left us with: Skrips is a hardened slave owner.

The General broke through: he spoke quickly, stopping me from objections with a dismissive movement of his hand:

"How quickly you judged, Colonel! You yourself have just been honoured with citizenship, but you are denying it to a state pioneer. You, like a lawyer, hook a gentleman, breaking the law. Colonel Turchin! You will return the horses to Skrips, forbid the officers to feed the soldiers at the expense of the farmers - otherwise we will all be on a powder keg. And blacks, blacks, blacks, "he raised his voice," they have no place in the regiment. " We must make a distinction between domestic life and the regiment! he said, alluding to more than just negroes.

He rode away as my enemy, touchy, unfair to my every order, enemy forever.

I dismounted, so as not to immediately return to the camp, and set off on my way back; with reins in hand, with angry eyes that do not see the forest, muttering curses that would not have been good for John Pope if he had heard and understood my Don verbs and adjectives. Fitzgerald Scrips stood guard on the forest road, towering in his saddle on the roadside, holding back an oak branch with his hand. By his stirrups, on either side, were Negroes naked to the waist: one in the hands of an expensive gun, the other a strange piece of iron. I would not have thought that the old man would gallop to his farm in half an hour, change the gig for a saddle, the gentleman's gray thin suit for breeches and a hunting jacket and make it to the oak forest,

"I branded my horses, Colonel," he said casually, looking down at me. - I have a rule: to brand all my living creatures, any property that could escape.

I tried to think about trifles, looked at the silver of his spurs, the gun, it must be not French, but a German gunsmith, with salting in chasing and inlays.

"Sometimes misers brand chickens," I remarked.

"When you live among thieves, you have to dirty your hands about them. Before, we lived quietly, and then weasels, traitors, vile plebs romana [16] got divorced. Where does the devil bring them from, colonel?

"Ask the black servants where they were stolen from?"

— They were born on my farm; I'm talking about white, unsolicited shit.

"It's just how the nose works: some gentlemen can't smell anything but shit."

Scrips held out his hand, and the Negro put a dark, twisted iron into it - I saw a hefty monogram, two letters "F" and "S" connected from iron - Fitzgerald Scrips.

"Here's my brand!" he boasted, holding the monogrammed iron against my eyes. - If I want, I'll put it on them. He pressed the monogram against the servant's black chest.

The faces of the Negroes remained the same: closed, with an expression of humility, as if petrified reverence.

"I can brand them and sell them, even if your Illinois attorney vomits sermons and proclamations in Washington. They are my dogs.

I jumped into the saddle, not hastily, without losing my dignity.

"In every nation there are those who see first, and there are those who are servile.

"Give me back my horses," he said hoarsely, grieving that he was letting me out alive. - I showed you the brand, do not try to change horses.

I picked up the reins and said to him in the end:

- Live in peace, stranger, and take care of the sheep from the company's boilers, I will not let the soldiers starve!

I heard curses, clicks of triggers - Skrips was testing me - waiting for shots; moments stretched along the entire length of the remaining road, and only the dull thud of hooves on the turf was heard in the oak forest. The road turned into a clearing, I spurred my horse and rushed to the camp cheerful, as if General Pope had treated me kindly and promised a brigade.

But I ran ahead and did not tell you the important thing: before Scrips, before General Pope, there was a meeting with Grant. It is necessary that on that one day when the volunteers were indignant, Colonel Grant appeared next to me. I did not mention: not the captain, but Colonel Grant. Even on the "Jenny Dennis", I heard from the ship-owner Shible that two allied officers, General Lyon and Colonel Grant, are fighting better than others in Missouri. Then I doubted whether he was talking about Grant - it turned out, about him; at the first detail I recognized Ulysses S. Grant. In Palmyra, he stood before me, emaciated against the month of June, wrinkled the suffering face of a skeptic in a fleeting smile, shook my hand, then put his tenacious, thin hands behind his back, under his unbuttoned uniform, which reached below his knees. His hat is old, turned up in front, and his uniform looked old and worn, and the same trousers worn over his boots; on the cuffs and on the chest remnants of cigar ash. If he wasn't wearing a brand new, not yet faded epaulette, I would have sworn that Grant had just stepped out of Governor Yeats' carriage at Camp Long.

My regiment came to Palmyra to relieve Grant and, with Hannibal and St. Joseph's reservists, guard the railroad from the Mississippi to the Kansas border. Grant had never won battles the world would hear of—the federal banner had never known such victories—but he dug his digger's shovel deep and expanded the Union lands mile after mile. Grant's success lay in himself: he was a worker, a clear head, will and military talent. It is not a sin to say about him that he was born to win the battles that he won; he did not stumble in someone else's field, like McClellan, but plowed his own.

It seemed to me that he did not come on duty, but out of kindness, sent the orderly with the horses and stayed with me in the tent. He didn't ask too much: intelligent eyes surveyed the inside of the tent, the second bunk, with women's shoes under it, a table with many papers, but without a damask and glasses, so familiar in the bivouac housing of a senior officer.

Grant accepted my congratulations with a long face and, pursing his lips, asked:

- Have you ever regretted taking the nineteenth?

"I am happy with the soldiers. But the last word belongs to the war.

- It's difficult here. Especially difficult without cavalry. And it is difficult for those who expect war according to the rules.

He was afraid to offend me, but he saw in me just such an officer, of European training.

- You did not require cavalry? I asked.

- Here's my advice to you, Turchin - avoid the generals. Take yourself as much independence as possible. Try not to write to them, leave your place a few hours before an officer is sent to you with a headquarters package.

— How to guess?! I threw up my hands.

- Here lies the talent of the commander: to smell not only the enemy, but also the headquarters package in time. In Missouri, special caution is needed, here you lose a busy city at once, when not all the wagon train has got out of it - so that even a wagon train cannot be without weapons. Don't judge by Hannibal or Palmyra, it's the breath of Illinois.

"Even in Palmyra there are not too many sympathetic eyes.

- Some want secession from the Union, others peace, so that the war is banned north of the thirty-sixth parallel. He suddenly asked, "Did the ex-commander, Bashrod Howard, leave the regiment?"

"His company is one of the best.

"It's hard to get food in Missouri," said Grant, preparing another cigar to replace a cigarette butt that had flown from the tent. It seemed to me that his thoughts were still in Camp Long. "Learn to take without making too many enemies, and half the battle is done. We have neglected the cavalry, and we will pay for it, and it costs nothing to feed a horse in Missouri. - He got up, - And yet you can live, since there is a fight, we will trample in the saddle to the end.

Before Grant had left, a band of Irish volunteers approached the tent rowdily and noisily. They vied with each other for justice, reproached me for indulging the Protestants in the regiment - these orders were brought by Adjutant Chauncey Miller and Quartermaster Weatherell - if the Protestant is out of time, he is escorted like a son, and the Catholic is escorted out like a beggar. They yelled badly, like at a drunken talker, especially two: Barney O'Mullen and Daffyd Keeler - a lawyer, a scrofulous desperate guy with small round glasses. Grant looked at the Irish with a frown of surprise.

"I release every volunteer myself," I said calmly. "And I make no distinction either by religion or by language.

"And you are being deceived, Colonel!" Barney shouted.

"Let the one who saw the overpaid dollar or even a cent in the hands of a Protestant come forward.

Keeler left.

"Can you see how a small coin falls into someone else's pocket," he grinned. - Black deeds are done at night.

"The Irish are at war, and the Protestants are in command!"

- And now name: which of the Catholics left without money?

— Miguel! Keeler recalled.

— Spaniard! Hispanic! Several people shouted at once.

So, you showed your lies! The Spaniard left suddenly, when boarding the ship, he did not want to go to Hannibal, where the ticket office was waiting for us. Are all Irish Catholics?

"A good Irishman is a Catholic!"

- The Irish - sow, but who harvests ?!

— The Irish! Irish! Grant didn't hesitate. - Listen to you, so the Irish saved the world, like those geese that did not let Rome perish! Fight first, get an Irish bullet in the ass and realize that there are enough Irish everywhere, both in Catholics and in Protestants, both in the North and in the South.

"How long has he been cramming into our regiment! Grant pissed them off.

- You quickly jumped to the colonel!

"They buy divisions from us, and even more so epaulettes!"

Behind Barney and Keeler, a couple of dozen screamers were inflamed, things were taking a bad turn.

Grant is a guest, but he steps towards them, almost lost among the tall volunteers.

- Regiments and divisions are bought from us, but what can you do; so, you started a conversation with money, money people equip soldiers. And I'll be a general, if they don't kill me, what's wrong with that?

"But it's bad that they will never assign me a general!" Barney shouted.

"There is no other regiment under the command of an infidel in the entire Union army," said Keeler; volunteers didn't see me at Conant's services.

"You are mistaken: the German regiment is commanded by Hecker," I corrected him. "And not only Gekker is like that.

"They say Lincoln was dragged into the church by force," Barney announced. "He approved the Bible to flaunt in the courts, but he does not honour God.

"I despise those who make distinctions between people by blood or by faith!"

"People are different from each other in some way," Keeler persisted.

— Conscience! Courage. Mind. And then at least believe in Allah.

- Let in Allah, if only with God in the heart.

"Will we be lucky if the regiments are against God!"

The entrance to the tent is open, two bunks are visible, a woman's saddle thrown on a straw bed.

"Perhaps they serve the devil here!" Barney pointed inside the tent. - From where the gentlemen were expelled, the devil is spacious there.

- Until now I am without a batman; we can't wait any longer, tomorrow we'll attack the rebel camp. "I spoke calmly, infected by Grant. "Come to me as orderlies, Barney, get accustomed to the demonic life.

- And what - I'll go! — the Irishman was not lost.

- Only an agreement: where I am with the devil, you and St. Patrick go there - even with bayonets and buckshot.

I saw off Grant, the colonel slyly stroking his face in a dull, sparsely gray beard.

"Don't worry, Turchin," he said, squeezing my hand. "Remember the motto on our coat of arms: E Pluribus Unum[17]. I don't know the work is more difficult than to do one out of many; this life is not enough.

- I would like a second fight as soon as possible! I sighed.

— And the first one? Grant was surprised.

The first one doesn't count. In the first, the green volunteer fights blindly. I need a second one, I need a second one.

General Grant's staff.

Chapter Seventeen

I sent scouts under Marshall's Mills, listened to the townspeople and farmers who took their products to the markets of Palmyra, Hannibal and Woodland, and formed a fairly complete picture of the rebels under Marshall's Mills, up to two regiments strong. They waited, fearing to attack Palmyra, but I did not wait, struck a blow with the whole regiment, and at the same time having one volunteer against two thugs who fired from the saddle and without stopping. I arranged the companies so that next to the newcomers, the Chicago Zouaves and all those who had already fought at Cairo in the spring, under the command of General Swift, would attack the enemy, participating in the march known as the big march through the mud. Grant's advice came at the right time: at the end of the night, before the speech, I sent a report to the headquarters that tomorrow morning I would attack the secessionist camp, I sent it with the confidence that General Hurlbut would open my package when the battle on the Fabius bank had already broken out.

Pickets had closed the Philadelphia post road since night: we approached Marshall's Mills secretly from three sides, and the rebels would have suffered severe losses if Fabius had been deeper and wider. The rebel camp was at the sandy fords; raised by our volleys from three sides, the enemy rushed to Fabius, stirring up the river with thousands of hooves and shooting back at a gallop. The rebels left us a tent canvas, fire pits, several kitchens, a frightened bunch of Missourians - servants herded into the camp - and three dead.

The Chicago companies were annoyed at the flight of the enemy, the taste of victory only touched their lips, and the newcomers rejoiced, rushed around the captured camp with burning eyes, entered Fabius, as if threatening the rebels with persecution, but in fact enjoying the coolness of the river, the clear water of the rivulet, which, soon uniting with another river, he will give himself to the Father of Rivers. This is the blessed hour when both grumblers and misanthropes rejoice, like children, in a happy life, a summer day, juicy coastal greenery; when they embrace, forgetting who is Protestant and who is Catholic; when even a negro is embraced by a friendly hand.

Success at Marshall's Mills did not deceive me, we only knocked the enemy out of the picturesque camp. The officers understood this too; standing on the reach of Fabius and looking out over the green plain beyond the river, they were bitterly annoyed that they had no horses for their soldiers.

I rode alongside Howard as the companies moved towards Fabius. Howard urged on his horse, as if he were thinking of breaking into the enemy line alone, leaving his foot soldiers to Tadeusz Drum and Lieutenant Maddison. I warned him: "Captain!" He didn't hear or didn't want to turn around. For a moment, a dark thought flashed through that Howard was jumping not on bullets and a bayonet, but on treason, rejoicing that he heard me behind him and could betray the regimental into the hands of the enemy. My horse's head reached Howard's knee, one day the captain turned with an expression of blind excitement, and the excitement immediately turned to hatred when he read suspicion in my eyes.

After landing at Hannibal, Howard was in disarray. The glory of his older brother almost overshadowed the black glory of Green and Harris, Howard Sr. surpassed them in cruelty from the first weeks of the war. We could run into him any day and for any mile - on the outskirts of Palmyra, near Taylor or Philadelphia, at Marshall's Mills on the banks of the Fabius. The figure of the elder brother—if rumor is to be believed—a giant clad in leather like armor, with a blond beard that he would not touch with scissors until he rode his horse up the marble steps of the White House—that figure haunted Howard night and day, he was afraid that the reckless brother would openly appear before him somewhere in the middle of the day. And Howard was silent, wandering about the neighborhood of the Palmyra camp, silent at the council of the senior officers of the regiment; any thought he had, prudent or overly daring, could be misinterpreted. Words of sympathy became impossible; a friend who dared to express his confidence aloud would become the same enemy of Howard as the one who did not hide a cautious look.

— Captain Howard! I shouted to him, fearing that we would crash into the midst of the rebels. He didn't want to stop, someone else's sabers would have cut him down, if someone else's bullets hadn't saved him: a horse fell under Howard. He immediately jumped to his feet, saw that I, too, was dismounting, hiding from the guns of the enemy, and began to pray that I would save him and give him the horse. "The enemy is gone, captain," I tried to sober him up, "you won't pursue two regiments alone!" - He looked across the river, at the horsemen galloping across the plain, without clouds of dust under the cavalry, looked and seemed to see nothing. The rebels left, and he remained here with his sleepless agony. "The enemy is leaving," I repeated, "the deed is done.

I did not yet know that he, too, was wounded in the shoulder; maybe Howard didn't feel pain yet. Having cooled down, asking my forgiveness for what he said was a stupid request, Howard hurried to his company. But as I stood with Nadine and Dr. Blake outside the hospital tent, I noticed Howard not far off; the captain's uniform hung on a branch, the undershirt was torn, thrown off his shoulders - bloody, it hung down to the captain's boots.

- Who is bandaging? I asked Blake. - Why not you?

"George Johnston," Nadine replied. "He is the son of a doctor and knows everything.

"Howard told me to go to hell," Blake said inoffensively. And it seems to be further away.

There was nothing to reproach him for; a lead barrier grew between us, I took the regiment from him, and when he was forced to stop at the arena in Chicago and his black-haired Elizabeth, lifting her son on her shoulder, clung to the stirrup, Nadine and I were next to the governor and the fathers of the city. We kept everything we had, he parted with everything that he loved, that was his life. The war blew up two families from within; Elizabeth was left alone against a large relative who despised Lincoln and his soldiers; Howard lost everything, even his home state, the trumpets were blown there only for the patriots of the South. I took the regiment, and the news of my brother robbed Howard of sleep and rest; a premonition of trouble tormented him.

"Don't go to him," Nadine said. "Johnston has everything he needs, he washed the wound.

She said that she had not seen Thomas since the signal of the attack, she was worried and wanted to find him.

We soon found Thomas: he was standing over one of the dead rebels, leaning on his gun. I read fatigue and devastation in Thomas's bent shoulders, in his drooping head and distant look. The dead Confederate fell on his back, in a suede caftan, in high boots with spurs, huge, with silver threads in his beard and curly hair. The mouth is bared, pink as if in a scream, and the eyes are open, brown and white, like two chestnuts that have fallen out of a prickly shell.

"I killed him, Mr. Turchin. It was I who killed him.

"If you know for sure that you killed him, then here is his Winchester.

Thomas did not take the Winchester, which I picked up from the ground.

"I... saw him fall..."

"But you can't know, Thomas," Nadine said, comforting him. Any bullet could have killed him.

- No, madam, I know ... I killed him, I think he was a brave man. He turned around, saw me and watched as I knelt down ... - Thomas knelt down next to the dead man. - How he aimed ... Then he took aim, and we both fired. The horse stood still, sniffed it and galloped after everyone. This is where he hit me, and I killed him ...

Thomas understood that his enemy lay on the ground, but he had not yet seen how life was interrupted, how it left.

"The eyes of the dead are closed, aren't they, madam?" Thomas asked.

He was still on his knees.

Do you want to close his eyes?

"I don't think I can do it, ma'am.

- I will help you.

Nadine squatted down, lowered the dead man's eyelids, and rested her hand on them.

"Thank you, madam," the young man whispered.

Get up, Thomas. I put my hand on his shoulder. - In real battles, it will also happen that you will not have time to close the eyes of a comrade.

- Isn't it real? he was amazed.

"Real, Thomas, but small.

Thomas is right, he killed the bearded Goliath, and I'm talking about some other, real battles.

Dinner on Fabius in foreign kitchens is our last quiet meal in northern Missouri. Since that morning, fresh beef has seldom entered our cauldrons; The rebels attacked us that same evening, they were met by my pickets, but they did not go far and made frequent sorties. I, too, had to, returning to Palmyra, scatter the companies throughout the entire area where the enemy was located; you don't have to stand still, you have to move on, and when you move, you understand that the dark water has closed behind you and, where you were just now, the rebellion is again prancing on horseback.

I gave greater independence to the companies, which means I took for myself that freedom that infuriated General Pope. I asked Hurlbut for fresh soldiers to rest my two or three companies in Palmyra or Quincy, I demanded horses, horses, horses, rifles, maps, cartridges, trousers, flour, linen, company wagons, money and bandages, and I knew that nothing will be sent to me, and everything that can be taken in the Theatre of war must be taken without violating the holy rule: 'no property of farmers loyal to the Union should be taken except by their good will.' We rarely got enemy wagons. Unfamiliar convoys followed the troops of Confederate generals Pillow and Price, south of the full-flowing Missouri, there was something to profit from, but there was no one to fight off the wagons - in the north of the state, the rebels themselves lived on the means of the country, ruining the right and the wrong, devouring friends and destroying the property of enemies. And no one wrote it to them on account! The generals did not reprimand the colonels, the regimental officers did not frighten the company officers, the company officers did not lock up the soldiers. A supporter of slavery, robbed by a gang related to him, kept a mournful silence, telling himself: endure, endure, and everything will be rewarded to you a hundredfold. The robbed farmer, loyal to the Union, secretly shed tears or left with a gun to us if he was left with his head on his shoulders. The mutiny generals praised their commanders for their economy and lessons for the damned abolitionists.

And we? Any piece not taken with a powder seal could forever get stuck in the throat, become the cause of infamy, expulsion from the army. If our regiments were made up of only saints, meek monks or law-abiding officials, then it would be impossible to wait for a person who has not eaten for a long day, excited by battle and good luck, to stay alive! - did not touch the bread of the one who generously fed and sheltered the mortal enemies of the republic from bad weather. Every robbery, every injury committed by the rebel cavalry, was basely attributed to us, even if our soldiers did not come closer than twenty miles to the estate.

I ordered the company commanders to feed the soldier; God will forgive you, but you will not please the general's epaulettes. Give the generals military successes, and I will turn my sides, maybe my ribs won't break. I never regretted it: none of my people took too much, did not stuff their knapsacks with other people's goods. We soon learned that both Harris and Green's thugs and Pope's staff officers honoured me with a wild Cossack. Well, let him be a Cossack, let him be a wild Cossack, but a volunteer needs to eat.

I didn't see Nadine every day anymore: she hurried to where the battle had died down, I to a new skirmish; she took the wounded to Palmyra - I left the saddle for the short hours of the night. Like a man charmed by a bullet, I galloped across the steppe and through the forest, from squad to squad, sometimes with several horsemen, with Chauncey Miller and Quartermaster Weatherell, sometimes accompanied by one Barney O'Mullen. The Missouri War did not linger, did not follow any rules; the echo of rifle salvos more reliably led to the war than the dispositions of the company. I was shy before Nadine after every short separation; laughed at her fears and yet was shy - is it possible to love me, who has come out with ten sweats under a uniform cloth, with matted hair on a balding head, hoarse from tobacco and roaring - can this woman expect me like that? Agitated, I listened to voices approaching: Blake and the fistula of paramedic Preston Bailhatch; they are about to leave from behind the trees, Nadia is with them, and if she is not there, then I will receive a note, a few words written in Russian: a complaint that she is bored, that it is bad without me - everything is in Russian, everything closed from prying eyes by the language of the motherland.

She jumped from the saddle to the ground, stood in rough boots, in a long skirt of light blue army cloth, sewn by a negro woman from Palmyra, in a caftan hemmed to the hips and a bright hat, which was not enough to hide her blond hair. At her feet is a heavy medical bag, a briefcase in her hand, and a nifty pistol at her waist, a weapon obtained for Madame by one of the volunteers near Emerson. The Missouri sun reached her face too; swarthy, thinner, and therefore big-eyed and young, she stood in front of me silently while I listened to the officers, examining me with a look: how did I live without her? In the eyes shaded by the hat, gray, but also taking in the greenish shadows of the Missouri forest, there was tenderness, severity and exaction and belonging to another life, where groans and pain, wounds and a long croak of the dying. What is more in the look that will prevail? I never knew this and became shy, my heart beat louder, but it happened that business kept me away from Nadia for a long time, and only in the middle of the night did I enter the tent. Sometimes she fell asleep without taking off her clothes, only unlacing her shoes on her swollen ankles, and I could stand over her endlessly, look at her swarthy face, on which shadows ran when the flame of the lamp flickered, or small butterflies touched the glass with their wings.

Do you remember Howard's wife? she asked one day when the darkness covered our faces. — Black-haired, with her son on her shoulder?

— Indian Madonna and Child. Elizabeth.

- If not for the war, I would have given birth to your son.

I squeezed her hand in a gesture of gratitude, but also a gesture of soothing deceit; because I knew that we would not have children.

"Before, I didn't believe... But here I believed.

So, we are united in this too: the feeling of youth returned to her. As if she was at that time when it was still smart to appear for the first time in public with spots on her face, in a dress raised on her stomach by a new, brewing life.

And she, emaciated, exhausted by the saddle, felt the strength of her body and that fullness of life, which prompts her to generously give, to give from herself everything that is in human power.

And Barney, the batman, I lost to Nadine. This happened later, in St. Louis, during a nightly meeting with the shipowner Shible, but I will tell here, by the way. Barney did not escape the common fate; he cherished Nadine's glance, her kind word. But the volunteers looked at her from afar and Barney every day: ask anything - and she will answer. Barney pestered her with questions, the volunteers only heard the cooing of the Irishman. This is their word - cooing; I found out about him late, having lost my batman. All my knights in soldier's uniforms, in top hats and kepis, in felt and bright hats, in holey boots, were equal before the Lord and Madame, and only I, Dr. Blake, the batman and the wounded sufferers were closer to her. All this was forgiven, but not the batman. They took revenge on him, calling him a cooing dove, a flatterer, a fox, a woman's saint, a sly one, a maid, and even, for some reason, a eunuch.

We were getting ready to board the "Jenny Dennis" in St. Louis, Nadine asked Barney for something, but he rushed over to me and said arrogantly that he wanted to take orders from the Colonel and not from his wife.

Is Mrs. Turchin abusing her position?

"But... I would like to receive orders from the regimental one!"

— What happened to you? Are you sober?

On a crowded pier, there is no hiding from prying ears, and Barney suffered: for the sake of a red word, he is ready for the gallows.

"I don't want to be pushed around by a woman!"

"Lady Turchina," I snapped, "could command a whole regiment of people like you. Go to your old company!

How unhappy he was at heart, this momentary winner.

I observed changes in people: even the chaplain changed, he more than once showed courage, forcing to silence those who used to amuse themselves by pinning sheets of paper or scraps of tattered envelopes on his back. He did not change in one thing - he was a zealous lawyer and guardian of another people's property. One day in early August, near Bird's Point, Augustus Conant caught up with me on horseback and demanded me to turn around, complaining that the soldiers were devastating the potato field.

"Why didn't you act on the soldiers yourself?"

— When there is anarchy in the regiment, the word of the shepherd loses its value.

"I notice under you is not a regimental horse," I said suddenly.

"I borrowed it from the owner of the farm, otherwise I won't be able to catch up with you.

The chaplain gave the name of the host, in the past he was famous for his brutality in suppressing the Kansas abolitionists, but since the beginning of this war he had been cautious.

"I wouldn't ride that scoundrel," I chided the chaplain. "It would burn my ass."

"From pride, Mr. Turchin," Conant replied calmly. - A good saddle, our guys would have such!

"Where do they have such saddles and bridles when they spare potatoes.

Behind the grove, a well-groomed field turned green, not entangled with dodder, with a strong overgrowth of tops. The farmer was waiting for us with a bunch of his Negro workers.

"We could buy potatoes for the soldiers," the chaplain objected to me. - Companies have fodder money, is it really appropriated by officers? Hard to believe.

- Difficult, you say! And for me, it's just impossible.

We approached so that the master heard us; Conant needed this.

- Then let them buy grain, and potatoes, and livestock.

"Don't you know, Mr. Conant, that these scoundrels refuse to sell to us?! "Now I have reined in the horse, let them hear. - They answer us: early, the potatoes have not gained weight. And we'll find last year's one, in the cellar, they'll say, we ourselves have just enough of it, the pigs don't have enough ... Guys! I shouted, as if I had covered the volunteers in a prank. - What are you doing here! "A good half of Raffen's company worked hard. Army shovels, knives, bayonets were used, and most of all quick hands, choosing large tubers. "Don't you know that foraging is forbidden?" — I waved my hat, giving a signal to everyone. - The local owner, not sparing the slaves, sowed the field, and you came to the finished one. Where is your commander?

- Pray to God for us!

It was William Christian, thin, tall, with large Anglo-Saxon teeth. The farmer was standing behind, I hoped that my horse would freshen the air by wagging his tail, but I did not want to look at him.

"Foraging is prohibited," I told Christian and those who stood closer. "If you don't stop digging potatoes, I'll post guards in this field in exactly two hours." In two hours! I repeated, and whipped the horse, breaking off into a gallop. After driving a little, I was convinced that the Chicago Zouaves understood me: the blue backs in the belts were bent again; they have an hour to dig potatoes.

Chapter Eighteen

From the banks of the Fabius, Thomas returned to Palmyra not as impetuous as before, now he was a soldier from head to toe and a comrade you could not wish for in trouble. "I can handle this gun," Thomas said as I handed him the Winchester of the dead rebel, "you won't regret giving it to me. And forgive me, Mr. Turchin, that I did not accept him then, at Fabius's…" I shook his hand. "Is it okay that I call you Mr. Turchin?" "Many people say so, and we are old friends." - "A letter has come from Mattoon, from my mother." "What news do they have?" - "O! Mr. Turchin," Thomas perked up, seeing my interest, "I thought that everyone was at the front; nothing happened. A beer hall was opened, next to the pharmacy, in a wasteland. The second mill has been set up, houses are being built … Don't you regret that you didn't take Editor Dawson into the regiment? "Everyone should mind their own business," I said firmly. "Give him your Winchester and he'll shoot half the farmers in Missouri." "Why are we fighting, Mr. Turchin, if the Negroes remain under their former masters, even such as Fitzgerald Scrips?" - "For me, even blacks should be given guns; that's how it will ever happen." "It won't help Skrips' Negroes; the Lord himself will not be able to define them in companies … "

Yes, Scrips' black slaves are dead, not excluding the two obedient giants who stood at his stirrups with their master's gun and an iron monogram in their hands. Scrips' son came with a gang to his parent's nest, burned down the farm of an abolitionist neighbor, scolded by Scrips with a ferret, exterminated the farmer's family, and discovering that his father's negroes had decided to run away, killed them and rode off with a gray-haired old man to the army of their relative General Pillow. Military fate again brought us to the farm, where Brigadier General Pope promised Scrips to return his branded horses;

blackened corpses lay in the yard and on the current: the sun turned black and white - and nothing blocked the eye of the horizon - everything burned down.

Two undersized boys from Presley Guthrie's company - James Fenton, nicknamed "Pony", and George Johnston, the company leader, whom the volunteers hid inside the line so that the inhabitants of Palmyra or Hannibal puzzled over where the voice of the flute was heard among the silent soldiers - even these two changed from pink chicks to soldiers. When Fenton first appeared at the Chicago Volunteer Depot, company commander Presley Guthrie looked down on him: "Look for a depot where a company of ponies is recruited, I have some tall horses." - "Sir! said Fenton, who was born in England on the shores of Lake Windermere, near the Scottish border. "God created ponies for this, so that draft horses could see how much mind can fit in a small body!" "How old are you, little one?" - "Eighteen". "Well, I'll risk it. If you are also a cunning Catholic, then I will regret more than once that I took you. Fenton replied calmly, "I am an Anglican, sir. But if I were a Catholic, like my stepmother, you would not have to regret: we are not going to mass, but to war.

Many soldiers approached the term; like summer labourers hired on a farm, they could put down their tools, turn in a bunch of cartridges, and trudge to the Mississippi. Across the river, the land of Illinois, their home, the smoke from the roofs, the beds, not yet completely weaned from their bodies, yearning wives; here - life from hand to mouth, on the march, under the bullets of mounted thugs. And in the midst of short breaks, at dawn, every now and then someone would get ready for the road, quietly pack their bag, look at their poncho, ponder whether to take the tear with them or leave it at the tent, hide their meager dollars away and go east.

The regiment was exhausted. Not a day without skirmishes and skirmishes, without crossings and futile pursuit of detachments of horsemen, but on top of hunger and wounds - powerlessness to cope with the planters, with the scum of slavery. An exhausted soldier, with lips torn by the heat, with a dry throat that had lost his speech,

approached the well and found a lock on it, a fresh oak cover on con-
stipation, purposely dated for the war, and could not, did not dare to
knock down the lock - the property of the planter. In the estate near
Warren, I found my soldiers at the well, the owner-farmer explained
to them that they could not take water, yesterday he was robbed by a
gang of abolitionists - not a company, but a gang - and poisoned the
well. He played the role well and, only seeing me, an officer in a high
rank, lost his tone, overdoing his courtesy. I ordered the children and
the farmer's wife to be called; his swarthy face turned green with fear,
but there was nothing to do - he called, then I invited the negroes,
raised a full bucket from the depths and asked him how he wished:
should I drink first, his wife, daughter or himself? I warned that if I
died, poisoned by his water, then he and his family would be executed.
He did not hesitate for a long time, shouted that he would die first, he
saw, really, saw how they poured yellow powder into the well! — it
would be better if he died and his children and the dear Colonel lived.
He. he drank, he vomited, but not from the powder, from vile fear, the
water in the well was clean.

I took from him flour for the company, maize, butter and two stacks
of hay, and gave him a receipt as a consolation. Since then, I have
become especially severe with the enemies of the Republic and the
Union - gentleness with them I would consider a betrayal and would
rather throw a sabre at Pope's feet, let it be broken over my head, than
retreat from justice. From the inhabitants of small settlements and
secluded planters, I began to take an oath - not to assist the rebels, not
to recognize flags with a white star on a blue field and the inscription
"Rights of the South". The planters took the oath not in the master's
living rooms, but on the estate, in public, so that both the servant -
black and white, and the whole servant, and my soldiers would hear
the words of renunciation of the rebellion. For some, the words of this
oath were like a momentary gag in the mouth, like someone else's
spitting on the tongue, like a poisoned piece of bread that you hold
in your teeth without chewing or swallowing, for death is in it: such,
without hesitation, will betray the oath, - but the oath is given, and the

fear of retribution has already settled in a low heart; and the servants know that the master is available for punishment, and for my volunteer - the triumph and rest of the heart, as in the capture of a small fort. Who knows how many more volunteers, recruited in April or May, we would have sent to Illinois if not for this sense of justice being done, the legitimate satisfaction of the citizen and the Republican.

And on Friday, July 26, 1861, at Hannibal's Wharf, we learned of an event that had already shaken the country. At the behest of the new commander of the entire Western District, Major General Fremont, we left northern Missouri; the regiment was baptized by small fire and was assigned to go down the Mississippi to St. Louis and south, in the direction of Cairo, towards the combat regiments of the Confederate army. The border of Arkansas breathed hatred nearby, the number of flying bands doubled, here the planters were ready to open fire on us from their ivy-and-rose-covered fortresses, and to the regular regiments of Pillow and Price to open with a bow the gates of estates, the doors of barns, cellars and funerals with a bow. From the banks of the Arkansas and the White River flocked here land servants of slavery, lovers of a quick tavern court and bloody massacres, they knew how important this land was, where the Mississippi receives the waters of the Missouri and Ohio, and together with the Ohio, Tennessee and Cumberland; whoever establishes his steamships, gunboats and amphibious assault boats here will be a big winner.

The regiment reached Hannibal, the companies proceeded along the spacious streets to the pier, to the echoing boards of the landing stage and the river slope, from where the sunset-gilded Illinois land is so clearly visible. We were waiting to be loaded onto a large steamer with two tall iron pipes, but before loading we saw the "Jenny Dennis"; the ship Shible came again from Quincy with soldiers. I was preoccupied; approaching the main headquarters of the West, I went to my execution; General Pope demanded my resignation, and the new commander Fremont was an original, a legend in the flesh, a man who does not change his mind, an idol of the nation, almost on a par with Lincoln.

Shible waved his cap at me from the upper deck, disappeared from view, and I saw him making his way along the crowded gangway: heated, he ran towards me.

- You argued, Colonel, but in fact how it turned out! he shouted. "Even I could not have imagined such a betrayal. They say there were good fellows who ran from the Bull Run to the Capitol without stepping on the ground - on blankets, on knapsacks, on overcoats, uniforms, on butts of abandoned guns, on boastful Washington newspapers ... - He was shaking, but I did not understand anything . - Holy Lord! Don't you know anything? Like all the valiant officers of our army, Colonel Turchin is doing his job, leaving the highest chips to sell the republic! Where are you taking them, - he pretended to be surprised, - who will guard the black property of the planters?

— Mr. Shible!

We were betrayed at Bull Run! He dropped his beard to his chest. The battle could have been won but lost. McDowell had ten thousand more soldiers than the rebel Beauregard, but the troops must be led, and this is more difficult than flaunting in the uniform of a general.

Schieble spoke about the defeat of the Potomac army, about the mistakes of McDowell and Patterson, about the panic and the shocked country.

"Now they will begin to draw troops to Washington; We must protect the unfinished Capitol! They will never understand that this is not a war, but a squabble over a piece of land, just because by chance it was called Washington or Richmond. If I believe in anyone, it's in Fremont, he is a warrior, he is not one of the metropolitan talkers. And General Grant brought fear to the rebels here ...

— General Grant? I didn't know about his promotion.

— Ulysses Simpson Grant!

Ulysses Grant climbed another step, climbed quickly: military fate is not always so fair. I did not envy him, but only thought how different our service was: Captain Grant became a brigadier general, and Colonel Turchin wakes up every day with the danger of being removed from the regiment.

Shible saw concern in me.

- Let's go up to him, Turchin, let's have a glass.

"I don't drink, captain.

You know who hasn't drunk yet? How many did you drink?

Many, I hope.

— John Brown! He is the only one I forgave sobriety.

"Forgive me too, Shible.

"But I will drink to your health; Brown wouldn't let that happen either.

For some reason we hugged. Shible's beard touched my cheek, I thought that he was not so severe, but his life was difficult.

I was standing at the gangway, loaded soldiers were walking past, carrying sacks and bales, rolled-up tents made of rubber-impregnated canvas, ammunition boxes onto the ship; three cannons were rolled in - a movable battery attached to the regiment in Hannibal, several wooden boxes with Colt and Mignet carbines, they promised to give cartridges for them in St. Louis. Already by the light of the torches, horses and mules entered the gangway, the riders carried in their arms a heavy harness. The company commanders lingered around me, and each inquired about the Bull Run. But human ways are inscrutable; the Bull Run did a good service to our regiment—about two hundred of my volunteers at Hannibal stepped on the ship as summer soldiers, overdue or at the end of their term, and on the wharf at Bird's Point they went out as volunteers of the whole war, and covered, dead and alive, with glory the regimental banner.

Hot August has arrived. The mutinous General Pillow gathered a large force in southern Missouri, around New Madrid, and we did not have to land in St. Louis, Fremont decided to draw as many troops as possible to Bird's Point, and for two weeks we did not leave the fierce vanguard fighting, pushing the regiments Pillow. Volunteers trained in independent combat in northern Missouri became an inconvenient nut to crack for the enemy. The rebels had cannons, but powerful ones, mine were shot with grapeshot; gunners let the enemy close - even the newly converted artilleryman Thomas tamed young impatience. The regiment did not change its rule of swearing in planters, and in the south of the state my actions caused such screams that if the White House were closer to the Mississippi, its columns would collapse from curses. The black list of our sins grew, but the regiment fought well, and the generals endured. And not only did they endure it: after Norfolk, we were sent to the most risky places, testing me and my soldiers, as if rewarding me with pokes: you are hot, so it's hotter for you; you are itching, so go ahead, go every day, fight and fight, so that there is not even an hour left for trifles, for training planters. They thought they would annoy us, but we are glad—isn't the soldier going to the war of the Right in order to fight! These days, at the robbery auction of the Knights of the South, the price of my head has doubled, but in order to get the jackpot, someone else had to get to me and separate the head from the body.

Intelligence reported that the left wing of Pillow's army was moving to Dallas and Jackson, preparing to strike at Ironton, and now our regiment was back at Bird's Point, loading onto ships, going up the Mississippi and hurrying from Sulphur Springs station by train to Ironton. And then a regiment in the vanguard, this time of General Prentiss' expeditionary force, behind us five regiments, an artillery battery and a horse reserve, we go in the direction of Dallas and Jackson, meeting the weak resistance of the rebels. Only once, near Jackson, did the regiment take on a long, uncomfortable battle, we were attacked from different sides, furiously, I lost men, the baggage train and guns were threatened, the rest of the Prentiss regiments moved on Cape Girardeau, making sure that Pillow neglected Ironton.

On that day, General Pope's auditors came to me: a food commissar with the rank of major and his assistant, a lymphatic youth with sad, watery eyes. Their appearance among the driven, fighting regiment was like a miracle: two officers in shining uniforms, on fresh horses - a major on a bay stallion, a lieutenant on a white filly - and not a speck on them, not a speck of dust, as if earthly vanity and dirt let them through, were heard to the sides, as once the sea was heard before Moses. The major interrogated me kindly, patiently waiting for me to finish urgent business.

His assistant's ledger lists everything: every gallon of beans, lard and corned beef, flour, every bag of crackers, a pound of coffee, sugar, candles, soap...

"Of course, you didn't tally up the grand total," said the major. - Regimental commanders are rarely concerned about the overall score.

- I leave you to make such an account.

"But you also took food from the farmers?"

- They took - a gift. Bought from those who were inclined to sell. Confiscated - in extreme need.

- It seems that the need did not leave you all the days.

"With your help, food commissar!" I didn't owe.

He was an imperturbable officer; not a shadow of discontent, the same gaze of intelligent, sympathetic eyes, an even voice reading our sins from the list - Fitzgerald Scrips horses, someone else's maize, butter and sugar, corned beef and molasses, a bull sent to the company boiler, a light potato field ... Everything is calculated as if a pack of scribes were on our trail.

"Order the lieutenant to record the new," I said, "everything taken at Birds Point, Norfolk, and Sulphur Springs.

"We'll do it when the companies come out of the fire.

"Then you will have to wait for the end of the war; See, they're jumping after me again.

Lieutenant Maddison was approaching, but on Howard's horse.

- What's wrong with the captain? I rushed to Maddison.

"It's all right, Colonel," Maddison smiled, my heart was relieved. "They rode out of the woods with a false federal banner. - He snatched a crumpled banner from under his unbuttoned uniform, - And here we have it, and I am without a horse.

Did you save the saddle? I asked Maddison. "Then unsaddle the stallion," I pointed to the bay, "and God bless!"

- Mister Colonel! The Major did not believe that this was possible.

Do it, Maddison! I'm confiscating it for the needs of the war.

The major looked furiously from his saddle thrown into the grass to the retreating lieutenant.

"A fair amount of savings must have accumulated in the companies," he said, showing remarkable restraint.

"The company money is not accountable to me.

But when it comes to abuse...

"Don't rush to declare war on my company officers," I interrupted him. "By nightfall, if they come together, throw your suspicions in their faces, but beware, I beg you.

"You are wrong to think of intimidating me, Colonel. He chuckled. "I'm an old soldier, I spent the Mexican war next to Grant. That war taught us to punish guilty soldiers and officers. When we returned to the serene cities of Illinois, Ohio, or Pennsylvania, the inhabitants found us cruel; we did not spare the marauders, we tied them to a gun carriage or sent them to walk around the city gagged, with their heads threaded through the broken bottom of a wine barrel ...

- Then you did not spare the enemy either: a Mexican, an Indian with a bow in his hands. And now you send clerks to count the losses of traitors.

"Now those who are close to each other, divided by delusions, are fighting. Here the former cruelty is inappropriate.

Why is the rebel so cruel? There is no meanness before which he would give in: the forgery of the flag, the killing of prisoners ...

"Even blood brothers are somewhat different: one is kind, the other is cunning; one is brave, the other is peaceful.

- And you give them courage, cunning, military prowess, and leave us the kindness of simpletons?

"You won't understand the Americans, Colonel Turchin!" The South is a special world. Daring, arrogant guys, funny old people who imagine God knows what about themselves.

- Why, if this is a nice domestic massacre, to call under the banner of tens of thousands of Irish, Germans, French, Italians ?! You give us the right to die for the Union, not allowing us to judge and think; then you yourself are a slave, an involuntary slave of the South.

- You took away the horse - I endured, but I will not allow dishonour!

- Ride away from us: no matter how I take the second horse.

Returning an hour later, I did not find the auditor: he left on a white filly and sent the lieutenant to the station on foot.

Prentiss's regiments safely reached Cape Girardo, and in the first days of September we had a hard time: from the vanguard of the expedition we turned into a rearguard and retreated under the harassment and bites of a pack of rabid dogs of the rebellion. We fought without knowing the number of enemy losses, without reporting on successes. We were pursued by flying units, Pillow's cavalry squadrons and gangs under the command of Jefferson Thompson, nicknamed the swamp fox, we met them with fire, ambushes, close buckshot and took away from them two more false federal banners - in the first military summer, the rebels, with the help of low deceit, even reached success in battles against the brave Lion and Siegel. Bands of rebels succeeded each other - there was no one to replace us: all the same weak Austrian guns, a meager ration of cartridges, legs buckling from fatigue, in torn boots and boots, the same chilled bodies under uniforms, ponchos and blankets damp from rain and dew .

We arrived in Cape Girardeau in the evening, shabby, happy, noisy, ready in joy to overturn both the log wharf and the cozy houses of the townsfolk into the deep Mississippi. The Chicago Zouaves yelled at the top of their throats, the other companies were not silent either, each wanted to make the residents of Cape Girardeau go to sleep happy with the song of their county - the landing stage and the cobbled marina square were swept over the camp. And when in the darkness a slender man in a black cloak with a gilded buckle pushed his way towards me and stared at Nadine and me with ungracious eyes set close in his narrow face, I mistook him for a steamship steward and prepared for a skirmish.

"If this is a regular regiment of our army," he said through his teeth, so that his mustache and flat beard with a noticeable gray on his chin hardly moved, "then it is not surprising that we lost one battle after another.

"You have a regiment of Illinois volunteers," I objected, "and not the worst in the army of the North. When this regiment fights, no one complains about it except the enemy.

And local farmers! he added quickly. — And food commissars!

"Did God send me an inspector in Cape Girardeau, too?"

"John Fremont," the stranger introduced himself. He held out his hand to me, the cloak opened slightly, I saw the uniform of a major general, a wide, gold-woven belt. "It's time for us to get to know each other, Colonel Turchin.

— John Basil Turchin. - I felt an energetic, not promising complacency squeezing.

— Miss Turchin? I am glad to welcome you to Cape Girardeau. There is alienation in the voice, official courtesy, as well as in the gesture of the hand touching the French felt hat.

Fremont was overtaken by the officers of his headquarters, around us became crowded and tense.

"I need you," Fremont turned to me again. "Just don't try to confiscate my horse, it's at the tavern hitching post. The colonel has taken the privilege of taking horses from officers whom he considers idlers," he explained to the crowd.

No one laughed: neither in the simplicity of his soul, nor obsequiously. I didn't feel like laughing either; I knew something about General Fremont (below).

Chapter Nineteen

We walked through the crowd: the general was in front, I was behind him, trying to keep my head high in front of my volunteers. The way seemed long to me, although John Fremont's flying home was in an iron warehouse; behind walls of bales, sacks and boxes, a large room was fenced off, decorated with carpets.

I did not expect mercy: five other regiments of Prentiss, covered by us, had already departed from Cape Girardeau, and we arrived late, battered and with losses. In the briefcase of the general are complaints against me - the food commissioner, General Pope, our chaplain. With me in the regiment is my wife, madam, the only one in the army of the United States. In the companies of Negroes, I entrusted them with both intelligence and guns, in the companies of Ruffen and James Guthrie. Everything is bad, everything is against custom and charter, I did not have time to acquire laurels, I did not take my Carthage or Troy.

Desperation has its own convenience: I was ready to break straight ahead. When I first looked at Fremont, the sight of him annoyed me: the wrong person I expected to meet under this big name. Few in the republic could match his fame, in the West he had no rivals at all: explorer, scientist, brave pioneer, conqueror of the Rocky Mountains and the Pacific coast. Senator is a high-profile title, and Fremont was not an ordinary senator: he had first conquered California, presented it to the States, along with the gold found there and the graves of the exterminated natives, and only then came from California to the Senate. Fremont is a topographical engineer, whose name is given to the mountain passes and peaks he discovered; Fremont is the Creuse of the Republic, the owner of boundless lands, heavy with golden sand; Fremont is a darling of fate, escaping from death, already leaning over him; Fremont is the first presidential candidate from the Green

Party back in 1856 of the Republican Party and Fremont is a breeder, an adventurer, an idol of the crowd, who has acquired laurels even in Europe. After Mexico, Lieutenant Colonel Fremont was court martialed for insubordination and liberties and expelled from the army, but the first treacherous volleys of the seceding South reached him in Paris, he offers his sword to Lincoln, and the delighted president dresses him in the highest army uniform of the Republic.

He threw the cloak over the batman's hands, gave him his hat, sat down and invited me to sit down "Did you really take the Major's horse?"

- I would take it again if it were all over again.

He was waiting for an explanation.

- A horse fell under the officer; there was a fight, and I confiscated a stallion from the major. If the major had a cannon, I would take it, but they don't carry cannons with them.

We were listened to by two officers leaning over a map, and someone else doing nothing; Headquarters officers caught our voices in the embrasures of the inner walls made of bales and boxes - everyone is flattered to hear how the general skins the "robber colonel".

"You took the horse," he came to my aid. Why didn't they bring her back?

— The Major can be proud of his stallion: he fell in battle.

The general jumped up, walked inaudibly along the earthen floor of the warehouse, turning his displeased, preoccupied face towards me, and began to carefully reprimand me, still not losing his strange interest in me. I mentioned Fremont's narrow face, but it was both proud and strained by the constant whim of choice. I listened to Fremont calmly until he mentioned Fitzgerald Scrips, saying that by extremes I pushed the old man to revolt.

- General! I stopped him. — I hope you didn't know Scrips?

"I believe Brigadier General Pope, and you had him at the Scrips estate.

"He was on the farm of Scrips' neighbor, an abolitionist exterminated with his whole family by Fitzgerald Scrips.

"You risk a lot, Colonel, even at the very beginning of your career.

- Only life; And who in the war did not risk his life.

I began to see the pattern in the fussy and silent movements of the officers of the retinue: the general was preparing to return to St. Louis, to his fortress and his Washington.

- You are Russian? I'm not looking for clues in blood," Fremont said to my nod. "Not knowing, I would take you for an Irishman.

— This is a common misconception: it does not offend me.

- With me at the headquarters of the Germans, the Irish, and more than others - the Hungarians. Hungarians are born warriors.

"I don't put blood before man either, General. Perhaps that is what made me an American citizen.

He was embarrassed by the lack of search, the desire to go ahead where there is a safer road.

"Still a career, Colonel. It is life expressed in society.

- I am ready to end the war with the rank of colonel, which I achieved back in Russia, if only the war would be decided fairly.

- Is there any other possible outcome, except for the restoration of the Union?

Raising my voice, Fremont invited the officers to listen to us, I cursed myself in my thoughts: step back, fence yourself off from the learned general with an evasive word.

- For the sake of the unity of the Union, I would not put on a military uniform.

The word has been said, the heavy cannon ball has fallen at the feet of the commander. What did I know about John Fremont: who is he in the depths of his heart? Ambitious, seduced by the highest army uniform, or a man who cannot bear the spectacle of slavery?

Is the unity of the Union not a high enough goal?

"On its own, it wouldn't have put me in line; I would let others fight for the old frontiers.

Why are you fighting at the head of a regiment?

- For a republic that will destroy the shame of slavery. You talked about my sins, general, here's another grave sin for you: I drove out the masters who came to the regiment for their negroes. And if they offered to buy off the blacks, I would throw money in their faces. There was a long silence, Fremont did not interrupt, and the crackle of two large lamps could be heard. "That is my goal, General, and the goal of the regiment, almost all of its soldiers.

"I know two Illinois regiments—both officers and men—average young Midwesterners. What is special about your soldiers?

"They are brave and trained to think.

And they owe it to you? - The thin nose of the proud man squeezed unkindly at the nostrils.

Republic, general. They are my first success since the day I landed on the American coast.

"You don't like our army," he said suddenly softly. - The nineteenth is good, but the whole army is bad.

- I like that in our country you can punish an officer only by dismissal from service, and there is nothing to reward him with. He does not expect orders on his chest, and even more so he wants rewards for his heart. But there is no true army yet, you know that.

He leaned over to the table, found some papers, looked around under the light-swept rafters of the warehouse, as if wondering why he was here, and not in the Bois de Boulogne, not in the opera box, under the glances of envious Parisians.

"On my way from New York, I saw McDowell's army after Bull Run. He turned to the officers, saddened by the events. "The soldiers were badly disposed of. Fremont looked at me with an absent-minded look, perhaps seeing not me, the scolding officer, but a tempting image of his future victories. "I know what needs to be done here on the Mississippi, and I won't let them confuse me.

That's when I felt that he was not surrounded by seekers and flatterers: the officers looked at Fremont with a willingness to share his fate.

Yes, I know, I've thought about it. He approached a large, well-drawn map, solemnly picked up the papers in his hands and tore them up. "These letters and reports were related to your sins, Turchin," he said, "let's forget about them. I like your way of fighting. It is a crime for a soldier to starve and not to take the cattle from the enemy just because you did not overtake him with a red-hot gun in your hands.

He spoke without pause to spare me the difficult duty of giving thanks.

"We won't get better," I said when he stopped, "until we make it our goal to destroy the rebel troops, and not push them into the depths of the Confederacy.

- He began to walk, in the power of his thoughts; fuss arose again, the officers took down the map, the orderly took away the commander's suitcases. "We dream of Richmond, as if this piece of land in itself means anything, as if it is difficult for the South to move the capital to another city, even in a wagon. And how many soldiers are idle on the Potomac, how many forces and weapons are chained there!

Fremont stopped abruptly and announced to his subordinates:

"And the president is demanding that my proclamation be rescinded!" To do it myself, with my own hand!

The cries of indignation and dissatisfaction with Washington showed that it was about something extremely serious.

"Washington is obsessed with Kentucky!" They are ready to lose the war, if only not to disturb Kentucky, not to alarm our mythical Kentuckian friends. He spoke loudly, as if not in the Cape Girardeau warehouse, but in the Washington State Capitol. — Colonel! How many friends of the Union did you find in the South?

"There are friends in Missouri, but they are waiting for our decisive action. What proclamation are you talking about?

Everything stopped and froze around. They looked at me as if I were a marvel, insulting the eye and mind.

- Here, if you please, fight when the regiment commander does not know the orders of the commander! "He seems to regret having torn up the denunciatory sheets. "This is a proclamation for the people of Missouri, for anyone who is hard of hearing, for a state governor playing a double game. For you, Colonel, this is an order. General Prentiss received it three days ago.

Frémont handed me the St. Louis Courier; casually and angrily, as if even one person in the universe who did not read his proclamations was a challenge to heaven. What a glorious, dapper man he was, in love with himself, with his star, a capricious, graying youth! If I had the right of friendship, I would have embraced the man who let me breathe with all my chest: Fremont extended military laws to the state of Missouri - the property of all citizens of the state, declared a proclamation, no matter what this property consisted of - in slaves or real estate, is confiscated if the owner is proven to be actively collaborating with the enemy on the battlefield, and his slaves are freed.

"But Washington cancels it, demands that I back down!" - He was angry, but also played, showed how unbearable his situation was and how difficult the victim was.

- In the opinion of the government, the proclamation could reveal a single flaw.

- What is it? Fremont asked jealously.

- The one that it affects one state of Missouri: the whole South, the whole Republic must live by this law.

"You don't know our politicians very well!" Fremont exclaimed. "Like old matchmakers, they still hope for marriage where there is only enmity left. They will come to the emancipation of black slaves, but too late, having paid for their cowardice with unheard-of blood. I will not make things easy for them; if the proclamation is canceled, then only by the president, let him take responsibility: he, and not me, will have to lose some of the fans.

The warehouse was empty; only the lamps swayed in the wind, he burst out of the darkness through the open doors. We were going to the pier: I - to the regiment, Fremont - he again donned a black cloak - was waiting for the steamer to St. Louis. The general asked about the tactics of the rebellious partisans, about the roads, but spoke more himself, complained about Lyon's recklessness, praised Milligan, but found mistakes in him, complained that the Irishman had gone too far, denounced Washington for the constant demands of the regiments to strengthen the army on the Potomac; for if the government keeps near Washington all the guns and guns that come from the factories of the East or bought in Europe, he will remain here unarmed and will not be able to create an offensive army. The storm had passed over me, I listened to him and noticed that Fremont rejected all disagreement; I had no success when I said that the war in Missouri was impossible without a strong cavalry reserve, or found it important to immediately support Milligan - about Lyon, left without support and now dead, I was silent, a matter of the past, and for Fremont it remained as open wound. With every step along the dark, sleepy pier, Fremont grew cold towards me - it was true that he was offended that I passed the tests of our conversation, without yielding to him in anything, without giving a real price to his disposition.

Our steamer on the Cairo had not yet docked, the Fremont's boat was making steam and winking lights in the night. "Your soldiers are asleep," Fremont said, stopping abruptly and farewell. "I can't be sure of their exclusivity. They haven't slept for three nights."

So, what is their talent? he repeated the previous question.

— In the ability to act independently and in their esprit de corps.

- O! You give the dorks and esprit de corps!

I could tell that there were more intelligent people in the 19th Illinois than in other regiments, but would he believe it?

"I want to give you advice, Turchin. My silence hurt him. — You carry your wife with you; in the entire army of the North, and indeed of the South, there is nothing like it.

"But here and there, there is no second such woman. Even in Sevastopol, under the hand of the king, women managed to distinguish themselves by saving the wounded.

- The army is guided by the charter, and the charter prohibits women, except as scribblers.

He became dry and official, there was only one thing left for me - disobedience.

- Nadine Turchin - paramedic. I won't do any harm in this.

Frémont took off his hat and shook his maned head.

"Among the sheets I tore up was one written by her hand. The chaplain sent it to the headquarters, convinced that it was a treacherous cryptography; Here is your harm and embarrassment. The chaplain did not know that the letter was tender.

"But you don't know Russian either, general!"

- Your compatriot, a Russian colonel, is here, and now a successful recruiter of soldiers. He read it and completely reassured the headquarters. Frémont held up his hands, palms facing me, in reconciliation. - We did not find treason in the regiment, even marital.

"If business is done, let me leave.

I trembled, not knowing what to attribute it to: a damp night on the river bank, rage, or sudden shame that our lines were touched by the pupils of the chaplain.

- What are you all of a sudden, for nothing ?! Fremont was surprised.

"If you could see with what fury the chaplain's heart will respond to your proclamation, you would never accept a piece of paper from his hands. They are everywhere - secret friends of the South, blind or traitors.

"I don't have them at my headquarters," Fremont said with all possible sincerity. - Europe gives us excellent people, but they come to America, rejected by their homeland, with all the passion of political dissatisfaction, they hurry us, whip the horses, ready to drive them, if only quickly. Farewell!

We shook hands, investing in a short handshake forgetfulness of grievances in the face of war.

... The regiment did not linger on the Mississippi. In Cairo, Grant was in command, subordinate to Frémont, but here, too, he achieved a marked independence. The stamp of his personality lay on army affairs at this most important point at the confluence of the Ohio with the Mississippi: the immediacy of decisions, the calculation, organization and sufficient food of the regiments spoke in his favor. With false movements of ships, he misled the enemy and secretly, in the holds, transported troops, gathering them for a possible strike towards Arkansas, or Tennessee, or down the Mississippi, to Memphis. This time I did not wait for Grant in Cairo, and he took care of us, having heard how exhausted the regiment was on a futile expedition to Ironton.

We were transported to the Kentucky coast of the Mississippi at Fort Holt, a newly built and poorly armed fortress. We had hardly rested for five days among the green hills and groves, by the embankments and fresh logs smelling of tar, we had barely allowed the wounded to throw away sticks and makeshift crutches and remove bloody bandages, when a new order came: having joined with the 17th Illinois Regiment, move downstream and take Ellicott's Mills. A week at Fort Holt and Ellicott's Mills gave me a glimpse into the life of the state for which the President demanded that Fremont revoke the proclamation of the emancipation of the Missouri slaves. In Kentucky, everyone was lurking, waiting for the art of war or the vicissitudes of fate to give an advantage to one side, when it would be safe to join the lucky banner.

In Kentucky, I had time to consider the Fremont Proclamation. What a humiliation of freedom! Washington is tearing up John Fremont's Missouri charter, which, after all, is a concession, a compromise, an involuntary admission of slavery. I couldn't say that to Fremont in Cape Girardo, happy that the word was spoken, and what a word: and their slaves were freed! But Fremont also declares freedom only to that slave whose master openly joined the ranks of the Confederates, risked his life in rebellion. And thousands of cowards who have remade the word "treason" into the convenient and evasive word "loyalty," thousands of prudent flatterers who excel in cruelty to a slave and those who have drawn a sword against us - all of them, even under the Fremont charter, retained the rights to black slaves. Did I know about the black sufferers when I drove up with Nadia to the Austrian border in order to close the doors of Russia behind me? I knew, I knew, but I am here among the healers of leprosy; a person needs to stand so close to evil in order to become a warrior from an observer. The horse of the republic must not slow down - the guns of the South will take the right aim at the slowed down target and shoot it. It would be better if Lincoln looked not in the mirror of Kentucky, but in the Missouri, split so that no concession could glue him together. What is in Missouri today will be in Kentucky tomorrow; silence will end, evasiveness will end, the hidden weapons of feigned loyalists will thunder with shots.

We were just settling into Ellicott's Mills when a new order came: the 19th Illinois immediately depart for Cairo, board the Illinois

Central cars, and proceed to Washington, under the command of General McClellan. On the steamboat, accompanied by Grant's courier, was Teddy Dawson, war correspondent for the Chicago Daily Tribune, Madill's and Ray's papers. The order to move the regiment delighted Dawson, alarmed me. We have adapted to the war, and the Potomac is calm. In Missouri, senior officers gave up on me; even Pope saw that the regiment was fighting bravely, but what would it be like in inactivity, in the nooks and crannies of headquarters, where even a capable officer loses a share of his mind?

Grant wrote to me that Washington again demanded 5,000 equipped men from the West to defend the capital. Fremont was unable to provide five regiments and decided to sacrifice the two best: mine and Hecker's German regiment. Grant wanted to keep me, but Frémont stood his ground; if he sends two regiments instead of five, then let the best ones go.

"You see, you are an impossible person," Dawson attacked me. "You have been honoured, and you grumble.

I kept quiet: an incorrigible breed of newspaper warriors. They are the first to see victory, where it does not smell; they are the first to draw the hasty lessons of war, and the first to betray these lessons at the slightest difficulty; eager to fight, but no further than a certain line; they advertise as a virtue what in reality is the misfortune of the nation.

"Don't bet on Fremont, Turchin," Teddy Dawson went on. - He's a broken card. They say in Washington that he surrounded himself with adventurers, created a military Versailles in St. Louis, distributes money from the treasury ...

— Nonsense, Dawson! I cut him off. Frémont is a good citizen, he declared war on slavery.

- And immediately backed off! Dawson triumphed. Lincoln ordered Frémont to change the proclamation, and Frémont complied.

After the first letter from the President, Fremont was determined, in Cape Girardeau he encouraged my determination, and here again the fetters on the hands and feet, the scope for enemies, the deceitful secular rules for the Union soldier, again he is an open target. I took off my hat and bowed low to the northeast of this country.

- Who are you, Turchin? Dawson took off his glasses: did another boat appear on the Mississippi?

"I mourn my hopes, Dawson.

Teddy Dawson was out of need for the first time, he didn't spend money on the voracious Courier, now he was getting paid. Teddy dressed, ate for two, and most importantly, was not the last of the patriots, joined the soldier, sharing his glory, but not sharing the mortal danger. Few things corrupt a person more than the glory of war and closeness to it, without the obligation to fight and die.

Outside the open window of the cabin, the water was splashing, we were going up, against the current of the Mississippi, shoveling hard with our plates, and our conversation was just as hard. Dawson laid out homegrown plans for the defeat of the Confederacy, and the capture of the rebellious Richmond turned out to be the crown of them. Nadine noticed that Napoleon did not enter Richmond, a town adjacent to Washington, but the capital of a huge state, that, settling in the midst of burning Moscow, he sent an offer to Emperor Alexander to begin negotiations, but the answer was: "The war has just begun!" Teddy assured us that we did not know the South, its adherence to idols and superstitions, that if Richmond fell, the whole cotton empire would collapse, we would not sneeze to death among the dust rising from the rubble. He was angry that Chicago had spoiled us, but we were different in Mattoon - he became a boring herald of common places, a zealous servant in the temple of the Republicans.

Chapter Twenty

We landed at the landing stage at Cairo, and Edwin Ramage limped up the ladder first. When the enraged slave-owners wrecked the Herald of Freedom's type-setting cash desks and machines, they did not forget the editor: Ramage paid with a limp, broken ribs and a scar that cut through a thick dark eyebrow, cheekbone and cheek. He brought me, Nadine, and Dawson back to the cabin and asked, not without mystery, how my relationship with Fremont had come about.

"We parted well," I replied. "But Teddy thinks Fremont is a lost horse."

Edwin's face darkened, the scar became dead gray, his olive, lively eyes flickered from me to Dawson, are we kidding?

"I was waiting for the steamer to see if things could be fixed?" - he said.

"We are not being exiled to New Caledonia, Ramage, we are going to the capital.

"I was at Fremont when Grant rode up and asked to keep Turchin's regiment in the Missouri army. He gave Ross's regiment and some other, but about yours, Turchin, he said that one such regiment is worth a brigade. Fremont protested that he had seen the regiment and found in it the most ordinary soldiers.

"Sleepers, Ramage," I said.

- Fremont answered Grant: "Maybe you are right, Ulysses, I did not see them in battle. But if they are like that, let them go to Washington, two regiments - Gekker and Turchin - are worth the five regiments that are required of us. It seemed to me that you baked him with something.

I bowed lightly, jokingly and silently.

You have done something and shut up. "This Turchin," Fremont said, "plus royaliste, que le roi[18]. He has an itch to teach senior officers. Let him enlighten Lincoln, inspire him with an ardent idea of the general abolition of slavery, here he has no one to argue with ... "

"What an excellent, what a noble court adviser! I said, not even to Nadia, who could understand me, but into space, into the gray fog of my irritation. I grabbed the door handle. Frémont is an extraordinary man, and his proclamation is the best I have read since Brown's court speeches. And what a pity that even the best people in our republic do not tolerate the truth.

Leaving the horses at the tavern, Nadine and I had a quick bite to eat in the garrulous hall; from the tavern we went to the station, to check on Weatherell, whether everything was fine with him with cars and steam locomotives. Nadine and Regimental Medical Officer Samuel Blake lingered in the medical supplies car, I stood by the tracks, the daily train to Sandoval, Cincinnati, and further east pulled up here. I was staring at a steam locomotive with a large conical pipe and rear wheels that were planted when I was accidentally pushed by a gentleman in a new-fashioned summer hat resembling a military cap.

Who do I see! Can happiness be trusted? he spoke, spreading his arms: in his right hand he held a cane, on his left a summer coat fluttered. - Ivan Vasilyevich! Dear!

Sergei Alexandrovich Saburov stood in front of me, and behind me stood a soldier with luggage in his hands. All this did not fit into the picture: the bivouac Cairo, the provisional officers and the impeccable civilian Saburov with a batman.

- Not happy with the countryman? - Saburov twisted his touchy lips; a dark purple plaid suit, a light waistcoat, a starched shirt, and a visor lowered over his forehead distinguished Saburov with a mixture of nobility and efficiency; one might have thought that he himself was being carried along the streets of Cairo by soldiers, so shiny were his boots with untouched varnish. "Don't be angry with me!"

— Hello, Saburov.

- Well, that's good and that way: briefly and angrily. Still dear - hello.

- I remember you in the uniform of a lieutenant; you were going to make the natives of the West happy.

"Alas, the night came when I abandoned my military dress in the Rio Grande; the uniform could have cost me my life.

- People of honour put on a uniform for this, in order to give their life for it if necessary.

"Well, let's suppose I have enough honour in me for ten Yankees!" Saburov objected, not offended. "They humiliated me with the uniform of a lieutenant; if I were a general, I wouldn't have to look for where to drop my military dress.

- You have become one of the worst Yankees; buying and selling is on your tongue.

I despised him without condescension, completely and completely.

- And you did not keep your word, returned to epaulettes when it became profitable: I know how profitable your position is.

He broke off, afraid of a change in me, but Nadine's appearance saved us.

"I'd kiss you the old fashioned way, on both cheeks, but he'll kill you, he'll kill you!" Saburov lamented. "He is waving his fists at me. What did I say? Yes, the position of colonel is profitable, no less than the California estate where they found gold, but you won't bend down for ready-made gold. Here is your gold! He pointed to Nadine with bitter envy.

We were silent.

- Push away, drive your brother! For half a year now, since I have been supplying soldiers to the North, I alone have given the army no less than a regiment, and all not for myself, without counting on epaulettes.

"So aren't you the Russian colonel who seduced Fremont?" Nadine asked.

"Here's my guilty head!" He leaned over. "If it weren't for that duel in the Caucasus, I would have become a colonel long ago, I served well ..." He noticed that we were retreating from him, and hurried up: "But all the lack of imagination, impoverishment of the mind: at the first question - who are you? - I lied: Colonel! I remembered you, Ivan Vasilyevich, your oath not to return to the army, and thought: I'll take his shell. I'm a bit of a Buddhist, I believe in the transmigration of souls. Then, in Missouri, I heard about you, but it's too late, you can't deny the colonel... Wait a minute! I want to tell you about Balashov...

We knew about Napoleon: he was sent to his homeland in chains. And what about Balashov?

- Bell hanged himself in hard labour, who would have thought? The court was merciful to Balashov, he received a short term, and the war began - he was allowed to serve. He accidentally wandered into my volunteer depot, and I anointed him as a warrior! - Saburov approached us, spoke quietly and quickly: - Emperor Alexander favours America. The new reign is happy and humane. Shouldn't we hurry back, gentlemen! Whoever wins - North, South, damn them, what difference does it make to us? Their whole anthill with the Capitol, with steam ships, with cannons and epaulettes, is not worth your head, Ivan Vasilievich, and your pen, dearest Nadine. We are strangers here...

And again he bowed his head with a movement of perfect reverence; Saburov believed in his charms and waited for Nadine's hand to reach out to him. And she stretched out with a sharp, backhanded movement and fell loudly on Saburov's cheek. The bowed face did not immediately rise, Saburov hesitated, sideways, from under his sideburns, looked around the square, how many witnesses were attracted by a slap in the face.

- The handle is heavy, isn't it? War, blame it.

Before her, the thoroughbred Saburov swayed in despair; his eyes filled with moisture, the desired martyrdom softened his features, he would fall to the merciless hand, apologizing and asking for forgiveness. I was afraid for Nadine; is she capable of such a thing, will not Saburov's astonishing sadness touch a woman's heart? Nadine turned away, and we walked away from this unfortunate place.

— Farewell! Farewell, friends! - we heard a dramatic whisper, but there was a cry of a locomotive, the clatter of running feet, the iron clang of a train. We looked around: Saburov jumped onto the footboard of the carriage and, forgiving us, waved his hand.

At the end of the day, the companies stretched across the Cheiro to the railroad. I was with Grant on the stone steps of City Hall. The general leaned wearily on the cast-iron railing, stood bareheaded. We almost did not speak during this meeting, looked at the companies passing in the lazy, late afternoon dust and thought our own. With the arrival of Howard's company, Grant perked up. I thought Howard struck him; instead of a fresh young man, we had a Don Quixote, with a face tanned in the sun, with thin cheeks and a thin, light beard, so that you can't immediately make out.

- And you deceived the commander! Smiling, Grant scratched his rough beard. - I know this horse, the third, under the lieutenant. A food major used to ride on it. "Grant knew and appreciated horses. The bay, however, still served Maddison. "Somehow you didn't please Fremont.

On the contrary, we agreed on the idea of abolishing slavery.

"The commander is close to that view," Grant remarked, but with a certain detachment, as if his business was war and he left philosophy to others.

"He took the slave from the rebel and left him to the Pharisee.

Grant shook hands with me, wished me good luck, apologized for not being at the station - they were waiting for him on a gunboat that would go up the Ohio at night.

In the evening, on loading, by the light of torches, Thomas found me: Captain Bushrod Howard had disappeared somewhere. The horse is in place, next to the saddle on the station fence hang his hard drive and a pistol on his belt, but the captain is not. Nothing about Howard promised treason, and yet, for the first time since the Fabius battle, I allowed it in my mind.

The square in front of the two trains blazed with torches, the lanterns suspended from the cars swayed, illuminating the volunteers with luggage on their backs, with rolls and knapsacks, rifle barrels, blades of wide bayonets and horses resting on the gangway. Nearby the Father of the Waters breathed: the swift, dark bed of the river in the September night cut the earth in two; here Missouri, on fire, in battles, in the treacherous blackness of autumn nights, with the possibility of surrendering to yesterday's enemy every hour, to come to the rebels as a prodigal son, to receive from them what the North refused Howard, preferring him a Russian colonel; on the other side, lurking Kentucky and loyal to the Union, Illinois, with his wife and son, with peace already beyond the strength of the shocked Howard. But why did he go to them without a horse and weapons? Isn't this the whole of Howard, with his chivalrous refusal to confiscate, to take someone else's, even in extreme circumstances?

I did not rush things: betrayal is bitter, but if it happened, I will not attach too much importance to it, I will patiently wait for Tadeusz Drama and Maddison. Nadine was asleep when I heard the quick foot-steps of the officers and stepped out of the car to join them. From their gloomy faces, I realized that their anxiety was about the same thing. But next to suspicion is resentment, shock and an underlying thought that such a betrayal is impossible.

- Did you expect this? asked Tadeusz Drum.

- Never! I answered quickly, "But here is the night, it speaks for itself. And the fact that he didn't take our weapons, isn't that Howard?

I sent them away, saying that the night is long, their company is in place, with them, the best company in the regiment. They slept in the wagons. The night was pitch black, with an inaudible drizzle as warm as the south wind from the Alabama hills. Illinois Central fell silent, forgot herself in a short sleep, the workers left, snooping between the trains until midnight, shining lanterns under the wheels and tapping on the iron with hammers. The fire in the lanterns was reduced, bats burst into unsteady circles of light, it seemed that the night was filled with fast creatures, there are thousands of them, aliens from the heavy, viscous darkness.

Leaving the train towards the station, where I suddenly thought of Howard's tall figure, I watched the loaded train from the side, the oily spots of wet roofs, the fading light of street lamps, the fleeting light of a match in the hands of a sentry. The cars stood deaf, defenceless before the impenetrability of the night, its weary, hostile length, before the betrayal and alienation of the world, and I was drawn back to the train. The night dragged on, I forgot about sleep, and it felt like if the darkness of the night went on and on and the wheel of the universe misfired, skipping the day, leaving us on the Mississippi the blackness of a double black night—if that would bring Howard back.

But dawn was approaching, the sky beyond the Mississippi turned from black to gray, the train appeared in the dawn twilight all in one contour, there was nothing to do, we had to return to our place - Nadine would wake up and hear the truth about Captain Howard.

After Drum, Maddison and Thomas, there is no person in the regiment for whom this news would be as difficult as it was for her. There is no friendship between them, there is no external understanding, but such is the property of a noble soul: to stubbornly maintain justice to those who are unfair to you. Howard might have hated us, and he was only aloof; up to this point we have regarded Howard as one of the better parts of humanity. Now he's gone, now treason, and Howard's petty nobility - the saddle and weapons of the North taken from the fence by Thomas - is nothing before the sin of betrayal.

Shivering in the dawn dampness, I turned to my car, got up, creaked the door and heard a voice behind me:

— Colonel!

I turned around and didn't immediately recognize Howard. He ran to the train, when I turned around, took a few more steps, holding out his sword to me in outstretched hands.

— Captain Howard! You left your weapon... betrayed him... Now you have come to return the sabre too?

He stood tall, light, almost incorporeal, with huge, mournful eyes in a bony face. It was anyone but Howard, and the voice in my ears was not Howard's, but hoarse, low, alien.

I descended to the ground, to the sabre stretched out in trembling hands.

"A family sabre, as far as I know, Captain Howard," I said angrily. "No one will blame her for you!"

Figures rushed towards us from the far car - Drama and Maddison - we noticed them at the same time; Howard lowered his head, I raised my hand - the officers stopped.

- Well, let's go. — I took the sabre. — You have decided to announce your resignation from service; it does you credit, Bashrod Howard, but treason tramples on that honour too.

- I haven't changed. Not by deed, not by thought... But I must be detached. I spent the night with the enemy of our banner... with my brother. How dare you, Howard! His hands are covered in blood! "Blood is inevitable in war.

- The blood of the enemy - but not the inhabitants, not their wives and children. The rebellious Howard—I don't know his name—had been a scare to the Missourians.

"I hope it's not true, Colonel," said Bashrod Howard with bruised pride. "He is a violent but noble man.

- And you were looking for this knight, yearned for closeness, and now - you found it!

- Every day I was waiting for a meeting, but in battle. A meeting after which our mother would mourn one of us ...

- Speak! I demanded.

We didn't move; a walk along the carriages would have given Howard relief, and he didn't deserve it.

In the evening, in the bustle of the station, a woman found the captain, she identified herself as a resident of Cairo, from a family well known to the Howards. A beautiful, middle-aged woman, by her nobility of manner, aroused confidence; she handed over a note from Howard Sr.; the handwriting left no doubt. He asked to come, wrote that the meeting should be at home, farewell to brothers who, perhaps, were never destined to meet: he already knew that the regiment was being sent to the Potomac to defend the capital.

- How do they know this?

'They know a lot,' said Howard. "Their people are also in places where we do not suppose. My brother promised to come unarmed and

asked me to do the same; I left the saber, in agitation, out of forgetfulness, and then I could not return. You are right, it is family and equally
belongs to me and to him. Howard paused, reading the condemnation
in my eyes. "I thought: if the officer of the Confederation, the senior in
uniform, is not afraid of meeting me, why should I, the captain of the
North, retreat? Wouldn't that be cowardly?

"He seems to know you well," I remarked dryly.

"No more than I do. In front of this woman, I could not retreat, show
the callousness of the soul or fear. "He understood that too.

- He loves me. The hardest…impossible thing in all this is that we
love each other. He persuaded me to go over to them, but if this was
not possible, to leave the service, to untie his hands so that they would
not tremble with every shot at dusk or at night. He begged, threatened
and begged again...

"Why didn't you invite him to us?" Not to a regiment—a regiment is
too small for him—to a brigade, to the army.

"I know my brother well enough not to bother with trifles.

But he knows you too!

- Older brothers are more often mistaken about younger ones.

"And you listened to his pleas all night?"

The moment came when he retreated. He said enough, Bashrod, not
a word about it; the war will last as long as god allows each of us to
hold a weapon. But now we will sit at the table with the hostess, as if
we had come home at night and mother feeds us.

"Damn you, Howard! Your friends are at the carriage, they didn't lie
down during the night, and you?! Did you have a good dinner, Howard?

- When we got up from the table, my brother turned to me with a
strange request. Howard looked at the train, already visible from end
to end, in bewilderment, as if he wanted to read something else on its
cars besides the initials of Illinois Central. He asked me not to get on
that train today. In any other, but not in this one. He assured us that
misfortune awaited our train...

"Is he planning to frighten the regiment through you?"

"I told him that neither Grant nor Fremont would have allowed the doomed train to leave Cairo. He replied that the generals in all armies are blind: it is not with their hands that the rails are laid and twisted. Not a single headquarters, he said, can do without a venal skin.

"And you're giving up your saber to avoid this train?" "I spoke unfair words, but he did not deserve fair ones.

"I brought you my guilt, Colonel. Dignity returned to Howard. - I will be proud to take a place on the train, which you indicate to me, and if the place is a volunteer, then I will be a volunteer.

The wagon door creaked behind me and called out to me. Nadine.

- Take it! I returned the saber to him and said in a quick whisper: "Go back to your company, Captain Howard, and don't tell anyone about your dinner." Come up with a love affair ... the most reckless.

We started off quietly, one train after another, with a small interval, without escorts and wires.

Civil War casualties.

Chapter Twenty-One

From a letter from N. Vladimirov to his father.

"... My letters are still floating to you, even the first one, with a list of Nadezhda Turchina's "History of Lieutenant T.", and I keep taking the packages to the Chicago Post Office. The official here already knows me and Virginia Fergus, he weighs the bag on his arm, winks at us (he thinks we are a couple) and puts the bag on the scale. "If there are stocks or bankrupts here," he once said, "then you let America go around the world!" "We'd rather go broke on postage," answered the practical Virgie.

When the story came to the war, the general vacated part of the table and allowed me to take notes. "It will be better that way," he said. "We have many hunters to confuse not only the numbers of regiments and divisions, but also honour with dishonour." On my way back to Chicago, I'll make a full note and whitewash the manuscript for you. We are insured against failure - I still have draft notebooks, - if only Turchin had the strength to finish it, - and what other word, so corresponds to the local life, like insurance! Everything is insured here - not only the life of a person or his house, but at least one day of life, or even several hours in a railway car. You bought a ticket, you have luggage with you, and now you are a game worthy of an insurance agent. Insure for thousands of dollars and pay cents. Wish, and you will insure everything: a night's sleep and a safe awakening, love, marital fidelity, even the failure of matchmaking can be flavored by an insurance premium. The General reminded me of this the other day. "After the war," he said, "insurance agents clung to the republic like nits. Where were they when my train approached Beaver Creek? That's where they would have to fork out, but who appreciates the life of a soldier? This is the coin that pays for everything ... "

It's been a week since the old man's story was interrupted. I arrived on the appointed day but found him confused. Once I advised him to take a walk in the neighborhood, breathe in the air of the Epiphany, he looked out the window with the eyes of a life prisoner and began to reproach me that I did not understand him, offering him trifles when it was his business that I was a doctor, but did not see, what exactly constitutes a person's life, what gives it strength and reality, and what is death during life and the wilderness of old age.

After Cairo and Vincennes, we did not move further east towards Washington, where McClellan was waiting for the regiments of Gekker and Turchin. Several times the general proceeded to continue, but stammered at Howard, recalling himself, how he, still a lieutenant, carried a wounded Pole past his posts. Captain Howard occupies Turchin's brain, he does not dare to follow the events until he fully expresses Howard's personality, his proud and unhappy eyes, his courageous and at the same time tender flesh, which burned in the fire of internecine war.

Virginia says that I contracted the Fergus family affliction - love for Turchin, even in the clinic I often think about him and about my duties as an executor. The other day, Johnston knocked on my apartment, he brought a bundle from the confectionery, finding that the general would take a present from my hands. To appease me, Johnston gave me some of his old letters to his father and mother, written from Turchin's regiment, at a time when the young man Johnston believed that a noble heart beats in the chest of every person, and was amazed at how the rebellious generals find soldiers for themselves.

Chapter Twenty Two

How can a regiment's movements be kept a secret when its departure is rumored on both sides of the Mississippi, so loudly that the Father of the Waters seems to carry the secret to New Orleans; if the telegraph machines of the Midwest chirp with the inviting, excited voices of my officers, inviting wives to a passing date; if the newspapers, railway employees, everything, everyone, not excluding the trading public, are notified about the movement of Hecker's regiment and my regiment? Our war has not yet known secrets: everywhere scouts, the ear of a spy, the criminal obsequiousness of an official, and on our side - carelessness, concession, a dashing soldier, like thirty-day recruits from the Brooklyn Regiment, who tied a piece of rope to the muzzle of grandfather's muskets, as a sign that in less than a month, either of them would lead a defeated Confederate on a leash.

James Guthrie's wife was the first to meet us, a young woman in rough farmer's shoes, a striped skirt on strong hips, and a felt hat with an ostrich plume; after her, in the blue uniformed crowd, the delicate flowers we had forgotten began to bloom: turquoise, lilac, fawn, walnut, pearl gray; petals-flounces, ruffles, frills, laces swayed. We have a bountiful harvest of silks and taffeta, velvets, tulle, mantillas and bows, scarves, counterfeit silver buckles, agramant rosettes, prunel slippers, gloves and ribbons. If a flock of pink flamingos, royal peacocks and tropical parrots sat on the black plowing, then even then the spectacle would not be so beautiful. But our birds had gentle, human voices and such eyes that any pair of them is enough for the joyful excitement of an entire company. Tadeusz Drama's little wife, a polka, a slender grasshopper in a flat green hat with a greenish veil, with mint-coloured eyes, in a light-green jacket with an emerald braided cord on a daring chest, in shoes with green lacquer socks and jasper buttons; Ruffin's vociferous wife, a blue-eyed Scotswoman with black hair, in a straw hat with pale roses; shy friend of my assistant Chauncey Miller, smiling at everyone because everyone is a fellow soldier of her husband, serves under him and supposedly honours and loves her handsome, boring pedant Chauncey Miller, and more, and others - how good they were in the middle of the blue army camp, tattered uniforms, bearded and bristly faces, dirty overcoats and army ponchos.

The wives of volunteers also boarded trains, furtively, for one stage, recognizing their inequality with the wives of officers. At the Shoals station, a family was waiting for our leader Johnston: father and mother, two sisters, an old black woman paired with a black coachman. Little Johnston disappeared behind puffy skirts, capes, guipure kanzu and wide-brimmed straw hats, and two negroes murmured something aloud, as if they were reading a prayer before a colourful, living altar.

The anxiety of the night was gone: the sun had driven it away over Indiana, the beautiful earth in the forests touched by yellowness, the clatter of wheels, the merriment in the carriages where a woman appeared, where she - a wife and mother - took off her hat with a move-ment inaccessible to a man, and shook her hair, allowing them to fall on shoulders and chest, and laughed in a way that even a very happy woman does not laugh at home. Dozens of eyes looked at her, and what! - a man, a soldier who could die, but is alive, alive and smells of iron, gunpowder, a fire, a forest bivouac, an indestructible soldier's sweat. But my poor soldier and officer, even the one to whom fate gave a meeting with his beloved, they had nowhere to hide. I would not be afraid of regimental tartuffes, I would close my eyes to what our chap-lain calls sin, but how can you close hundreds of other people's eyes.

We passed south of Illinois and measured Indian miles in the direc-tion of Ohio and Pennsylvania, to the shores of the Potomac. Why did the insurgent brother frighten Bashrod Howard? What was he waiting for? Desertions? How little we know people, even those close to us, and how highly we value our talent as tempters. Bashrod Howard is with us, like after the yellow fever, but he does not shun people, but looks for them, wants to feel again and again his retained, unbroken connection with the regiment.

"You didn't telegraph your wife?" I asked Howard.

"I didn't have time," he answered frankly, returning us to the night conversation; brother took away an expensive watch, and Howard did not have time to call his wife - that's punishment and execution.

We would be glad to have her. Madame often thinks of your wife.

- I think that Elizabeth would appreciate Mrs. Turchin if she got to know her closely. - He remained honest here too, did not lie that his wife liked Nadine.

Soon the women who had remained to escort us to Cincinnati converged in Samuel Blake's carriage at the lint bags, flasks, and boxes. The regimental infirmary occupied a third of the carriage, in the center of the women's circle was Madame, in a dark blue uniform, in a dress of the same cloth, with the end of a blue-white scarf slung over her back. Opa was like a sly mother-confessor or a cheerful abbess among young laywomen who flocked to ask about the rules of a carefree monastic life. Nadine called Thomas into the car, he became a medical target on which Nadine and Blake showed dressing - quick, field, and obs.

The resentment that the Western District had sacrificed us faded away: from a distance, Fremont's decision to give us to McClellan began to comfort my pride. Washington demanded five thousand bayonets, they sent us, two regiments, mine and the 24th, Gekker, - that means that this is the price for our volunteer - one he goes for three, or even for four. The capital is in trouble, and who does it get to help? Gekker's Germans and my companies, where an Irishman fraternally shares the night with a born Yankee, a Frenchman carries a wounded Scot from the battlefield, where the arrogant Tadeusz Drum is ready to boast of anything but his own blood, where even the Negro is safe from oppression. In this alone I found the doom of rebellion, the doom of those who consider us strangers, alien blood, the trash of the Atlantic. I came to the republic from the redoubts of Sevastopol, Gekker - from the defeated barricades of Germany - and here we are side by side, we are equal and called to serve freedom.

At Shoals station we waited for an oncoming train to move on to the 48 bridge over Beaver Creek and Mitchel station. The immigrant train arrived, several irregular carriages filled with the prodigal children of Europe. The war did not stop the flow of hungry seekers, Castle Garden regularly supplied passengers to the northern railroads and cheap hands to the lands of Kansas, Nebraska and Iowa. Passengers, looking at us from all windows, arrived in Shoals through Philadelphia and Pittsburgh, they did not see the war and the marching regiments, we were the first on their way - reckless gang, tattered uniforms, songs, bugle sounds and, unexpected in the regiment, festive women. There was a reason to get off the cars on the ground, closer to the soldier.

Barney O'Mullen noticed in the crowd a fellow countryman, a green guy, one of those who have everything written on their face, without concealment. A rustic, in a furious scattering of freckles on pale skin, as if he had holed up in a cellar, stretched up along with watery, blind potato sprouts and came into the light faded, but like a peasant strong, with a sharp cut of nostrils and a tender, half-childish chin. Barney began to call him to the regiment. The desire to get straight from the immigrant train into soldiers, into the elite, as Barney promised, dismantled the young Irishman, but it was also fearful: were these guys in darned uniforms, godlessly smoking tobacco, really good?

"You yourself are without guns ..." said the young man.

— Guns in the wagons. And you will receive; and then they will put it to the gun.

"There are cannons in the fortresses, you can't take her away.

- You seem to be a scientist; and we have guns on wheels.

- Well, you're lying!

- Ask him. Barney tugged at the sleeve of Thomas, who was as young as the Irishman. - He's a gunner.

— Yes, horse-drawn guns; we tighten with a quadruple.

- Beat the horses! - A man came out. Is a cannon worth four horses?

"And it's worth ten," Thomas replied seriously. "Maybe two dozen.

The young Irishman could not understand why a cast iron pipe costs more than two dozen horses, if a run-down pair would be enough for their family to stay on the land of their ancestors and live in abundance. The peace account did not agree with the military one.

"We have people being killed, boy!" someone shouted.

"They kill, but not with you!" Wow you are so funny.

"We're really happy," said Barney, and just in case, he took off his cap, in memory of those who would not return to the companies. "The chaplain talks about us with God himself, and when God has no time, then with St. Patrick.

— Catholic? the young man asked Thomas.

Barney shouted, pointing at the ground crowded with soldiers:

"He's our black sheep, and they're all Irish!" Hey Irish brothers, show yourself where you are!

And everyone who stood close, without distinction of country or blood, yelled, raised their top hats, caps and hats into the air, frightening the immigrants. The young man glanced timidly at the carriage:

My father won't give me money.

- What are they to you? Barney was surprised.

- What about the uniform? For food? In America, you can't spit without money!

- Well, fellow countryman, you are completely green. Barney whistled. - You will still be paid. Twelve dollars a month, on ready-made grubs, with a free uniform ... We will dress you in an instant.

Barney put a cassette cap on him, someone, playing, held out a belt with a copper badge. But the young man did not play, he was preparing to jump over the abyss that separated the two trains.

From the train of immigrants descended to the ground a dark-faced Irishman with a belt whip in his hands, followed by a woman who was not difficult to recognize as the mother of a young man, and two younger boys, and in the door of the carriage two girls were crying loudly.

- Come on, in the car, but quickly! shouted the Irishman.

"They pay money, father.

We don't need their money! Wanted a whip?!

The son did not move, was silent.

- Well! - After a pause, the father swung and whipped angrily, with a pull, knocked Barney's cap off his son's head, cut his ear.

The young man rushed into the crowd of soldiers.

- Get on the wagon! the father ordered. - You're jumping.

"We will not allow a soldier to be beaten!" shouted Pony Fenton. - At the front, you can shoot for this.

- I'm his father! - Great was the temptation of the future farmer to knock down the little talker with his fist or put a meth with the same whistling belt.

The Irishman asked for a commander. I stood nearby, told him that I had heard everything and wanted to know what he wanted from me.

Give me back the guy.

A father who hits his son with a whip loses his rights.

He looked closely to see if I really was a senior officer if I said this, but the uniform and age left no doubt.

- I raised him and feed him; who should teach if not me.

"There shouldn't be a father or a teacher with a whip. That's why we fight.

- So, fight! And we still don't know who the truth is.

- Sit on the ground - you will soon find out.

Passengers hurried to help the Irishman: the solidarity of immigrants often dozed off, lulled by self-interest, fear of falling into someone else's misfortune, but it happened that it woke up.

"It seems that you were not beaten in your youth, but in vain ..." said the gray-haired old man with the air of a patriarch.

I remained silent and thus reassured them.

- The guy's ass won't fall off!

- Someone else's three skins are being pulled, but your own father - do not touch?

"So, it happened," I answered peacefully. - At first, the fathers made domestic executions a rule, and following their example, the highest despotism arose: butchery, slavery, whips.

"You're all about the slaves, lieutenant!" - A bitter gentleman with poor English speech, a Frenchman or Italian, but with the stamp of an

American, old vagrancy, stepped forward. The soldiers shouted that I was a colonel, - he grinned unkindly. - I am a timid person, even with a lieutenant my soul goes to my heels.

"You don't seem to be a fresh immigrant?" I asked.

- I'm wearing cast-offs - not the first; how much someone else's, abandoned goodness suffocated me. Adverb your bastard accepted. And if the Lord would allow it to turn black from muzzle to ass, I would, with a dear soul, become a slave.

- Was it worth it to swim across the ocean for the place of a slave? Starvation is worse than slavery.

- And for me, it's better to die than to serve as a dog for the master.

- And breathe! So, no: you will first raise confusion, you will begin to exterminate others. What are you staring at? he shouted to the volunteers. - Dollars are paid for a slave, and we are a free commodity. A slave is fed, if he happens not to work for a week, a month, he will not be left without bread. And we? We are hungry when we work, but without work we die. As soon as a person dies, not a single creature will be able to do this: the Lord gave him reason for misfortune. The creature dies overnight, and we year after year.

"Although you are rude to me," I answered, "I will not dispute your words: everything is right. Just remember: as long as there is a complete slave, do not expect the scoundrels to get off your neck.

My peacefulness deceived the Irishman, and he said humbly:

"We can't manage the land without a son, we, mother, can't manage without him. Look how many mouths I have.

"I see," I said. "It's a sin not to give up one soldier from such a family.

"It's good for you to say, you damned Yankee!" the boy's mother screamed. "You won't climb into the fire; my boy will go for you!" You are luring Christians from all over the world to death and torment!

She, apparently, boarded the immigrant ship not of her own free will, according to her, it would be better to live, albeit in black need, but at home; she was one of those who know their villagers, their people and their faith, and humanity is closed from them by darkness; somewhere it is there, but it is scary in it, scary and scary. Such people, even the hungry, can remain happy until they leave the land of their fathers.

- There are Yankees in the regiment, we have a lot of them, but not me, a woman. My home is away from Ireland. "I took pity on her. - You have not insulted your son, your right to ask him to return.

Opa rushed to her boy, but the engine gave a voice, shook the cars, and the soldiers rushed to the doors, dragging the young man along with them. He stood at the door of the carriage and, without moving, looked at the huddled family, stunned no less than if he ascended to heaven or suddenly went out in hellish flames.

I spent the day in the second train, where the staff car, the infirmary and half of the companies. With the onset of twilight, I wanted to be in the lead train, to be the first to cross log bridges, to enter dark forest clearings, where the trunks came close to the embankment, and where they touched the windows and the roof of the cars - where we will pass in safety, nothing threatens the other half of the regiment . There were not many officer wives left with us by night: four on the first train and Tadeusz Drama's wife on the second, with Nadine. They were united by a special interest: a young Polish woman was born in the States, her parents fled overseas after the blood of 1831, she spoke Polish, but did not know Poland, Warsaw, Krakow, everything that became Nadine's second love, after St. Petersburg and the Caucasus. To the surprise of the soldiers and officers, they kept switching to Polish, and when two beautiful women speak Polish, you hear music, sometimes sad, sometimes careless, and the heart of the rudest volunteer is embarrassed.

So, we drove through the dark earth, with plowed up in some places, with bushes mysterious by night, in black forest clearings, I with the company commanders in the first car of the lead train, Nadine and the green grasshopper Drama - in the penultimate, staff car of another train. Two locomotives gave voice to each other, called to each other in the September night.

At about ten o'clock in the evening the locomotive screamed, demanding the road, but the road was not given, and we began, blowing steam. We lowered the windows: the light of the torches pulled out of the darkness a rare shrub and an embankment. Looking out the window, I saw a locomotive driver leaning out of the booth, the fire of torches and people walking towards us. In the moving flame it was difficult to see clothes and faces: but it seems that they were road workers approaching.

- What happened there? the driver asked.

"Forty-Eight Bridge over Beaver Creek. The supports are washed out.

- From what? Second month without rain.

- It's pouring somewhere, but our water is rising. It's like with a war: somewhere they fight, but we don't have peace at night.

- Who's being smart there? - The driver shone his lantern, we saw a bunch of road workers. The man in front was independent, his quick, urgent eyes surveying the train with interest.

The engine driver hid the lantern, the torches moved away to the river, darkness shrouded the earth, the soldiers who had jumped into the lead train at Shoals station went downstairs to wait for their cars.

"Thank you, grubby..." came the unkind voice again. "Or you should have crashed into Beaver Creek." You won't drown in it, but you can break your neck. Show us your handsome general.

"They don't have a general, Colonel. Headquarters in the second train. Don't move without a signal. And you will go quietly.

The eye could make out a faint glow ahead, where a fold of earth hid the bed of Beaver Creek, an unknown rivulet - it did not give a voice, did not splash, did not rattle on the riffles; there was silence, and below, by the water, a fire was burning. I did not attach any importance to the hitch: wooden bridges were always the concern of the owners of roads, the republic was in a hurry to lay new railroad tracks, logs served as the main support even in bridges raised over rapids. We were given a sign with a lantern, the locomotive answered with a grateful cry and set off slowly, at a human step. Our iron, machine steps resounded on the high bridge; in the darkness it seemed higher than it was, but it was high; the water below it shone with barrels, its narrow bed, in steep banks, was guessed. The first wagons passed the bridge, the earth responded to us with a soft voice, like a sigh, then the sound of the last wheels on the bridge subsided. The eyes of a second engine flashed behind us, as if searching for us in the black night of Indiana, and there was a short voice questioning him; I and Raffen were still standing at the open window, and if I could, I would have called out to my comrades: Drive without fear, we have just passed this way.

And then we heard a rumble and a rumble, a dry crack, we saw how the windows of the cars that had just been glowing were disappearing, falling through, and a flash of fire, and a gray ghostly cloud of steam, and a moment later screams were heard.

We experienced everything during the long war: the enemy hidden in the thicket of the forest, the surprise of the cavalry reserve, his blow on the flank, the inequality of battle, the blood of the Chickamauga, death on the steepness of the Missionary Ridge, it also happened that Nadine in the saddle, replacing the fallen officer, carried away like a soldier into an almost lost battle - that's what the war is for - now hundreds have suddenly gone into the abyss, into the crevice of the earth, life has gone. We ran to the river, where the bridge had just darkened; somewhere there was Nadine and hundreds of people dear to me; Howard and Thomas, Drum and Maddison, Guthrie, Blake, Chauncey Miller, Tadeusz Drum's wife, and Nadine, Nadine, Nadine, always with me, and now, in death, alone.

We ran past a locomotive with a wagon; a black, short stump of a train, the hoarse breathing of a boiler, and darkness ahead, cries for help, groans and the other side of Beaver Creek, with the flat, lifeless end of a stopped car. How many of them were kept there - three, two cars, or only the last one, and the staff one was also below, crushed? Lights flickered on the rocks of Beaver Creek; something flared up above the water itself. I saw broken logs sticking out at random and at random, broken boards and pierced walls of wagons, failures of windows with hands grabbing air, lifted up by the visor and flying off the roofs of wagons, bent iron, half-drowned people and crawling blindly, not seeing the shore. I ordered large fires to be lit at the top, near the scree, and rushed down, not looking for a way, colliding with others. Who arrived in time before me, was already working on the fateful harvest; torn boards cracked, the remains of window glass rang, the soldiers dragged the unfortunate ashore without waiting for a stretcher, not knowing whether they were carrying a cooling corpse or a future cripple. I stumbled on the stones, fell into the water up to my waist and chest, not recognizing the voices around. On the other bank, as I grabbed hold of a bush to crawl under the dark bottom of the car, which was leaning a fathom towards the abyss, Quartermaster Weatherell called to me. He said that three carriages were intact, but with the one with the locomotive, four.

Nadine is in the headquarters, Nadine is alive, - as soon as I understood this, the pain for those who died on the stones of Beaver Creek multiplied tenfold: this is how personal grief, unbearable when you ask for a quick death for yourself, leaves, leaving you for another pain — long and painful. Fires flared on the banks, the wreckage of four wagons: the factory tree, even taken out of the water, burned brightly, with a crackle, as if the guerillas of Howard Sr. continued to shoot my regiment.

Everything is against us: shallow water, islands and stones; four wrecked wagons lay in a heap, now the unfortunate was taken out of it, carried ashore, laid on spread overcoats and blankets.

On the shallows in the middle of Beaver Creek, I saw Tadeusz Drum, mad, on his knees, with the body of his wife in his hands. I wanted to help him get up, not yet realizing that the woman was dead—I saw only his bloodied hands and the body of his wife in dark green clothes, as if she were covered in river mud and algae—but Tadeusz stubbornly shook his head and did not notice when I look, he has a gun.

- She was with madam ... With madam, in the staff car. And I took Jadwiga away, she followed me ... then I let her go ahead, but I myself did not have time to enter our car ...

Maddison came up with the soldiers, and I hurried on. By the piles someone was lying face down, and it was not clear whether he was struggling to get up or whether the current was swaying a dead man, not a soldier, but, probably, one of the road workers, in heavy boots, in a wet, pulled up jacket. I lifted the heavy body: it was an Irish youth, a prodigal son, with an unrecognizable, broken face and lively, tortured eyes. Fate purposely showed me vain sacrifices - a broken stem in the hands of Tadeusz, a young man who sailed from across the ocean for bread.

Are you alive, son? He lowered his eyelids and raised them.

Are your hands intact? - That the legs are broken, I saw myself. "If you can, wrap your arms around my neck." He held on to me as I carried him to the shore, while I spoke to him, panting, touching his head with my head: - Well, son ... you wanted to be a soldier ... for the war ... she met you ... You landed right into battle ...

On the shore, I handed it over to the volunteers. Close fires illuminated the sand and wet stones, I could make out the crouching figures of Samuel Blake and his assistant Balehatch, and just as I thought of Nadine, I heard her voice.

Slipping her hands under Bashrod Howard's neck and back, Nadine carefully, as if made of fragile glass, shifted Howard's body onto the stretcher. At the sight of his dead legs and arms, I was horrified: the spine was broken, life was almost cut short, its young, strong threads were interrupted; maybe he does not experience pain, does not suffer, but after all, pain also distinguishes the living from the inanimate, and it happens that you pray for pain, you beg for it from fate.

They carried Howard, I walked beside him, Nadine touched his head with her hand, removed her hair from her forehead; she only glanced at me, silently asking me to forget about her for the time being, to leave all feelings, ties of kinship, the joy that we see each other, to save only what connects the sufferer and the healer. Howard lay flat, dead, and his face was petrified, without asking for anything, only his eyes looked up into the black sky, as if before them was the hot black of Elizabeth's hair and, somewhere behind their dark, thick feather, the saving eyes of his wife.

Howard's lips moved, Nadine stopped the stretcher: death could come at any moment.

"God has been good to me... My brother killed me... It's better than if I killed him... I didn't call Elizabeth... if it wasn't for that night... I would send a telegram..." He spoke softly, amid groans and screams and loud orders, but his every word reached our ears, and what seemed to others to be nonsense, was revealed to me with all the burning depth. Howard closed his eyes. "Close my eyes... madam..."

When, coming ashore, Nadine took her hand from Howard's eyes, he was dead; he was carried to a row where seventeen privates from Howard's company alone were lying, not counting the other soldiers.

I walked along the banks of Beaver Creek, between fire lines and volunteers. Hundreds of the living stood like a gloomy wall, many wept without shame; it was no easier to look into the petrified faces of those who could not weep. At the feet of the living, on straw, on wagon boards and blankets, lay the wounded, the dead, and those who were yet to die in Cincinnati hospitals. There were not enough bandages, Nadine, having opened the suitcases, tore our linen for bandages. She didn't give herself a break; sometimes I turned away from the sight of extreme suffering, and Samuel Blake, a doctor from the Mexican war, sometimes became unbearable, he stepped back to breathe in the river air, to see before him the dark bed of the river instead of blood and white, torn flesh - Nadine was looking for herself difficult. Thirty dead and more than a hundred wounded, thrown to the ground, with no hope of a speedy recovery; even if a division had stood behind us, the heart would have also contracted and suffered along with the victims of Beaver Creek, but the regiment did not number eight hundred people, and many could hardly keep on their feet due to illness and old, unhealed Missouri wounds; what was it like for me, a regimental officer, on this night of September 17, 1861! I ordered to start repairing the bridge, soon trains with Gekker's soldiers will come from Shoals station; the road is necessary for our country and our war.

The young Irishman did not die. Blake took his feet into railcar tires, wax fingers protruding from the bandaged wood. I stood over him; but he did not remember who lifted him out of the water and carried him to the shore, he was waiting for something from me, perhaps the decision of his fate. I called Adjutant Chauncey Miller and ordered the Irishman to be enrolled in the regiment, in the company of Tadeusz

Drum, which Howard had commanded until that night; to put on him the uniform of a soldier, and, upon arrival in Cincinnati, to determine him in a military hospital.

Someone called out to me: Barney O'Mullen was sitting on the straw, his left leg was bandaged and stretched forward, his hand was in a linen bandage, his head was bruised, but the red-haired Barney bared his teeth, his eyes were lively and mocking.

"Wherever they're handing out cuffs—Barney won't be late," he said. - If manna from heaven from silver dollars had fallen on our train, I would not have been there, and in case of trouble, I'm right there. I would have stayed in orderly, I would have sat in the staff car and did not know grief.

"I didn't push you, Barney.

- Still would! he exclaimed repentantly. "Neither you, nor madam…" I waved my hand to him, he called after him: "I will find our regiment, heal and find it!"

It dawned, we did not notice traces of high water on the banks, but the soldiers, restoring the bridge, saw sawn logs and a tree burned by a powder explosion. There was a murder, someone was waiting for us and set fire to a short cord: a weak explosion merged with the roar of a collapse, but it did its job.

A small procession wanders towards me: I recognize the wife of James Guthrie, the black-haired Scots wife of Ruffin, and two more hunched over women who have lost their recent brightness; Together with Maddison and Thomas, they accompany Tadeusz Drama, who carries his wife and love to the car assigned for the dead. Nadine can't join them while everyone else is moaning, bleeding, dying. For long hours she will not be with me, not with Tadeusz, but with those who can still be helped, until she identifies them in the Cincinnati hospital, and then, returning to our car, she will fall face down on the bed.

The funeral train, the infirmary train, the train of living mourning, moved across the land of Indiana, crossed the Ohio border and reached Cincinnati. The rumour of misfortune was ahead of us, foreign soldiers, road employees, onlookers gathered at the stations, involuntarily taking off their hats in front of a carriage with drawn curtains.

Two runs to Cincinnati, Elizabeth Howard rushed into the car, past the sentry, rushed in as if Howard was still alive, he still had to say his last words to her, and if she did, perhaps she would not let him die. And this collision of a dead, cooled body with the desperate energy of a loving woman, her refusal to imagine her husband inanimate, disappeared was unbearable; it seemed that moments would pass, and something would happen inside the car, the iron-bound walls of the car would explode. The train started moving, Elizabeth remained inside.

I could tell her whose hand threw Howard and 29 other people into the grave. Could, but didn't say. I didn't tell her, I didn't tell Nadine, I didn't report to the senior officers, so as not to cause wry smiles in the back, that, they say, the colonel was delirious with scouts of the rebellion, looking for a conspiracy where the misfortune happened.

We only saw Elizabeth in Cincinnati. At the deserted station at dawn, the mayor of Galina, the city from which Howard's company had been recruited, was waiting for us. The mayor did not arrive alone, with him were employees of the mayor's office and relatives of the victims. The coffins floated on the hands of the soldiers from the car to the wagons waiting outside the station fence. No speeches, no orchestra, silence and the creak of heavily loaded soles on gravel. Elizabeth also appeared at the door of the car - the woman we remembered from Chicago, with her wild and majestic beauty, that woman and not the same anymore, the woman of one love for life, with eyes burned with grief and with a parched, bloodless mouth. She asked Mayor Galin to bury the captain along with her soldiers; Howard commanded them in his lifetime, may he stay with them forever. If only she could feel how her decision lightened my soul! After all, the hour will come, Howard Sr. will also die, the child-loving Howards will find his body in any grave and bring it to the family crypt, and Elizabeth's decision will not allow the murderer and his victim to unite even after death.

I did not dare to approach her: there are moments when everything is given to the will of a suffering person. Elizabeth came up to me.

"My husband wrote to me about you, Colonel," she said. - About you and about madam, he forgot all insults. Farewell!

Before me stood a strong woman; I would like Nadine to be with us, but she was delayed in the hospital.

"I am proud of his friendship, Elizabeth," I said, showing that I was not alien to her name. "I've met few people worthy to stand next to Captain Howard.

This amazing woman laid her head on my chest, only for a moment - one mournful touch - and I felt the devastation of her grief; if her spirit had not been so great, she would have fallen to the ground and no one could have lifted her.

General Oliver Otis Howard.

Chapter Twenty Three

We were forgotten.

The defeated, even without guilt, should not wait for relieving words - there is no one to say them to. Some are in battle, saving mournful words for nearby graves, while others are so far from the danger of perishing throughout the war that they condemn the death of others as the fault of the dead themselves. For them, war is numbers, and no matter how great the number of losses, it will immediately be overshadowed by another number, huge, reconciling with any misfortune.

The Western District was silent, the Missouri doors closed behind us, Cairo's landing stages forgot about the holey soles of the Illinois. Frémont loaned us to McClellan, Hecker and I were rushing east, Missouri brawlers, guns hot on the right bank of the Mississippi. Now we were out of breath, had to pick up a crutch, bury the dead, the name of the 19th of Illinois was surrounded by a halo of misfortune, and on the Potomac, they were afraid of the smell of bad luck. In the state of Ohio, we are passers-by, burdens, sufferers; on the city cobblestones of Cincinnati you can't pick up even a handful of grain, you can't dig up potatoes in a company boiler. I won't blaspheme Cincinnati: the orchestral brass rattled regularly at the funeral, the wounded got places in two hospitals in the city, and the townspeople took off their hats in front of us. Our train and two Gekker trains were taken to sidings, we went into quarantine, whose name is not plague or yellow fever, but bad luck. I knew McClellan closely, a kind man in ordinary life, and hundreds of miles later I could clearly see his well-groomed, trim figure in the Washington office and how he was suffering, precisely suffering, from the inability to call such glorious colonels as this worthy Russian to the capital. and no less worthy German; how interestingly he thinks about the necessary change in the plans of the campaign and about the fate unfair to us; he does not think secretly, but aloud, logically, roundly, gracefully, as roundly and gracefully as his movements; and his smile is coloured with sadness, which makes it smarter and thinner; he thinks, thinks, seeks advice from his interlocutors, looks up at them with hope; but there is no way out, there is no other way out than the one he found, and now he speaks firmly, almost dictates, in a low, clear, quick voice, announces that Fremont was late with his regiments and now he, McClellan, guarantees the security of Washington on his own, although he does not have two hundred thousand bayonets at his disposal ... McClellan did not answer my telegrams, and the Secretary of War did not respond either - the Cincinnati telegraphers began to fear me.

At the end of the second day, locomotives were sent to the trains, a military agent announced that the regiments were ordered to depart for Camp Dennison near Cincinnati. Camp Dennison did not change our situation: the same life, albeit in a field camp, but half-starved, without kitchens and company wagons, without clarity - but Cincinnati got rid of our lean faces.

I turned to Lincoln: the president, who tolerates too slow generals, takes on the everyday worries of the war. Here is the telegram, she left Cincinnati before noon on September 23, 1861.

"... I have one hundred and fifty (150) people injured in a railway accident, about a hundred (100) sick and incapable of serving, and five hundred (500) remaining in the ranks. Uniforms, shirts, shoes were worn out. People haven't been paid for two (2) months and our equipment from St. Louis has been sent to Washington. The regiment is ordered to proceed to Louisville. I telegraphed the adjutant general twice, asking which way we should go, but received no answer.

It is true: my two telegrams to Lorenzo Thomas, Adjutant General of the Union Army, received no reply. I did not flatter myself with the hope that the name of John Turchin would make Lincoln tremble with gratitude, but stubbornly waited for an answer. Nadine and I rode from the camp in Cincinnati, although if Lincoln's telegram had arrived in the city, she would have been rushed to Camp Dennison in no time.

I never heard the President's response. A quarter of a century later, having been admitted to the military archive, I also learned about the fate of my telegram along the way. On September 23, 1861, Lincoln wrote on my telegram: *"To the general ad-tu, please give an answer to this or order an answer."* Below is a handwritten note by Assistant Adjutant General Absalom Baird on September 24: *"The President is respectfully informed that yesterday a telegram was sent to this officer to 'do his general's orders'."* Nicely they danced the departmental square dance, went full circle to the music of clerical quills - I alone did not receive a word in answer, and General Robertson had not even had time to look at the regiment before we were transferred under the command of W. T. Sherman. On September 25 we set out for the Kentucky border town of Louisville, on the Ohio coast, and by nightfall arrived by riverboat at Lebanon Junction, 35 miles south of Louisville. Here we set up camp, from here we began a new military campaign.

In Louisville we had another test; I don't want to blame Hecker in him, a defeated knight of the German revolution, he imagined fast, victorious wings behind his back when there was a soldier's satchel, a gun and a roll. Hecker's regiment was made up exclusively of Germans: the regimental regiment was German, the Germans were officers, the German pastor and the soldiers were Germans. The uniformity of the regiment accelerated order, discipline, external unity - I was also aware of the burgher poetry of such a formation, but I rejected its idea. The apostles of this idea speak of the wisdom of nature, of pomegranate seeds under one peel, that one can never find a banana fruit in potato nests, just like potatoes in banana clusters. But if nature cared only about separation, it would not have created a person, the only living being that will come to a community, solve this problem or disappear.

We arrived in Louisville together. German was as common in this town as Yankee. Bakery, sausage or beer trade was not enough for the Germans who settled in Louisville, and they penetrated into all crafts, making up not one middle class, but several - from farm labourers and wage labourers to manufacturers and co-owners of offices and banks. No matter how corporate the German, the Louisville community was torn apart by discordant passions, and the German judge, indignantly, imprisoned the impoverished German thief. The arrival of Hecker's regiment provided a welcome occasion for the manifestation of the unity of the German community, and with it the burgher arrogance that flourished under the protection of foreign bayonets.

The regiments went down to the pier and moved towards the town square. We ended up on one side of the square, and Hecker's companies were stationed at the city hall and the Louisville post office, having received new uniforms, boots and winter overcoats back in Missouri. Hecker's soldiers were surrounded by hundreds of Louisville Germans; well-dressed women with treats in their hands; gentlemen in black and fashionable plaid suits opened their arms and boxes of cigars to the soldiers; tables were taken out of nearby houses for convenience; there were speeches, psalms and songs, and shots of champagne to boot. Where can we remember us - a crowd of military vagabonds on the other side of the square. The regiment wore itself out, asked for a change of uniform back at Ellicott's Mills - Beaver Creek did not adorn us. We tightened the straps on our collapsed bellies; and there, at the mayor's office, without finishing, they threw meat of turkeys and chickens under their feet, pieces of fried pork, throwing their heads back, drinking from bottles, as if looking through telescopes, looking for a lucky German star in the blue sky. I sent a message to Maddison's City Hall to have them arrange for us to be escorted to the station. Gekker's officers rushed off somewhere in a crowd, and running up to me, ringing with spurs and holding his sabre in his hand, the adjutant of the regiment, to call the Turchins and Lieutenant Colonel Scott - we are talking about him ahead - for a German dinner. Nadine beat me to it by saying that it is not proper for officers to eat when soldiers are hungry, but if they call the whole regiment, then we are not averse.

- It's impossible! the adjutant was upset. "This is an officer's... noble dinner.

It turned out to be not far from the railway, but the dispatch was delayed until the arrival of the German regiment. Gekker found me in the carriage; he came with a belated friendly alarm: why didn't we share their holiday?

"I am not accustomed to feasting when the regiment is not fed.

Our alienated glances brought Hecker back to the square, to the disunited regiments.

"You offended us," he sighed forgivingly. "Someday I will take revenge on you, I won't even accept a mug of beer from you!"

"You and I have someone to take revenge on, Colonel Gekker," objected Nadine.

Gekker pondered, looking out the window at the bustle of the station.

"If you only knew how great the suffering of my people is!" he exclaimed softly, with the passion of the chosen one, who took upon himself the crown of thorns of the nation. "Standing one step away from victory, already touching it with your hand and suddenly losing everything, everything... If only you could measure the depth of our suffering!"

"In suffering, our people will not yield to anyone!"

Nadine said it: in such cases, she stepped forward to prevent me from going to extremes.

"But vast Russia is dark and mute, and from our barricades light could be shed for all of Europe.

— Who knows what Russia will give the world; she has already thrown off her serfdom and will not stand still. We are outcasts of the monarchy, why should we boast to each other with chains.

We are here for brotherhood, not isolation. Do you know what I discovered here in the five years of my life here? There is no other freedom. No!

He shook my hand and held it in his.

"You are a noble man, Colonel. But..." He leaned his head on his shoulder. Forgive us a little German holiday.

How many concessions were made for the sake of Kentucky, for the sake of its dubious loyalty to the Union! Like an Easter egg, politicians rolled in the palms of Kentucky - just not to drop it, not to break it, not to let it fall into the wrong hands. And here it is, carefully, exploded louder than the Missouri bomb. Rebellious General Buckner's troops marched on Louisville, Tennessee and Virginia, condottieri *(Noun: a troop of mercenaries especially in Italy)* marched into Kentucky, and thousands of those who yesterday were considered friends of the Union threw off their masks, hoisting a blue banner with the motto "Rights of the South". As soon as there was a smell of gunpowder and Morgan's black gangs loomed on well-fed horses, my regiment was again the same, and the companies fought as if they had not shed terrible blood on Beaver Creek. After the fighting, we returned to Camp Lebanon Junction, here, then fresh volunteers came to Elizabethtown, a few of the crippled on Beaver Creek returned, and Barney O'Mullen was among them.

The "Elizabethtown Democrat" was printed in the city, one of the most vociferous insurrectionary leaflets. Demanding insatiably our blood, the publishers of the newspaper valued their own and fled the city, leaving us the printing house. And I had experienced people: there were reporters in the companies - among them Redfield from the Chicago Evening Journal, and compositors - Haney and William Christian; in mid-November we published the first issue of our Zouave Gazette. I printed in it an article about General Fremont; he had just fallen victim to his enemies in Washington and had been removed from the army when the Kentucky mutiny told the whole country that Fremont was right. I devoted most of the newspaper page to military work, printed articles on rifle training, on the guard service, on company tactics and battalion tactics, and every literate soldier received a Zouave Gazette free of charge. I could also become an engraver for the Zouave Gazette, but I didn't have to, Balashov did this work. Yes, our kind and dangerous friend Balashov!

Soldiers, thirteen in number, were sent from the Cincinnati Volunteer Depot to Howard's former company, and a day later twelve of them deserted. The thirteenth volunteer was absent overnight, came in the morning and announced that he wanted to see the colonel, was this the Turchin he knew from New York?

Tadeusz Dram brought a volunteer to me, and I recognized Balashov, with a sad, driven look. He took off his hat, revealing thinning hair and flat, sunken temples.

- Saburov told me that you are already in the regiment, fighting. Are you still wearing your frock coat?

At the name of Saburov, he drooped in fright.

Do they cut the tongues of counterfeiters in America? It can be seen that you, Balashov, are Russians like us, strangers, you better deal with Saburov.

That's how I lightened him to the point of madness, to a scream.

— I hate Saburov! This is my fourth regiment...

- Can't stand the fight? I despised him. I haven't been in combat.

Are you deserting?

- Three times I ran away without getting a weapon ... Once with a gun.

I didn't understand anything.

— I have no free time, Balashov. If you don't want to fight, why are you climbing into a volunteer's shoes?

- I'm afraid of Saburov; I can't get away from him.

What else have you done?

- Saburov says that I was released from prison under his guarantee, and if I stop obeying, he will report that I embezzled eight thousand roubles by false power of attorney, which were sent to

Mrs. Turchina from Russia to Philadelphia.

"You'll go crazy with you, Balashov!" What are the thousands more?

"Someone wrote to Mrs. Turchina's father that she was in Philadelphia, abandoned, in need; he sent money, and I allegedly appropriated it.

I remembered Saburov's carriage in Philadelphia, the London debt returned to me in dollars. Did he slip me a small part of Nadine's embezzled money, exhausted by the nobility, generosity and breadth of his own nature? I did not regret the stolen money: had they come to us, we would have sent them to Russia; the blood pounded in my temples at the thought of the torment that the suffering father had taken.

- But if you become an American soldier, what do you care about Saburov?

"I'm in his herd," Balashov said lostly.

Here's what he told me.

He met Saburov in one of the volunteer bureaus in New York, from his hands he received, like two dozen other volunteers, a few dollars in advance, then Saburov got everyone drunk, and Balashov woke up far from New York and from the office where he put his signature . Saburov put together more than one gang of false volunteers, they were recruited from rogues and fallen, weak-willed people. Having deserted once, they are already in the trap of Saburov, in his herd. The recruiter receives a bonus for them and, having assigned them a distant meeting place, waits for them to appear for new recruitment. In New York, Saburov was suspected, and he moved his enterprise to the West, to Illinois and Ohio. Now he was waiting for horses in Hamilton, on the Indiana-Ohio border. I would, without delay, go to Hamilton with Balashov and put the thief before the court; alas, the duties of a commander do not give such freedom of action - I learned from Balashov the names of some "horses", the places of assembly and sent a report to the adjutant general of the state of Ohio.

My life brightened outwardly. The Zouave Gazette established itself not only in our regiment; the newspaper was also read by the neighbours, even by Hecker's officers, who usually did not expect the truth from other people's tablets. One thing sometimes covered the horizon with a gray cloud: Illinois did not hear me when I wrote about the promotion of officers, asked for new appointments or the cancellation of unfair appointments. This is how our army was built: if an Illinois regiment were even in New Orleans or in Washington itself, I could only write about the production of officers to my overlords - Governor Yeats and Adjutant General Fuller. You would find dozens of my letters in the state archives: I asked, threatened to leave the regiment, begged to cancel the appointment of people who were completely unfit, and invariably heard from Yeats: *"I will not cancel the assignment."*

Among the papers there is a letter sent to Governor Yeats from Elizabethtown, it will tell you more than what I said today: "If you, Your Excellency, appoint and promote officers and other people of my regiment without asking me about their business qualities, then you are doing me injustice. If my recommendations are not considered worthy of attention, then I have nothing to do in this regiment. Your appointments undermine discipline and offend really good officers and sergeants. I beg you to cancel these appointments, or I will be forced, against my will, to leave the regiment.

Written in extremes - I beg you, it burns my pride - written with a threat, it was painful for those whom I tested in the war; but I believed that I would achieve justice and did not leave the regiment.

In the meantime, the Ohio armies received a new commander - Don Carlos Buell showed up in Elizabethtown to check on our regiment as well. Buell is the true talent of the immobile army. If the goal of the army was not war, but formation, recruitment, accountancy, parade grounds and dinners, you would not find a better commander. The mind is direct, aspiring, the eye of a pedant; patience with flattery and the courage to do without it even for a whole day; a kind heart, compassionate not only for the lives of eminent rebels, but also for their sacred property; slowness, caution, prudence - the totality of talents necessary for the commander of immobile armies. A velvet voice, a velvet heart and a hand in a velvet glove - his enemies were in a better position, Buell's chivalrous nobility did not allow him to harm them.

I realized this too late and paid for my blindness, but in November, when Don Carlos Buell replaced Sherman at the head of the Ohio armies, we were pleased with each other at the review in Elizabethtown.

"I have never seen such excellent training of people as in your regiment!" Buell admitted.

He did not leave Elizabethtown that day. He remained in the camp until dead darkness, praised the company commanders, memorized their names and singled out the best among ten good companies - and not out of a whim of the eye, but after weighing and evaluating everything; we have never met such a leisurely, sharp-sighted senior officer, and he put us above other regiments of the Ohio army. Buell even changed his gait; in the morning he got off his horse like a bulky father-commander, who had no time for smiles and politeness - the regimental review rejuvenated him, the silver voice of his general's spurs became louder. Buell is the exact opposite of both Fremont and Grant; without the artistry and poignancy of the first, and without Grant's manly, grumpy directness, but it's true that Don Carlos Buell was exactly what I needed then. During that day, I saw in him the business-like and not alien enthusiasm of an officer. At the review, Buell often hit his left hand with gloves, clasped in his right, and slightly twitched his foot, spurring not a horse, but people, regiments, armies, events.

At such moments, he must have seen all the regiments of the Ohio army like ours, saw them advancing south, through Kentucky, through Tennessee and Alabama, saw the enemy fleeing in fear before them, scattering into smoke.

We mutually rejoiced at our acquaintance, instinctively holding ourselves back from secrecy and deepening into the soul. Instinct told each of us: this is all I can give you and take from you, further the road is unknown. He saw my blacks, asked whose they were, where they were from, if they had been taken away from loyal owners, but did not listen to the answer, he wanted to teach me, an exemplary regimental officer, also a lesson in wisdom, and in his thoughts not admitting that we might not see eye to eye. So did Buell and Nadine—it never occurred to him that she was in the regiment; he liked that we ate in the tent, although the outskirts of Elizabethtown were within easy reach; that the November wind strikes the canvas; that a woman is in charge of the table, sharing with her husband the cares of a military lull; that educated people crept around, at least speak French - you won't stay in the desert ... Buell noticed the violin, I had to play, and I played, badly, played with my soul taken out, then I understood why, but then I didn't understand, I got angry and so tormented the strings with a bow that Nadine's face changed. And all because the violin deteriorates in the company of such people. Judges puzzle over whether the defendant is guilty or not; jailers are looking for ways - and what! - find out the truth, but entrust this to the violin, blindfold the musician, bring the test subject, and the violin will tell you whether you are a criminal or a saint.

I played in front of Buell, thrice damned I played in front of him: for forty years after that my fingers held the bow, but this piece played by him died to me.

"I'll give you a brigade, Colonel Turchin," Buell said as he said goodbye. "I need people like that at the head of the brigades. - In the darkness of the night, he moved closer to me on a tall, black-gray stallion and said with the grouchiness of a friend: - In vain did you print about Fremont: he is ready to give the country to the mercy of passions; expelled from the army, he wants to raise a new rebellion in favour of his name and profits.

And he recoiled, disappeared into the darkness, leaving me forever with a lump of unspoken words in my throat. On Buell's orders, I took over a brigade of three regiments. Among them was the 19th Illinois.

INTER-CHAPTER THIRD

Excerpts from Letters written by D. Medill and Charles. H. Ray to Allan K. Fuller, Adjutant- General of Illinois.

"In my humble opinion, Colonel Turchin is the best colonel in the entire western military district, and he has the best trained regiment from Illinois. This colonel deserves to be a brigadier general."
Joseph Medill November 15, 1861

"The 19th Illinois Colonel Turchin is my favourite, and I really want him to fill a place worthy of his very capable commander."

Charles H. Ray, November 15, 1861

"If, as I have heard, the promotion of Colonel Turchin is supposed, then I consider it my duty to inform you that he does not deserve it at all. I appointed him as a brigade commander, and now I see that he must be relieved of this post, since he clearly did not cope with the task of establishing and maintaining discipline in his troops.

Major-General Buell, to Secretary of War Stanton, June 29, 1862.

From a letter from volunteer Johnston.

Athens Alabama, May 1862

"Dear Mom,

We all arrived at this place two weeks ago. We left our tents in Huntsville, but a week later they caught up with us. The place is healthy, the water is clean, and it is possible that here, if not elsewhere, our summer camp will be. That is, we will stay here if we are not driven out of here or, on the contrary, we do not push the enemy. At present, on this side of the Tennessee River, the enemy has only Scott's Louisiana cavalry, Morgan's horsemen, and other units that have left the Confederate camp in Corinth due to lack of fodder. However, getting food for horses is difficult even here. I have heard from the townspeople here that the available provisions will not last even two weeks, and if they are not brought from Nashville, people will have to starve. This state of affairs is more or less characteristic of the entire South. The weather shifts from warm to hot, and the hazy blues of spring are gradually replaced in the sky by the coppery reds of summer.

Athens is a place of tranquillity, peace and contentment. Athens to this day is a beautiful and cosy corner, a real garden city, where, wherever you go, you everywhere feel the delicate aroma of roses and all kinds of other flowers. You walk along the streets, and from each garden you can hear its own, special smell, depending on how many flowers are planted there. Whichever way you go, you will see planters' houses everywhere - and what houses: huge, with many rooms, comfortable, many of them really beautiful, and all with gardens and broken lawns on each side. Around each house there are large dense trees for shade, and in many cases, you can hardly see the house behind them. The fences are rough-hewn, mostly unpainted, some are made of rails, and some even the most beautiful houses are surrounded by a simple fence, only painted white; this gives the area an appearance that, as it were, says: "Here we are all at home, in our everyday clothes, and we don't care that our neighbours see us, because they are the same as us, permanent residents, and not visiting guests".

A little away from the city there is a fair square, where races are also held, surrounded by the same kind of knocked together fence, with elegant-looking (even though the places on them are cheap) stands and a very calm general view, as if there were no races here and it doesn't happen, but people just come, sit down and watch what they are shown, without the hustle and bustle that happens at hippodromes in the North or in other countries. The city also has a School for Noble Maidens, an elegant brick building with many intricate ledges, surrounded by wide lawns and shady trees. The wooden fence around the school grounds is painted white. The only business center in the city has the same look as in most southern cities where I have been. This is a square in the city center, on which the courthouse rises, and all the shops are located. There are no other shops in the city, they are only on this square. All in all, this Athens is the most picturesque and charming little town I have ever seen. The sale of liquor in the city itself and within a radius of 3 miles from it is prohibited, neither gambling houses nor billiard rooms are allowed to be maintained.

Until the fall of Fort Donelson, the inhabitants of Athens were only glad that their town was the quietest place in the South, and that federal troops would occupy it, it never occurred to them ...

Colonel Scott [19] fell off his horse again. Half of his face was bruised and bruised, but in a day or two he was in the saddle. I haven't seen Mott Temple for some time now. Say hello to everyone at home from your loving son.

J. S. Johnston.

From a letter from N. Vladimirov to his father

"... Virgie is sick for the second week. She is consumed by an unexplained fever. Widow Fergus in fear and constant prayers; Virgie's condition is so bad that sometimes I'm offended by you, why didn't you teach me prayers - as a doctor I can't help her. All these days I have not left Chicago and would not leave even if the General called me by telegram.

I was afraid of an invitation from the general, and here was a letter yesterday, from a direction I did not expect. Under it is the signature: N. Turchina. The lines are polite, but also angry. She writes that she will call me soon, reproaches that I rush Turchin, do not spare his old age, that everyone knows his habit of "talking without caring about his own interests", and do I, a doctor, do not understand that he will have nothing to live with, when he finishes the story...

I will see Nadezhda Turchaninova!

Remember I wrote you about the cigar, and liquor businessman Johnston? It is hard to imagine, after reading a letter from Athens, that this red-faced, gray-haired gentleman and the young man who described Athens are one person ... "

Ruins of the Richmond Bridge, following the Richmond Bread Riots, 1863.

Chapter Twenty Four

The regiment was stationed near Athens, and the road to the court-house ran past the picturesque estates and wooden stands of the hip-podrome, lined with landaus and carriages of the surrounding planta-tion owners. When the orderly in the camp began to saddle the horses, a gang of regimental blacks, led by Abraham, approached us; he had joined us in Elizabethtown and turned out to be, an experienced type-setter and a quick-witted scout.

"They'll kill you, Mistah Turchin," Abraham said. — It would be better for you to go in a closed carriage; we'll get one ready.

"I'm not used to hiding, Abraham.

"They'll shoot you," he insisted, over the disturbed hum of the Negroes. – "They shoot from the attics; in the morning they go, as if to a fair.

The rumour that the rioters in Richmond had outlawed me and put $50,000 on my head reached every ear, and anyone could guess how much their money would have improved from a well-placed shot.

"If they are coming to Athens to seek my disgrace, then they will wait a while to shoot me.

I will put Nadine in the saddle."

— Ride quick, ma'am! Abraham begged; my reasons did not reas-sure him. Don't give them time to take aim.

By the bridge over the stream we passed a herd of black pigs and tall wagons full of old amber ears of corn. With our appearance, the streets of Athens fell silent, the soft step of two horses along the unpaved streets became audible. On the stone porches of the mansions, to the gates of the estates, girls in heaps of ribbons, dressed up as for church, and children ran out. Nadine was wearing her dark blue Union uniform, one that had been washed more than once from the blood, dirt and dust of Missouri, Kentucky, Tennessee and Alabama; back in camp, I had a cowardly desire to show her to the Athenian ladies in her best dress, so that envy would be added to their hatred, but I kept silent - in court, they will talk about her, a woman forbidden by the Regulations, let them see that this woman serves and the soldier's ration belongs to her by right.

The fear of the good citizens of Athens has passed away, and how much of it was unrighteous fear? There was the outward sanctity of the bigots, the puritanical self-satisfaction that a dirty Yankee's foot would never set foot on the bridge over the Elk River; there was a pretense at the first capture of Athens by my brigade, hasty oaths of loyalty, and then frenzied, vicious vindictiveness. None of their houses have been burned down, not even the robbers' nests, from which they fired at my soldiers, have been destroyed; even those who tied the prisoners and the wounded to the saddles and let the horses run, escaped the court, like the Athenian matrons, those who, having lifted their skirts, protecting silks from blood, rushed about in front of the city hall, among the wounded soldiers thrown to the ground, spat in their faces and showering them with curses. How can there be fear if General Mitchell's officers write in the office books the false claims of Athenian shopkeepers, accusations of robbery, theft and Sodomy sins, if a committee of citizens is assembled under the high hand of the General of the North, whose goal is to prove how vilely the rights of the noble South are violated!

The courtroom is spacious, with six tall windows overlooking the town square; Athens did not recognize the Theatre, did not let comedians and circus booths, but they wanted to enjoy the spectacle of someone else's vice and the humiliation of fellow citizens in court in full and with comfort. Alcoholic beverages were not sold in Athens, each estate had them in the cellar; here it was not allowed to open either billiard or gambling houses - billiard and card tables caressed the eye with green cloth in rich houses.

We drove up to the crowd at the entrance to the court. A few were allowed inside - senior officers, reporters, eminent citizens of Athens;

the mayor of the city stood at the entrance, together with a newcomer lieutenant colonel. Joseph Medill shouted at us that Chicago would not let a regiment offend, that Illinois had put more regiments under arms than any other state, and the 19th Illinois was the best of the best. To the cries of Medill, we dismounted, went to the porch, and from the quivering pupils of the lieutenant colonel, I realized that he was ordered not to let Nadine into court. But she walked imperturbably, Medill in front, me behind, forbidding and unceremonious, walking in her army uniform, and the lieutenant colonel hesitated.

In the hall I found friends - Teddy Dawson and Ramage, officers of the brigade; in the far corner, the regimental chaplain, head bowed, was talking to a woman in mourning - the Athenians occupied two long pews - soon the court appeared: Brigadier General James Garfield, six colonels and a secretary of the court with the rank of lieutenant.

You haven't heard from me about Lieutenant Colonel Joseph Scott. He appeared after the regimental election at Camp Long, flat-chested, with thin cheeks and a sallow face, but strong, well-trained and hardy. Apparently, the adjutant general of the state doubted whether I could master the regiment of volunteers, gave me a campaigner to help me, and Joseph Scott began to serve without further ado, not timid in battle, but not out of line; his lack of ambition harmed this officer. I chose Scott as a target for the eyes while the referee's lieutenant read the pages of the charge; a closed look, restless, rejecting all points of accusation from himself, rough hands, now crossed on a sunken chest, now lying on his knees, palms up. I was accused of three crimes: negligence in my duties to the detriment of order and discipline, conduct unbecoming an officer and gentleman, and disobedience to orders. The lieutenant read and read, announcing the names of the victims, the price of the looted, the size of the losses and illegal forage, the numbers and paragraphs of the violated orders and paragraphs of the army regulations, my sins and my outrageous orders, everything I did in the name of the Russian military idea, with the secret goal of turning the friends of the Union into his enemies. From the beginning of May, the newspapers of the South wrote about the sack of Athens, *"about the destruction of the lovely, peaceful city, which deserved the name of southern Eden."* They also named the amount of the robbery: 50,000 dollars, broken walls, ruins, smoking ruins - the American Pompeii, mortified not by the lava of Vesuvius, but by the mud of the North, by the barbarism of the "wild Cossack" and his hordes. The numbers are strange! $50,000 is also the price Richmond put on my head; hand over my head to the rebels and Athens will drop the suit. There were quills in the North ready to repeat the slander: especially Brick Pomeroy's La Crosse paper, even the Chicago Times picked up the lie about the "plunder of Athens."

I prepared to answer: a peaceful square outside the open windows of the court, shops without traces of war are my allies. How naive I was: the shopkeeper hastened to fix his altar so as not to lose money, but, having hidden the insignificant traces of the war, he could now exaggerate the damage a hundredfold. Until we came to Alabama, these people came from the plantations to the squares of such towns and there, in the vigilance committees, they judged and punished. The tavern served as a playground for them, groups of madmen, clustered at the tavern office, judged their fellow citizens; the accused, not yet opening his mouth, saw how a rope was attached to a tree or a gun was loaded with lead; shout only after the man - he is an abolitionist! he is anti-slavery! - and the tavern justice will not spare him. Athens deceived everyone; from the ancients they took a noble name; Tartuffe has hypocrisy; at the church - outward holiness; the slaves have their blood and the future of their children; the sun has fertility; at the earth - space; the devil has everything black, deceitful and cruel that seethed secretly and came out when one of my regiments retreated, leaving wounded soldiers at the mercy of Athens. Thirty-seven soldiers of my brigade remained on the green grass near the stands of the Athens hippodrome, among roses and heliotropes; we found one alive, in a dump of dead bodies, on the edge of a dug grave.

The number of robberies was over, and terrible words fell into the hall, more unexpected than if thunder suddenly struck in the dry, orange, cloudless sky of Athens. The soldiers were accused of murder and violence. Girls from the boarding school of noble maidens fell victim to the Illinois rapists; their flesh was abused, but honour and name must remain pure, the court cannot demand their return to Athens while rapists roam in the military crowd. The soldiers also abused a Negro girl; her body was not found, she lived far away, on the banks of the Elk River, the criminals drove her father and mother into the bushes behind the house, then in the darkness of the night there was a heavy splash of water and the laughter of fleeing soldiers.

My friends sat petrified, but how animated those who had had tired of chewing the cud of the pillaged commercial Athens and were waiting for more salty news! An individual can be defended against the accusation of violence, but not hundreds of soldiers, heated by battle, forced to break into houses from which they are fired upon; how to deflect impersonal, slippery lies? I was informed more than once about deserters from the mouths of both the North and the South, they hid in abandoned estates, drunk, lay on the master's featherbeds, put on uniforms from the time of the war for independence and did not budge, as long as supplies remained in the cellars, until they heard close gunfight. If violence is on their conscience, how to remove the stain from innocent soldiers? And the stooped lieutenant slowly moved along the bloody, ploughed road of my crimes - and how could I not plough it, if on any yard of the earth an Athenian saint could be discovered, killed and hidden by my thugs, the body of a dishonoured girl, a crucifix stolen in a church, pieces of cloth or the precious stones of the Alabama ladies, with which the poor things redeemed their honour. And I, the Antichrist, a wild Cossack, stood over the disorderly soldiers, stood with my bloodied hands upraised, and repeated to my people: "For two hours I close my eyes!"

— Colonel Turchin, do you plead guilty?

Get up, Colonel! Judge Garfield said. - In civilized countries, the defendants, answering, stand before the judges.

"In these countries, General, such a formality is also carried out as a consequence. No one will be put before the court and the crowd without first hearing his explanations.

"Didn't you get a full indictment beforehand?" Garfield was a serious man, but the judge was inexperienced and not a pedant. We started talking while sitting: Garfield in a high-backed chair, I'm on a bench, I wouldn't be torn off from it even with a bayonet.

"In the course of May, General Mitchell brought me six orders regarding the conduct of the brigade: all of them related to Athens," I pointed out the window, "barbarically destroyed Athens. Mitchell demanded that the seized property be handed over to the quartermaster ...

- Did you do this?

"The hams were eaten, and the silver spurs for Major Grosvenor were taken from a shopkeeper who deserved to be executed.

Were Mitchell's orders the same indictment?

"If that were the case, the divisional commander would have had to remove me from the brigade and from the regiment, and General Mitchell appealed to my vigilance.

What did you do in the remainder of May and in June?

"As Mitchel demanded, he wrote reports on excesses, real and imagined by the shopkeepers. General Mitchell demanded that the heads of the thieves be shaved, which I did not.

Why, Colonel?

— I didn't find them in the regiment. And you, General, will not find it. They are not here.

Garfield looked through the documents before him.

"Mitchell's first order was sent to you on the third of May; Major General Buell's order to remove you from the brigade was issued on the second of July. You have had two months to establish strict discipline.

- I don't need downtrodden soldiers, but men who are brave and think for themselves. What did we do for two full months, you ask? If you please, we fought, we occupied many cities, and among them are Fayetteville, Jasper, Salem, Chattanooga, Stevenson, Huntsville. The Eighth Brigade, along with others, that promoted Brigadier General Mitchell to Major General ...

"It is inadmissible to attribute to oneself the promotion of a commanding officer!" exclaimed the staff colonel at the judging table.

I responded with restraint, not to his zeal, but to Garfield's gaze.

- Generals can lose a battle, but it takes soldiers to win it. In the gold of our uniforms and their prowess.

"Isn't that why you dismissed the soldiers like that!"

"I am proud of the independent spirit of our volunteers: this is the best of everything that I have found so far in my new homeland. Those who emulate formal discipline will never become popular commanders in a volunteer army.

"We're not in the military to be popular, Colonel," Garfield pointed out.

- But we are waiting for at least new epaulettes - the volunteer does not even have this! When the alarm sounded, he rushed to the weapon; A naked, terrible truth opened up to his eyes, but he does not run, he fights to the last breath. And they want me to shave his head! From mid-February until Buell's dismissal of me, the Eighth Brigade took more than thirty cities and towns of Tennessee and Alabama, losing four people, and noble, unfortunate Athens killed my wounded and executed prisoners, about forty people! Now they want to take away our honour as well; Is not this too much! - Garfield did not answer, and I managed to finish: - I do not recognize the trial of the regiment, nor these false-witnesses!

"But you showed up at the court session," the judge objected.

"I came so that my absence would not be interpreted as an admission of guilt. I appeared - Colonel Turchin, my Volunteers are not here, they are above suspicion.

"But you wanted to avoid trial, Turchin!" said the Colonel from General Mitchell's staff. - They tried to resign with an offended mien! I request that the Colonel's letter to the Chief of Staff, James Fry, be read out.

Here it is, a letter sent to Buell's HQ from Bridgeport on July 5 - I kept the copy, I felt that there were still decades of slander ahead. I wrote about the affairs of the brigade, that in Kentucky, in Bowling Green, we captured provisions that would be enough to feed the division; we captured Huntsville and a 137-mile section of the railway, capturing 16 locomotives from the enemy, about a hundred wagons, depots, bridges, warehouses, food - the cost of these trophies alone is two million dollars. He wrote about the cities taken, about crossings on longboats and scows, about the movement of regiments and about the courage of volunteers.

"*I always and everywhere went ahead of my brigade,*" I concluded the letter, "*in the performance of my military duty. In official reports and dispatches, no one has yet spoken badly about me or my brigade. Now, instead of thanks, I receive insults, therefore, as a colonel and commander of the Nineteenth Regiment of the Illinois Volunteers, I unconditionally resign and respectfully request that it be accepted immediately.*"

"Isn't this an escape?" the Colonel laughed. "That's not what an officer with confidence does.

"That's the only way a commander who hasn't lost his honour should have acted!" General Buell sought my exile; it was he who demanded reports from Mitchell and finally received a false, evasive report.

"Permit me, here is General Mitchell's report!" the Colonel answered without hesitation; Garfield and the other judges relied on court discussion, the Colonel studied the papers and knew all about my service. — "*The pillages committed in Athens by the troops under the command of Colonel Turchin have become just a byword,*" he read Mitchell's lines. "*At my request, a committee of citizens of this city reviewed the claims of the victims, and according to sworn testimony, the total damages exceed $50,000. I ordered to search the satchels and backpacks of all the privates and sergeants of the brigade. The officers presented reports, in the proper form, that they allegedly did not find any things in excess of what was required by the charter from the soldiers. Colonel Turchin always assured that he did everything possible to prevent robbery and other violations of discipline by his troops. But he didn't seem to be able to.*" Here is the true opinion of you as a divisional commander.

- And according to such a report, the brigade commander is dishonoured! - I controlled myself: the reckless, as if from an ambush, look of the staff colonel, not prepared for the trial of the judge - everything required caution. "General Mitchell does not trust me or my officers, but he trusts the secret rebels, from whom he asks for papers against my volunteers. The knapsacks were searched—a shameful, vain humiliation—but the knapsacks are empty, in them, in all of them, there was not even a hundred dollars' worth of goods, but where are goods worth fifty thousand! Where are the silver trays, furs, silks, pieces of linen and velvet, ladies' gloves, armchairs, even the harmonium - where is the good that is put on our account? Ask at the post office - not a single Illinois volunteer sent mail in May - where is it all, then, this pillaged property? I made no promises to Mitchell to prevent pillaging; I stated that such a thing was impossible in my regiment."

Medill inappropriately shouted: "Bravo! This is the best Illinois regiment!" - and began to clap his hands.

"I will remove from the courtroom anyone who tries to interfere with the proceedings. Garfield warned the hall.

Medill loved me devotedly and disinterestedly: I was his favourite officer. To have your favourite regiment, to know its companies, to share their success, even if only on the pages of a newspaper – If it is possible to stand closer to the people when there is a war going on? If it is possible, then only in one way: to enter the regiment, but Medill was a publisher and patron of the arts.

"Persons not involved in the army will now leave the court," Garfield said, "we will summon witnesses as necessary. Reporters may stay."

The Committee of Citizens and the victims of our robberies, cleanly dressed, but with deliberate orphan poverty, as if my soldiers had emptied their wardrobes in their homes - all these birds of prey humbly left the court.

"That woman must leave, too," the Colonel told Garfield.

Nadine hesitated - is it possible to ignore the will of the judges?

"Mrs. Turchin will remain. "The slanderers, the Confederate spies, were gone, and I got up, it didn't matter to me now whether I was sitting or standing in front of a court martial. "She belongs to the regiment, as much as any of its soldiers.

"That is one of the charges against you, John Turchin." Garfield was getting strict with me.

- Mrs Turchin is wearing a military uniform.

- It's misappropriated! the divisional colonel intervened.

- That cloth is legal on Turchina, it has absorbed the blood of soldiers.

"I would prefer that Mrs. Turchin withdrew." Bowing his heavy head, Garfield looked at Nadine with respectful curiosity; he seems to have seen in her a woman whose tenderness is wounded by the trial of a loved one, her determination and confusion. "I have been told that you are a noblewoman, here there will be things told that are unfit for your ears.

Nadine got up. Before the judges stood a simple woman, for whom no one did even the dirtiest work, and at the same time independent, not knowing someone else's power over herself.

"The reason why we chose a republic is to throw away the privileges of birth," she said.

"I don't want you to be touched by any of the details of the case," Garfield insisted.

"Women don't belong here!" The chaplain rose, holding a black hat in his outstretched hands. Transparent blue eyes in a tanned, rustic face blazed with prophetic fire. - She cannot be called to witness, she does not believe in the Lord and will not lay her hand on the Bible."

"I will tell the truth even without the Bible, while others will lie with the name of God on their lips.

The judges were silent in confusion.

"If you remove your wife, I will leave too. You will have to take me into custody.

Are you looking for excesses?

- I want to avoid them: My wife is revered in the regiment, but the regiment is insulted - all, from the commander to the last volunteer."

The square was noisy outside the windows. From my position on top of the sycamore leaves, I could see the shops at the back of the square, the excited crowd, the faces turned gloatingly in our direction, as if the townspeople were waiting for the court to throw my body out of the windows to be torn to pieces.

"I hope," Garfield said, "you don't pay your wife's military pay?"

— I pay a soldier's salary to blacks: the Volunteers think it's fair. But Turchina does not receive money, in the war even a woman has nothing to spend it on.

"Why waste money when you can take it arbitrarily!" The chaplain still didn't sit down.

Our roles had changed - I groped for the right tone, recognized my judges in the face, Augustus Conant more and more opened his heart hostile to the regiment.

The judges didn't touch Nadine anymore; they bowed their heads over the table - strong heads, without gray hair and baldness, young, quick heads of the young American army - and decided by a majority not to exclude her from the judicial process. This connivance cost Garfield dearly - from the first day the newspapers of the rebels did not know leniency to his name.

Garfield turned to me again.

"John Turchin, do you admit yourself guilty of actions and crimes against law and honour ?"

— In what I heard here, there are two points that apply to me. I stood in front of Garfield.

"The first is what my accusers called the Russian idea of war. Inaccurately said; this idea was guided by the ancients. I confess it and, therefore, I am guilty in the face of evasive politicians.

"Before the law," Garfield corrected.

— And before an evasive law. There is a woman in the regiment: If this is a crime before the law, then, despising such a law, I recognize myself as a violator.

- There are no women's uniforms in the army warehouses of the Federation.

"Only, there's too many things lacking there, General!"

"Women are not forbidden to visit rear hospitals in the rear, or to serve as sisters of mercy.

"Women would not buy regiments and brigades for themselves, they would be brought to us by the patriotism of the Republicans.

- The army charter laid the border, and you crossed it.

— That I admit; but I did not secretly bring a woman into the regiment, even in Camp Long I warned the adjutant general of the state.

"Fuller hoped that you would think again and come to your senses, Turchin.

- No, he was expecting something else: that Mrs. Turchina would be unable to master the war.

"You didn't disobey orders?"

Thank God I didn't get any!

- The orders forbade foraging, and you took provisions.

- I paid for them.

- As always.

— I paid those who agreed to sell. I took them from those who supplied the Rebel bands but slammed their doors in front of us.

"Isn't it the right of a citizen to sell to whomever they want?"

- War puts an end to such a right. Bread or bull meat is as important as gun cartridges.

"So, you've arbitrarily stopped the operation of the law." And doesn't that mean turning friends into enemies! A staff colonel intervened in the interrogation. "You push a man into the Rebels' arms just because he is afraid of the Rebels vengeance.

"I am pushing him not to rebellion, but to a choice.

Suddenly Joseph Scott rose from the bench.

"Interrogate me, General Garfield!" he said unkindly. — I am Colonel Turchin's second in command."

"Your turn will come, Lieutenant-Colonel."

"You're all talking to Turchin, to Colonel Turchin," Scott said. "There will be no end to this.

I didn't know what was more surprising: the rebellion of the pedantic campaigner or the annoyance with which he said my name twice. What's on his mind? For some reason, he is not judged, although the regiment entered Athens under the command of Scott, and not Turchin. Isn't that why he's on a free bench because he's been assigned the role of witness for the prosecution? During May and June, General Mitchell and Major-General Buell's commissioners were zealous in trying to split the regiment, to divide the officers into innocent and guilty, but the regiment was not attracted by the smiles of army Machiavelli. Only Conant was with judges. But Scott: why is he avoiding my eyes?

— Joseph Scott! Cried Garfield irritably. "Don't make me think that in the regiment even the senior officers have forgotten the meaning of discipline.

Lieutenant-Colonel Scott stepped towards me, made a turn in all his uniform and sat down on the bench. Scott is awkward, with elephantine legs in huge boots, with a flat, sunken chest, shaggy eyebrows drawn together over small, dull eyes.

"You are sitting on the defendant's bench, Scott! Of course, in his excitement, he went to the wrong place.

"For half a year, from November to July, I, Joseph Scott, commanded the 19th Illinois Volunteers Regiment . He got up slowly. I do not see anything in the actions of the regiment, for which it could be so insulted and sent to the rear in the face of the enemy. To my regiment –"

He turned to me, - Forgive me, Turchin, for calling the regiment mine, at this moment justice requires it. On my regiment a stigma has been unjustly cast. I was commander when the regiment entered Athens, and remained so for another two months, and I demand that I share the dock with John Turchin.

Before Garfield have time to answer, as in the square, under the windows of the court, a cry was heard: "Mr. Turchin! Oh! Mr. Turchin!" Loud, lively and desperate. I rushed to the window, worried that I had once heard that voice; something familiar, even close, broke through despair and mortal anguish. If the shout had been: "Vanya! Vanya! " I would have thought that this was my brother Sergei or my desperate father. The voice called out to me again: "Mr. Turchin!" - and choked, drowned in the roar of the crowd.

A fight surged over the square. Yellow dust rose, as if raised by a tornado, and hid the fighters. I saw black backs, curly sculptured black heads, brandished fists: it seemed that a crowd of blacks attacked their brother and he stood alone against everyone, but it was impossible to see him in the mad tumult. The Negroes fought in a circle of whites;

Alabama farmers also strove to hit the Negro, to get him with a boot, but their main concern was the soldiers, the farmers did not let the Volunteers pass and retreated from the court, and the whole fighting crowd surged away from the courthouse, in the direction of the stores. Again I heard the call again: Mr. Turchin! - muffled, as if from the underground, a terrible cry of the victim and two rifle shots.

- They killed him! It's him, it's him, Vanya! Cried Nadine and rushed out of the room.

Everyone froze at the open windows, stupefied. James Garfield was next to me. The judge and the defendant at the same window - he, wearily leaning against the frame by his upraised arm, with bitterness and reproach on his face, and I, shaken from the fact that I remained a powerless witness to a crime. The crowd stopped, swallowed the killers and hid them. A Negro remained in the open space. He was on all fours; then his hands slipped through the dust, and he collapsed sideways. Nadine ran to him, sank down beside him, took his hand, bent over her chest; then the Volunteers, at her sign, lifted the big black body on their arms.

She looked back at the windows and, with a gesture of despair, of a personal grief that was still incomprehensible to me, covered her face with her hands.

"Excuse me, gentlemen!" I heard a polite voice; standing in the doorway was the Mayor of Athens, appointed by Mitchell to head the Citizens' Committee. — - A nigger stole from other niggers and wanted to run away. "

"But he called me!"

"I believe you were called by one of the soldiers to protect the black thief ... It's a local nigger, he lived on the Elk River."

- What's his name?

- I do not know. The mayor turned his neat gray head on its thin neck. "Hey there, Adams, do you know the name of the dead thief?"

"Sure I do!" replied a cheerful voice. "You won't forget such a name: his name was Napoleon."

Chapter Twenty Five

Death loomed by the bed; not yielding to surrender Napoleon's to our entreaties, Samuel Blake's potions, the psalms or the prayers and psalms of the regimental blacks. Three black slaves also came from the plantation to which Napoleon belonged. I needed Abraham, but they couldn't find him, he and two young Negroes vanished; their soldier's uniforms were lying in the tent, as if the Negroes had fled in their underwear, afraid to show themselves on the soil of Alabama in the uniforms of the Northerners. Strange Negroes peered into the staff tent; Who were they? Napoleon's friends or enemy scouts? Had their hands held the knives which inflicted five wounds in the chest and shoulders of Napoleon and a fatal one in his belly? I called them to the dying man, to take leave of him; they stood respectfully silent, and in the eyes of Napoleon I did not notice the enmity.

He was dying, and I sent for the chaplain.

This is what Napoleon told Nadine, while she was bringing him to the camp, in the first spring carriage that she came across at the hippodrome; he muttered, regretting that the blood was staining the pale suede of the upholstery. His masters did not punished him when he was taken from a New York prison to a plantation near the Kusa River, not far from Gadsden. The old planter died, the young one did not remember Napoleon and accepted the Negro as a baptismal gift. The old planter's widow would have surely tormented him to death; every black who had angered her husband, who died from a rush of blood to the head, she regarded as his killer, she would have plagued the life out of him, but by that time the daughter-in-law had firmly taken the reins. Her son and his wife managed the plantation in a scientific way, bought machinery, drove water from the Kusa River to corn and cotton plantations, they fed the slaves and even treated their illnesses, like a good master feeds and tends his livestock. They got rich, bought a city house not in Gadsden, but in Athens, and a lot of land on the Elk River and on the right bank of the full-flowing Tennessee. To pay his debt to a merciful heaven, Napoleon, got married and finally took on his shoulders the care of a widow with five children.

The eldest girl, Judy, was taken to the Athenian manor house, she was 14 years old, and there was no more beautiful girl for a hundred miles. Napoleon became a slave again; He recalled widows with big families in Philadelphia and New York, free and hungry, and was sorry for them. -they would have the long winter, the damp cellars, the search for bread; they would not have the Alabama sun, the bright water of the Tennessee and the generous earth which warm's one's feet without shoes. But war was approaching, and the devil took possession of the young master, as if he and his wife had made money for this in order to spend money on army horses, saddles and wagons, on Austrian and English guns in overseas boxes, from New Orleans, on gunpowder and old swords purchased from all over the state. He left the Paris landau to gather dust in the coach house, opened a volunteer depot in Gadsden, and recruited a regiment of horsemen. From then on the plantations on the Elk River and on the banks of the Tennessee now knew only cavalry posts, soldiers' revelry and ruin. The Negroes lived in fear, the honour of a slave, already miserable, and his very life depended no longer on the masters, but on the whim of these condottieri. And the master, a newly minted major, remembered that Napoleon was a fugitive, and Napoleon cursed the day when he brought a widow to the Methodist church; now neither she nor her children knew no mercy. Judy alone lived with her mistress among flowers and had all she needed.

The major and his cavalry lost Athens to the 18th Ohio Regiment of Turchin's brigade, saw how the blacks met the northerners, and, galloping to his place on the Elk River, drove the slaves into a barn. This house arrest dragged on for a month, then blacks began to be taken to work under guard. Napoleon learned about the second capture of Athens by the northerners and that the soldiers were commanded by

Colonel John Turchin. Once he saw me up close, I was returning to Athens with other officers, and a handful of slaves were hiding behind the dogwood bushes, fearing to give themselves away. Napoleon recognized me, but he would not have ventured to run away, even now a cook from Athens found him, she said that Judy had been taken away from the town and was to be taken to the courthouse; where she must confirm that during the second capture of Athens she was raped by northern soldiers. If she refuses, she will be handed over to the Rebel cavalry. Napoleon planned to look for me in Athens, but they had caught him in front of the courthouse.

Nadine changed the wet napkin on Napoleon's forehead, and moistened his lips, but they were parched again at once. The plantation had restored the tautness of his muscles, his chest rose and fell with convulsive breathing, there was strength in his arms extended along the body, in his long legs - Scott's bed was too short for them, his rough, scarred feet hung over the end with weight - only a black, distended belly spoke of the approaching end. Pain silently lived inside his flesh, in the redness of his rolling up eyes, in his lips stretched out, as if before a scream, his jaws clenched to a creak.

- Who are these blacks? I sat down on a folding chair.

— Dey's jest People... Mr. Turchin...

"Don't waste your energy, saying my name every time.

- Yes, Mr. Turchin.

Why didn't you wait for me at the camp?

I was afraid to be late...

- Why did Negroes want to kill? Not whites - Negroes!

He turned his head with an effort, looked, am I the same Turchin?

- The master gave orders, and the slave will kill the slave. Oh-oh! If our masters were not overcome by arrogance, if they would give the niggers free, the South would receive a black army ...

They're afraid to give you guns.

"God blinded your enemies and took away their minds...

I often thought about it after we entered Tennessee and Alabama. Among the thousands and thousands of blacks, not only our friends were revealed, but they were also cautious, prudent, gloomy people; they hid when we appeared. A well-trained Negro, if released by the South and placed under the command of a former master, will become more terrible for us than a white soldier in hilly and marshy lands. He will endure any heat, he will crawl like a snake through thorny bushes, and if yesterday's master convinced him that the Yankees are encroaching on the lands of his free children, then he would meet us as enemies. He will look at the Negroes of the North with compassion, as fools, and think that they fled to the North because they were slaves, but if they were free, they would not betray the South and their own race. I reached this depressing thought and, suffering, rejected it. And now suddenly the same sadness on the lips of a slave who knew secret freedom and slavery; in my confusion, I complained about Abraham's flight.

"Abraham is unbearable among us," Napoleon said sadly, he thought I was complaining that the biblical Abraham turned away from us. "He has returned to his shepherds and is sitting by the fire... Will he now take us with him to the land of Canaan, to his black tents? Mebbe he'd take the widows and children, dey ain't done nothin' wrong."

A chaplain approached along the street of tents, his narrow shoulders and round hat danced against the night-crimson sky, he walked quickly, his black robes clattering against his boots.

"Here comes the priest." I bent towards the Negro, caught the unkind smell of decay, and straightened the cross which had slipped down under his arm. "Tell him about your daughter, about Judy. It'll be like making your testament.

He rolled up his eyes to see Nadine at the head of the bed.

"Ain't got nothin' to leave the children, but love ..."

"Most people leave money or debts," Nadine said. "There aren't many who leave loved to their families."

"I tried not to make debts." He paused. "You called Judy my daughter, Mr. Turchin ... thanks. But a black gal should not be born so beautiful."

Augustus Conant slowed his pace; and his dark figure took on majesty. After the skirmishes in court, only a determined man or one

287

of a fanatic faith could come so calmly towards Nadine and me. When we had fought near Cairo, Grant and I were confronted by the Rebel General Polk, the Right Honourable Leonidas Polk, Protestant Bishop of Louisiana, at one time a West Point alumnus. Now, as major-general, Polk retained the episcopal rank, and tales of his courage tormented the ambition of Augustus Conant. Increasingly he discovered military talent in himself, and ceased to be afraid of firing.

"I have been called to a dying man," said Conant.

I lit the lamp so that he could better see the Negro, and Napoleon could see him in priestly vestments. The chaplain's eyes wandered distractedly, uncomposed, at everyone at once and at no one in particular. Something irritated him: the prayers of the Negroes, or the instantaneously swooping moths and other insects kept flying in and beat on the lamp glass with a dry tapping of hard wings.

"Are you sure the nigga needs a priest?"

Blake raised his hands in a gesture of complete helplessness.

"Medicine is powerless! "he said in Latin.

"Does he believe in God? Has he been christened?

"He's a Christian," Nadine said. "He's worn a cross since he was born."

The chaplain demanded that he be left with the dying man. We remained by the entrance to the tent, and in the dusk I recognized many: Christian, Barney, "Pony" Fenton, Tadeusz Drama, Johnston and other ardent supporters of freeing the slaves. Barney O'Mullen did not recognize Washington's caution. "Negro regiments?" he argued passionately. – "And why not! Why the hell don't they give them weapons! They have as much right to fight for themselves as we have to fight for them!" But when I told Barney that we don't need regiments that are separated from others by the colour of their skin, let the Negro join any regiment, the rascal bowed his head and screwed up his eyes. "Well, if the Bashkiy Negra, like our Abraham, gets promoted to officer?" "It's inevitable," I confirmed. "Let's stop right here! Otherwise, the white soldier may fall into submission to the black captain: it's the end of the world!

The Negroes sang at the food warehouse, their melancholy psalms did not drown out the voices behind us.

- What's your name, black?

- They christen me the name Bingham.

Why are you called Napoleon, then?

- I was given this name when I ran away from master.

- Leaving your homeland is a worldly sin. Denying the name, you were given at baptism

is a sin before God.

The chaplain raised his voice; although he taught the flock not to throw stones at the sinner, now the stones were assigned to me too; I left my homeland, instead of Ivan I had become John.

- I'm Bingham-Napoleon ... Folk like the name Napoleon more.

- And do you?

- Yes. I think he was a saint.

- He was a robber, a bandit!

"But Jesus also forgave the robbers who were crucified next to him. God is waiting for me in heaven, Father.

"How will you appear before him with the hands of a thief?"

Napoleon did not understand the chaplain; he did not know about the slander.

"You stole from the poor, from your brothers!"

"I once saved a widow and her children from want. Was that a sin?

- Why did you run away? If you're being honest, why did you run?

I was a-running to my death...

"No one knows where death awaits him; it's pride speaking in you, black man.

"Bad times has come, Father, death awaits a nigger under every tree ...

"But it was Negroes that rushed at you, Negroes, not whites." You ran away from your master again. Did you want to be a soldier?

- No, father: it's a white folks war.

"You're quite right. A white men's war. But whites are fighting because of you blacks. And they die because of you.

"The Masters in the South understand that but the Yankees don't, Yankees think they are fighting out of pride: to get Richmond and New Orleans to bow down to Washington. He dropped his voice to a whisper. "Father, Jesus placed you with a cross and a Bible between two armies, between hate and hate, between white and white…

I shuddered, there was such sorrow and compassionate feeling for the suffering of others in the voice of the doomed Napoleon.

Why did you run away, why did these men kill you?

"Come here close, Father," said Napoleon.

A heavy, dark-skinned book in dark leather lay on his heaving chest, Napoleon's hands covered the Bible. The chaplain bent over the dying man. Napoleon spoke so softly, that we could hardly hear odd words here and there; Judy's name, my name, someone's ranch or other, Elk River, violence. His strength was leaving him, Antonov's fire (n. *Gangrenous emphysema*) was getting the better of his mighty nature, or maybe Napoleon did not want outsiders to hear even about Judy's pretended dishonour.

Lieutenant-Colonel Scott avoided me that evening, as if he regretted taking the strand he had in court and giving up his staff tent to a dying Negro. I don't remember Scott talking to the Negroes, but he was reserved and silent with others as well, his gaze darting past the negroes, as if he hadn't been able to notice the black colour in nature. Yesterday Scott made a bold move; perhaps when he came to the courthouse, not knowing that he would be on the guilty bench, and only after hearing the accusation did he decide to share my fate. I did not expect this from Scott and breathed more freely. But did he not now repent of his impulse? Fate gave him a warning: black passions stopped the proceedings, the trial was postponed until the morning, he had time to think, to return to the witness bench, leaving the impression of a noble officer. The honour of the regiment is precious. Scott tried to defend her, but did he want to get involved in the affairs of slaves?

In the morning, when we entered the courtroom, we saw Scott in the defendant's bench and I was ashamed of my thoughts of yesterday. Perhaps he was avoiding a meeting, he spared me, saved me from the need to thank him. It was he who had the care of the regiment on his shoulders; I was returned to the camp as a disgraced commander of a disgraced regiment, not to serve, but to await the verdict; I stayed all night with Napoleon, Scott with the regiment, with its indignation. If we had a mutiny the blame would fall on me and on Scott, meanwhile Buell was doing everything possible to cause an explosion among the Volunteers. They were rightly proud of having pushed farther south than any other northern regiment in the summer campaign of 1862, cutting off the roads by which the Rebel armies of Missouri, Arkansas, Louisiana, and Mississippi communicated eastward with Richmond and the ports of the Atlantic.

The volunteers believed in their star, the Illinois men marched at the head of the brigade and at the head of Mitchell's entire division. I have not forgotten how the eyes of my soldiers had glowed when the Staff colonel, the one who now sits at the judges' table, announced Mitchell's order to them, threw praises at the lined-up companies: *"You dealt the enemy blow after blow with unparalleled speed. Stevenson fell, sixty miles to the east of Huntsville. Dicater and Tuscumbia have been in the same way seized and are now occupied. In three days of operations you have expanded your front by more than 120 miles; and this morning the thunder of your guns at Tuscambia could be heard by your comrades-in-arms, who glorified themselves with a victory at Corinth!"*

Then all of a sudden, collapse. Like the bridge over Beaver Creek. The thunder of the cannons ceased, speed was replaced by immobility; the Volunteer was stopped by a blow to the chest, spat upon, declared a thief and a rapist. No one is accused individually, no one was named, no one exposed, which meant that everyone is suspected; henceforth we are pariahs, outcasts and renegades. Buell ordered our companies to be withdrawn to the rear; Mitchell hesitated, detained the regiment near Athens, where hundreds of hating eyes were at every step making sentence and execution. The companies were seething in the camp across the Elk River, the appearance of soldiers in the streets of Athens was like front-line reconnaissance, when only extreme circumstances keep the parties from shooting and bloodshed. The planters and merchants did not dare to hurt my soldiers, but the contemptuous looks, the sneering smiles, the sly jeers, and a mare's snorts cast after the tattered army uniforms were no sweeter than constantly repeated slap on the face. At the end of May, Mitchell sent the regiment to Fayetteville, Sweden's Cove, Chattanooga, Jasper, Crow Creek, Belfontaine, Larkinsville, and Bridgeport; and God, how my guys fought, insulted, spat upon, exhausted among Athenian roses, heliotropes and nasturtiums!

A month ago, Mitchell wrote in an order about *unparalleled rapidity*; now he could write about the flight of the hawk. But no, his clerks used different ink now: while the Volunteers captured stations and trains, depots and food warehouses, while they defeated Rebel companies and bands of cavalrymen, Mitchel's clerks dedicated their pens to the Rebel sympathisers in Athens. We multiplied the score of victories; while behind our backs, the false account of robberies, violence and rape grew. And now here we are, the whole of our regiment, haled before this petty court, we are dirty in body and thoughts, rejected by the generous commander Don Carlos Buell, whose orders reliably protect the stores and warehouses of the enemy and of the rights to own slaves ...

One might well think that it was not I who spent the sleepless night, but Judge Garfield; who had spent a sleepless night. He was already annoyed at the beginning of the stuffy Alabama day, he was angry with the restless, plotting crowd in the square, the horsemen in hats pulled down over their foreheads, now and then looming near the shops, the sullen Joseph Scott, who challenged the court.

"So, Scott, you don't find fault with the regiment?"

The lieutenant-colonel stood at attention before the judges: it was a matter of a career and the only way he knew how to feed his family.

"Let this High Court decide whether we are innocent or guilty."

"Your men are accused of pillage and rape, Joseph Scott.

The criminals must be called by name. Yesterday, I did not hear a single name.

Once again, the colonel from Mitchell's staff put his word in.

"You allowed the companies to forage; when did it begin?

"On 10th of March, the regiment, at the head of the 8th Brigade, set out from Fayetteville, taking with them a two-day ration. Scott was speaking of a time when he became a regimental commander. "We advanced on Huntsville on back roads, over steep rocky hills, and struggled through swamps. "Your Honour!" Scott exclaimed, encouraged by the memory. "To drag wagons through the swamps or up a mountain, we had to harness mules from two or three teams and haul the guns up ourselves. We walked to Huntsville for sixteen days, and for those sixteen days - a two-day ration! We bought food or took it from the enemies so that in the morning the soldier would have the strength to get up from the ground ..."

"What about Huntsville? What did you take in Huntsville on the twenty-seventh of April?"

"Confederate Major Kavanaugh, six captains, three lieutenants, a mass of soldiers and a lot of booty. The hungry Volunteers carried themselves like gentleman in Huntsville; the enemy in such cases does not stand on ceremony. John Turchin appointed Colonel Gazley as Chief of the Huntstville Military Police, and calm was restored..."

"So, it's had previously been disturbed!"

"Sure it had!" Scott shrugged. "Shots, a gallop of cavalrymen, the thunder of Captain Symondson's cannons, awakened residents, women fainting ... And a two-faced mayor: who bows to us and tells the townspeople that he will send for the Confederate cavalry to drive us out of the city. Yes, the war disturbed the peace.

"But when Mitchell's division received a hundred thousand daily rations, did you continue to take from the population?"

- General Halik sent the barges carelessly; the rations had to be destroyed so that they would not fall into the hands of the enemy.

Garfield said, looking at his papers, "Your men have confirmed that after Huntsville, in the Cumberland mountains, they burned down farmers' houses and killed pigs.

"We came down from the Hills against Chattanooga and repulsed the continuous attacks of mounted Rebels. Turchin had a list of several leaders - we burned their houses. In log houses along the way we found women and children and did no harm to anyone.

"And the pigs that were shot!"

- The northern soldier didn't know that these were domestic pigs." Scott smiled for the only time in the trial. "They are semi-wild, and razor-backed, your honour, and roam far away from habitation... I dare say we have brought more gain than harm to this land. - The judges were waiting to hear what kind of economic good can a soldier bring? "We repaired culverts and road pipes, we saved bridges tarred and lined with cotton ready for burning, and with quick raids saved the estates of loyal planters doomed to destruction.

And the destruction of Athens?

- When we occupied the city for the second time and I saw the corpses of tortured soldiers,

I said to myself: This town ought to be burned down! Let them feel that there is a war going on...

"And you ordered the soldiers to do so?"

- I'm sorry to say I didn't ! Look out the window - no!

— But Turchin? He shouted: "For two hours I close my eyes"?

Joseph Scott looked back at me.

- No. John Turchin is not shy about speaking bluntly... But why isn't Colonel Stanley called here? In Athens, his people were also martyred, general.

"Staff officers questioned Colonel Stanley," Garfield said. "There's his testimony in the file."

Colonel Stanley would rather forget about Athens; if Joseph Scott had burned this little town, to the ground, Stanley, like a good Christian, would have shuddered, but he would have breathed a sigh of relief. I had left him in Athens, among the roses and subdued citizens, with a combat regiment and entrusted him to the care of the wounded soldiers of the brigade; among them was Balashov, who had fought well, received bullet wounds to his arm and chest.

Stanley camped at the hippodrome, the mayor's speeches put him to sleep, the silence, the nocturnal chorus of cicadas and the peaceful herds deceived the colonel's vigilance; he did not bother to put up pickets - Then Rebels attacked Athens. The dawn mist, the fire of two mountain howitzers from the nearby hills, the snorting of horses bursting into the racetrack, and Stanley's regiment, which I had rated as highly in the brigade no less than the 19th Illinois, fled, losing wagons and more wounded.

Stanley retreated to Huntsville, confident that he would had succumbed to a superior enemy; the regiment was overtaken by the stragglers, they fled in the backyards and spoke in fear about the large Rebel forces. When we regained Athens and yesterday's voluntary executioners again rushed to meet us, ready to wipe the dust from our boots with their bare hands, the truth about the Ohio men's flight became known. This was what Colonel Stanley should have answered for to the court.

He circumvented the military circumstances of retreating from Athens; however, he testified that *the owners of the houses were compelled by force* to give up their attics to the Rebel shooters by force: also that the women hooted after the northern soldiers, that the noble ladies spat on the wounded and an excited crowd threw dirt and rotten vegetables on the unfortunate men, that the horse insurgents tied the northerners with hemp ropes to the saddle and drove the horses at a gallop, shouted at the Athenian ladies to take the Yankees to the kennels. These circumstances, Stanley concluded, explained and partly justified the excesses that arose during the second capture of the city. Colonel Stanley expressed the hope that the merits of the brigade should outweigh the involuntary sins of the soldiers, the looting of the hardware store, food shops and "other unworthy acts quickly suppressed" on the scales of justice.

Nobody in the courtroom were pleased with Stanley's evasiveness: the officers of the brigade remembered only too well the colonel's oversight, which served as the loss of Athens. But Stanley did not please the accusers either: the hostility of Athens was so undeniable that the retribution did not seem criminal.

"Did you intend to protest Stanley?" — The testimony of the Ohio colonel offended General Garfield with evasiveness of the Ohio Colonel's evidence.

"The Ohio cavalry entered the city first," Scott said. "We were in a real bad mood when we occupied Athens... We didn't like the city very much, General.

"But you easily finished the Rebels off easily, Turchin," the judge turned to me. You captured the city, with practically no casualties.

"Scott means townspeople; It was they who were shooting at us.

"Such persons should be brought before a military court.".

"Got to find them first," Scott protested. "Two of them were killed in the shootout, but their bodies were not found either."

"None of the townspeople died," the lieutenant reported to the court. - They showed us the parish books: there we no funerals in those days."

"But it was then that two graves were added to the Athenian cemetery! This came from Tadeusz Drum as he rose, he developed a habit of visiting the cemeteries of small towns."

Garfield sent for the mayor, who explained that on the eve of the surrender of the town, two people who died of a fever.

"I wish I knew what calibre the fever was !" Snorted Thaddeus Drum as he sank onto the bench.

Among us all the mayor was the only white-headed old man, with gray wings of eyebrows over mournful eyes, his mouth constantly moving, chewing, or opening, as if preparing to speak.

For a long time, he did not take his sad eyes off James Garfield, probing into the face of young America - educated, intelligent, with good manners - seeking understanding, protection from the rough Yankee boot, from the rootless immigrant, ready to reduce a foreign country to ashes, a country belonging to others. Yet the Southern newspapers were dragging out all of the thirty-year-old General Garfield's dirty linen to the public view, declared it an ugly product of a puritan father and a Huguenot mother, they wrote that he had become hardened in early orphanage, in the rough life of a frontier settler, that he had to face the former pea-jacket of a helmsman and mechanic on the Ohio Canal, and not the uniform of a general, even a northern one.

"What I am about to say must remain in your papers' records..." Joseph Scott paused, to let the clerk prepare. – "The prisoners were shot. These murders had happened before, but in Athens the murder was made a public spectacle, allowing the inhabitants to be present.

"The town is slandered, your honour!" said the mayor resentfully. - there were armed troops, in Athens, this is their court and was their judgement and sentence."

"It was your sons and your brothers who were here," This came from Nadine's voice; the audience had grown accustomed to her silence, and now the judges were taken by surprise. "The troops you called in for help you".

"This woman does not know our life," the mayor said regretfully. "She does not know the suffering and kindness of the inhabitants of my country."

"Joseph Scott! What evidence do you offer in support of your accusation? Where are your witnesses? Name the names."

"I cannot hand people over to the bloody vengeance of the Rebels."

"We will protect them."

"God himself will not do it when we leave Athens.

"We're not leaving here, Joseph Scott," said the staff colonel. "North Alabama has been cleared, the road to Birmingham and Montgomery is open all the way to the Gulf of Mexico.

"Fortunes of War are fickle," Scott observed."

"Mr. Scott is right," the mayor said unexpectedly. "Confederate troops must take back Alabama and Tennessee, the armies of East and West must communicate...

"That is just why we shall not leave the bank of the Tennessee!"

"Let's go! Well, if we don't run as Stanley did from Athens." - I could not resist, all the bitterness of renunciation suddenly came out. "What else can an army do but retreat if it keeps combat regiments in the rear!"

"You complicate your situation, Turchin. Garfield regretted my intemperance. "You are in no position at the moment to criticize senior officers."

"The Republic gave me this right."

"But the court withdraws it."

"Colonel Stanley spoke of soldiers tied behind horses," Scott continued. "They were dashed to death on the stones, only one soldier survived. Stanley believes that Rebel soldiers fired from the houses, but where did they go?! Colonel Helm's mounted Rebels galloping off from Athens; then who were the people who fired from attics for a long time? On the square, in front of the hardware store, artilleryman Morgan was killed. There were none of Helm soldiers in the hardware store, but the lead came from there that smashed Thomas Morgan head blown in."

"Who did you find in the shop?"

"The owner. And a gun still warm."

"And what did you do with the owner?"

"Let him go! I shouted. "We had the murderer in our hands and we let him go, instead of shooting him in front of the regiment."

"Shoot him without trial?" Without proof of guilt? He wasn't a soldier."

"No, General, he was a murderer... A hardware store murderer entitled to trial by jury." Sudden fatigue approached me, the consciousness that we are separated by a fortress wall; the court would not hear me and my tears for Thomas. The rumour about the murder of Thomas quickly flew around Athens then; the town was ours again, the loyalty masquerade had already begun, the May thundering downpour poured over Tennessee and the Elk River, washing away the blood, then suddenly an almost inaudible shot, a blow of lead from the depths of the shop, and Thomas's body on the ground, as if stuck in rain beaten dust, an old poncho thrown over his hideously disfigured head, and Dr. Blake holds Nadine back from the dead man. And traces of the murderer, a few steps from the corpse; the imprint of long narrow soles, a backward step towards the store; it seemed unbelievable that the murderer would hide from the crowd of Volunteers, but he had vanished, disappeared, evaporated, no one heard even the clatter of hooves.

"No! Trial by jury is not for those who kill soldiers from ambush. That's how it all started, General; the soldiers smashed the murderer's hardware den, and now for that store we were also billed to us for a considerable sum - six thousand dollars. Who submitted this bill? The Murderer? His wife? His Son? Let the plaintiff appear in court! But he won't come...

"Your Honour, Mr. Eddie is in town," the mayor said, his hand outstretched toward the door, as if the hardware dealer I'd slandered was behind it. "Mr. Cornelius Eddy, is a highly distinguished citizen of this town, interrogate him."

The court adjourned, while waiting for the arrival of Mr. Eddy. Outside the windows the square was noisy, but with no singing. There was something menacing in the subdued passions. No one approached the open windows. Now, in a moment, a murderer will enter in the hall: not in handcuffs, but with his head up, and we will not be able to take his life.

"What about the Negro, which was caught in the square yesterday?" Garfield was sick of the waiting.

"Just a common story, your honour," the mayor replied. - A petty theft and a brutal reprisal!".

"Who's his master?

"Well you see, his master's Major....one of Colonel Helm's officers, he has estates in Gadsden, on Tennessee, and on the Elk River. So it's difficult to establish where these Negroes come from.

"Is this black man alive?"

"Turchin's people took him away. If he is dead, then he died without a confessor.

The regimental chaplain raised his voice - stern, annoyed by everything taking place in the courtroom.

"He was a Christian and died a Christian. I was beside him."

Chapter Twenty Six

Witnesses did not have to wait; Barney, followed by the hardware dealer Cornelius Eddy, an empty-eyed man with a gliding gait, a vile piece of tinplate as light as fish scales, in a gray top hat, with tufts of dusty gray hair above the lip and at the cheekbones, and pale eyes, in a dove-gray face. Could this be killed with a bullet? It seems that she will click, One would expect it to clang, to pierce several sheets of thin iron, finding neither heart nor mortal flesh inside.

Barney described how the murder had taken place. Three men entered the hardware store: Barney himself, Private Fenton, and Lieutenant Ball. In the depths, in the twilight, at the back the owner stood, thumbs hooked under his armpits, staring past the soldiers, out onto the square. "I wish you'd open your shutters, good sir." Ball said. "A man can break his leg on your metal ware" - "I like it the way it is," was the answer, "there's nothing for you to do here..." And so, he said.

Mr. Eddie nodded, interested and proud of himself.

The lieutenant ordered Fenton to open the shutters, and we saw everything in the store: Hammers, saws, axes, wheel rims, metal barrels, sabres, spurs, even Mexican War artillery helmets, cherry-coloured, with pom-poms on top. While Fenton was opening the shutters some Ohio soldiers also entered the shop: someone saw the silver spurs and took them from the window. "Go on, go on, take them, take, guys," said the owner, "go on, steal - you Yankee bums, that's why you and the Yankees are hungry." Then I noticed a shotgun in the corner and handed it to the lieutenant. "It'd say by the smell, that it was fired not so long ago." Then Ball allowed the soldiers to take a pair of spurs, and one, an Ohio man, took two pairs, the second for Major Grosvenor. And that was the end of it, we took the gun and left.

Then I met Thomas, and it's hard to believe how much he loved those toy rattles. "Show me the store, I'll buy myself some spurs," Thomas said, and off we went. I see the shutters are closed again, but the door is half open. "I've never seen a silver spur," Thomas said, "now I won't let this one slip." He asked me for a spur, tried it on his heels, and was delighted to find that they fitted. "Now I will buy them... I have long wanted to get these, and I never had luck. Now I'll buy them." So we walked along and Thomas was saying: "I only saw such spurs on one person ... Even on the Fabius River ... He was lying on the ground, with brand-new spurs were on his boots." "Dead? Well, I'd have taken them!"

I said. He answered "From the dead?!" You can only take weapons from the dead." "Well, what about a purse?" I still laughed. "Money shoots better than a gun." He stopped, If I'm not kidding, and says: "The vile and despicable one who will reach for a wallet in the pocket of the murdered. Turchin would not keep such a man like that in the brigade." I even got angry and said: "You should be singing in a church choir, not fighting!" "So that's your tune, is it," he says. "Then I'll pay not only for my spurs, but also for these ones." He held my spurs in his hand and took out a purse with his other hand. "If you're such a fool, then you can pay for all the seven pairs that we took. The devil himself can't keep a fool from doing a stupid thing" ... We were already approaching the store by then, and Thomas's skinny purset, he left himself only two dollars a month, the rest of his pay he sent to his mothers in Chicago...

"In Mattoon," I corrected the man. "Thomas' mother lives in Mattoon."

"So, Thomas said, I don't know how much they cost, Barney," "but it would be the right thing to pay, they consider us all beggars. Did you notice how the girls here are looking at us?!" That is what tormented him, so I thought I'd get under his skin a bit: "That's during the daytime," I said, "then they turn their noses up and have such proud eyes, but you ought to take a look at nights ..." He stopped, he wanted to say something, but he didn't have time, he fell, he was out of the store shot by a bullet in the head."

Barney's words resurrected that terrible day, the dark, black blood seeping through the old poncho, the hostile to the square of the Athenian stores, Nadine's cry of motherly horror. I could not tear my eyes never left the dry, grey, upright lizard standing there erect - you can't kill this one, it will grow not only a tail, but a head, and what, took the place of its heart. Mr. Eddie listened to Barney with the critical judiciousness of an arbiter, murderer that he was! and I made a mistake.

"James Garfield!" I cried. "You must know that I would kill this man even now, if I had a pistol!" So, let this murder be added to my account."

"The court can't sue for a crime that didn't happen," said Garfield with all possible composure. "But your words tell against you, Colonel", He hurriedly turned to Barney: "When you entered the store after the shot, what did you see there?"

"I didn't enter, general, I ran in, burst in!"

"And what did you find inside?"

- That jackal!

"But you took his gun away.

" There was another standing there; exactly the same, a double bore."

"Is that true, Mr. Eddie?"

"Honest truth: my son's shotgun. He had gone to New Orleans and, been delayed there,

thank God; if he had brought goods, I would have lost them too, Your Honour."

"Why was the gun there in the store?"

"Maybe, my black servant Jeremiah could have been brought it in," Eddie tried to help the inquiry. "When the thieves left the store, taking the gun and seven pairs of expensive spurs ..." He stopped, meekly apologizing, and looked at me. "I was told, Your Honour, that one pair of my spurs went to Colonel Turchin; I would like to see the riding boots of that officer...

"Mr. Eddie, why would the black servant bring a second gun?"

"I was in such a state when the thieves left, that I went to the living room for a glass of something... This is a dry town, Your honour, but there are moments ... The mayor will excuse me." He made a gesture towards the mayor, who looked at him unkindly. "I told my servant: 'Jeremiah, this town's full of thieves, stay in the store and guard the goods, these Yankees are friends of the blacks, they will not touch you.' But he was shaking; he got real panicky with fear when a man's like that it doesn't take much to make him shoot. Those Blacks, Your Honour, are like children: he won't come to his senses as his finger pulls the trigger before he knew what he was doing."

"Thomas Morgan was killed with a square piece of lead; Why was there lead in the shotgun?" Garfield asked.

"It's my son's gun, Your Honour, and why he loaded it with lead I can't imagine."

"But if your Negro fired,, you must've heard the shot."

"There was firing every day, it got so a man didn't know if they were really shooting or just thought he did."

"Did you question Jeremiah?"

"He's gone and vanished, neither hide nor hair of him! What do you think – a nigger who'd shot the white man! Not only the Yankees would have boiled him alive in a company boiler, I'd have pulled down three black hides from him.

"Why did you run away, Mr. Eddie?"

"Who wants to die, Your Honour?!" exclaimed Cornelius Eddy innocently. "As I looked into their faces, I imagined my own intestines on their bayonets. And when he caught his breath, he found out that the store had been smashed. I've been robbed all around, Your Honour, plundered everywhere" he said mournfully, "by Yankees in Athens, and by my son in New Orleans... Spent all my money."

" But where did the hardware goods go?" Garfield asked Barney.

"We smashed what we could but can you can't break much of that, it's all metal! So we got ourselves bruises and turned everything over and left."

"Colonel Turchin, did you hear the soldier's confession?

"This soldier you can trust, General. And please take note that no one in Athens became a victim of lynching, no one's estate was burned, not a single heap of ash, yet we wanted to do all that, and badly! — I pointed at Cornelius Eddy. "Those like that deserve an hour of lawlessness.

The court dealt with the claims of the storekeepers. They swore that both the counters and warehouses at the stores were emptied by my soldiers, but they could not explain where the goods worth Fifty thousand dollars had gone or how they managed to continue a brisk trade without delivering new goods; the whole thing degenerated into a tangle and insoluble. "Pony" Fenton was also accused of looting, as if he had taken part in the Athenian robberies.

"In Athens, your honour, only two stores suffered: the hardware store and a provision one. There they took ham, bacon and chickens - I know this from the stories. And there was also a requisition on a plantation near Athens"

"Who owned the state?"

"Jack Harris's. He is in the Rebel army, he took his men with him and left an old Negro as caretaker. I was sent with two soldiers to get cornmeal and a barrel of New Orleans molasses; if I came across some meat, the lieutenant ordered to take only chopped carcasses and provided me with an authorisation certificate. And when we opened the doors of Jack Harris' smokehouse, none of us have ever seen anything like it! Hams, shoulder blades, chopped carcasses, sausages in cotton sacks, smoked pigs' heads—"Pony" Fenton's eyes lit up at the memory. "All the slaves, how many of them were hiding in the hazel and dogwood thickets, came running under the platen trees to the smokehouse, as if we had opened the gates of paradise. They asked for meat and I gave them meat; it was they who watched pigs and bulls, sowed corn, drove cattle out to pastures. I gave half a pig's head to one girl of about seven years old and you should have seen how she put it on her curly hair and carried it home, with the whole family as a guard of honour."

"Do you admit that you took part in the illegal confiscation?"

"No," Fenton answered after a lingering silence, "I thought about that then, and I've thought again now. No!"

"Your whole story confirms looting."

"I took the meat from the smokehouse of the enemy."

"You couldn't have known that this planter was a Rebel.

"What about Jack Harris? In these places everyone knows him: he is a major in Helm's cavalry regiment, it was his horsemen who dragged our soldiers over the stones. Jack Harris should have been hooked alive in the ribs and hung up in his own smokehouse.

The cries of the sack of Athens did their job - to the inhabitants of Richmond, Atlanta or New Orleans, the city seemed to be in ruins, crucified and given over to Sodomite violence. But through the open windows of the judges' hall, the truth reached without difficulty - fragrant flower beds, mansions covered with greenery, undamaged roofs of tiles and iron, the rustle of dense sheets of plane tree and magnolias, the peaceful cry of a rooster and a stubborn mule.

"Not a single soldier of Europe and America would have been so merciful to Athens as our volunteer," I took advantage of a moment of confusion. "We have the right to be proud of them, not to bring them to justice."

"They are not being judged, Turchin," my enemy from Mitchell's headquarters called out, "they're uncouth and guided by instincts. The blame is on you: you pushed them to banditry."

I did not dignify him with an answer; I wanted to understand the young general, not one of those who buys epaulettes.

"General," I said to Garfield, as if there were only the two of us left in the room, "to me, war is a battle, and I would like to die with my boots on." The Rebel leaders have done me the honour of placing a price on my head. Now my head is put up for auction by the North: let's see what the new price is.

We dined in the rooms of Chicago friends in their hotel rooms; exhausted by required silence, Joseph Medill orated while Nadine and I walked through the crowd, as if guarded by Dawson, Ramage, Scott, Tadeusz Drum, and other officers of the regiment. Mutual hatred, the proximity of an explosion, could be felt in the crowd, in its deliberate movements, in the men holding onto their saddled horses, in the watchful eyes from under the pulled down hats. In the hotel lobby, the air of hostility hardened even more: if silk top hats, bowler hats and frock coats could have fired, if watches and charms exploded like shrapnel, and umbrellas and canes acted like bayonets, blood would be shed here too. At dinner too, Medill talked without stopping: the court could not confirm the sack of Athens; wasn't that not a reason for cheers, for the belief that the court would end with my acquittal and the punishment of the slanderers! How badly noble enthusiasm harmonizes with the viscous, gray matter of life.

Returning to the courtroom, the Court proceeded to the most difficult paragraph: outrages against women's honour. The voices fell, the sounds died down, and Garfield once more appealed once more to Nadine's prudence.

"Mistress Turchin," he said with a touch of request, "you are the only woman among us, I would prefer to protect you from this".

Nadine rose. Her eyes expressed gratitude to Garfield. How well and decisively her face showed enmity or friendship - not with a grimace or a contemptuous break of the lips, but with one light of her eyes, now deaf and hostile, now open to the interlocutor.

"General, I'm a regimental doctor's paramedic."

"But there is moral impurity, madam."

"There's too much of that in Athens, and I want to make sure that our Volunteers do not become like an beasts.

The General looked at Augustus Conant.

"Mrs. Turchin is not frightened by the spectacle of sin," said the chaplain. "You could have kept her out of the courtroom, just as Fuller had to keep the lady out of the regiment.

Soon Nadine was not the only woman in the courtroom: the hostess of the boarding school for noble maidens, Miss Swingley, the headmistress of the Young Ladies' Seminary was called and the wife of the planter, who owned Napoleon's stepdaughter Judy, was called. The boarding house, the prosecution said, was raided by Chicago volunteers, and four girls were caught in the bedrooms and violated. In the courtroom, the headmistress was poorly heard; her broken, strangled voice was directed towards the judges, as was her confused face and shocked, avoidant eyes. Her words were clumsy, confused; Nature gave this woman all the outward features of honesty and provincial nobility; towards the end of her story Garfield was glaring angrily at my people.

"'Pluvia defit, causa Christiani! '" I rose from my seat, in a state of agitation I had never felt before. The Roman rabble shouted; "There is no rain, that means the Christians are to blame!" And in Athens, the northern Volunteers is to blamed for everything!"

And then the chaplain spoke up; he hurried to the judges table and stood next to the woman.

"Your Honour! I was not then next to this unfortunate woman, and I have no right to testify to the violation, but it could have happened, it could! A senior officer breaks the law, brings a woman into the regiment, isn't this a disastrous example for soldiers! There are not only gentlemen in the companies; the dirty sediment of Chicago is also in the regiment; the debauchery of the fiery Zouaves is proverbial. Women were allowed on the train when we moved from Missouri to Ohio...

"Wives!" Nadine said.

"Wives! Sweethearts! Women! the chaplain shouted angrily. "And what happened, your honour? - the train of sinners was punished, the terrible catastrophe at Beaver Creek ...

Tadeusz Drum was behind the chaplain's back, turned his face around and shook him so that Conant's black Soutane seemed to creak.

"You... Augustus Conant..." he stammered, forcing out the words, "You are unworthy to your Church. I challenge you to shoot it out with me ... - An unfortunate thought suddenly dawned on the captain. "Mr. General, allow me to fight a duel with him!"

"I'll arrest you, lieutenant," Garfield replied.

"Murder is not my thing, I wouldn't shoot a peaceful mule," Drum said to Garfield. "But Father. Conant is a lion with the heart of a hyena!..."

Tadeusz was taken away.

Scott explained that the lieutenant had lost his wife at Beaver Creek and deserved leniency. The Headmistress of the Seminary wept quietly, her shoulders shaking.

"What's wrong with you?" Garfield asked sympathetically.

"Life is so hard ... I'll never get any more pupils... This place is accursed... but it's not my fault..."

Captain Presley Guthrie's report disappointed the judges with insensitive brevity. He and his soldiers entered the Seminary; finding here the Headmistress and one terribly frightened girl there, he had advised them to leave and had given them two escorts. Nothing in the Seminary had been touched, neither living nor dead. I understood Presley Guthrie; it was not a coldness which spoke in him, but insulted dignity, but yet he lost next to the oppressed Miss Swingley. She, was an old maid, heiress to a vast mansion converted into a seminary; she did not recognize either Presley Guthrie, or the sergeant with a soldier, summoned by the court. Guthrie squinted scornfully as Swingley shook her head at him, but the sergeant and soldier were speechless with surprise.

"Calm yourself,", the sergeant said to Miss Swingley. 'turn your back and close your eyes, and remember everything as it was, and then look at us. Remember how on the banks of the Elk River you hugged your girl and shouted: 'Saved! ...'"

But Miss Swingley remembered nothing of it: no regimental horses, no dinner under the cedars, and no conspicuous face of the sergeant.

"I'm very sunburnt," the sergeant said, excusing her. "Even my own mother would hardly recognize me. Better to call the girl, she was smart enough.

Garfield asked about the victims, Swingley repeated that there were four of them, but she would not name names, the girls had their lives ahead of them.

"Mr. Mayor, do you know the names of the victims?"

"I know them." The gray head shook woefully, often in response to Miss Swingley's tears. – They are too well known, I will not name them either before the court or in confession.

"Did the Athenian doctors certify rape?"

"We have such a paper, General."

"Does it have names on it? "The doctor had to flee from Athens for the mere fact that he was involved in the investigation. The girls' parents are rednecks, they are rough and straightforward, and the war's hardened them. Here's the paper."

Garfield said that the anonymity robbed the document of meaning.

"I can't guarantee that the girls weren't taken to another state," the mayor said. "This misfortune forces parents to sell their estates, incur losses, and move away.

The accusation is shaken, but the battlefield was left to the mayor; the judges could not believe that Miss Swingley, oppressed by the imperfection of the world, could lie and pretend like that, and the nobility of the mayor was a cruel, no-lose game. Still, Garfield remarked dryly:

"You care about the dignity of your citizens, but they were forced, too, to write down the mayor's answer do not want to spare the honour of officers and soldiers."

The Chicago reporters' pencils ran faster, but they also had to write down the mayor's response:

"I did not offend them with false suspicions."

"Miss Swingley," I said, "could you identify any of the soldiers?" I hope the court will allow the company to line up.

I came up with a plan to build any company other than Presley Guthrie and expose Swingley's lies when she points to any of the soldiers. But behind Swingley's apparent meekness and dismay lay the nightly vigilance of an owl.

— Oh, no, no! she pleaded. "Spare me that! I was hiding behind the window drapes... I heard screams... the clatter of feet..."

The trial stopped again, the threads of the inquiry were interrupted, the charge was left hanging like a rumour, not confirmed and not completely rejected.

The court turned to the ruined Judy.

Judy's mistress, the wife of Major Jack Harris, a little woman in mourning, spoke of the rape against Judy. I was astonished when, throwing back her black veil, she walked to the judges' table: on a tender and nervous face were burned crazy light-violet eyes, all of the same tone, as if without pupils. A violet fire seemed to bubble up inside her and burst out through the slits of her eyes.

"I quite understand the inconvenience of my presence in this court," she said, "I am the wife of Major Harris, your enemy. Before you stands a woman forced in this unfortunate war to choose a flag and a capital, and I chose not Washington, but Richmond. Can such a witness be trusted?

She fell silent in a proud and intelligent readiness to leave the court; Garfield was frowningly silent.

"We moved from Gadsden to Athens because in Gadsden we were considered almost abolitionists. When our fugitive slave Napoleon, formerly known as Bingham, was brought from New York, the husband forgave the black man. Bingham undertook the care of a widow and her children. The counterfeiter, the New York villain, the runaway again became pious and pure; that's what the South is doing to them! I talk about Bingham because Judy is his stepdaughter; on learning about the violence and dishonour, he lost his reason and again became a thief.

"He died that night," Garfield said.

"God, take his soul!" She looked up at the stucco ceiling. "He believed the North was fighting against slavery."

"We are fighting for the unity of the Union, against seceding states," the staff colonel corrected her.

"The Negroes don't understand that: they're children. I don't understand a lot either: they say it's a war between brothers, then why do I so often hear Irishmen, of Germans on the streets of Athens, and why is a Russian commanding a regiment? What is the life of Athens to him?"

"Madam!" Garfield interrupted her. "I shall not allow you to speak ill of Federation officers."

"Without lowering her eyes, like Miss Swingley, Harris's wife told how soldiers entered her house in the middle of the night, pushed her into the bedroom, posted a sentry at the door and abused the beautiful Judy, who was probably a mulatto; "Napoleon was her stepfather, but it is rumoured that Judy had been adopted by her mother from a young white officer."

"Why don't you protect her good name, as you spared the others here?"

"Judy is a servant, General!" she exclaimed with a measure of contempt. "Who in Athens does not know of her disgrace!"

Arrogance of this kind, it seems, was not tolerated by Garfield either.

"Rape is reckless; why did the rapists neglect you, young and attractive lady?

At that moment, Harris hated Garfield more than all the generals in the North.

"Gentlemen don't ask questions like that, ladies don't answer them.

"When a gentleman becomes a judge, he has more troubles, and among them the most unpleasant duty is to achieve the truth hidden behind a wall of lies. Were they sober or drunk, those people who locked you in the bedroom?"

Mrs Harris was arrogantly silent.

"Where is Judy? "Garfield asked the mayor.

"The   servant   was   taken   to   the   estate,   she   had to   recover,   because   there   were   many   of   them   ... But we will present her to the court."

"Judy has been stolen! "cried Mrs Harris, gloating. "She was stolen from during the night from the Elk River estate."

Harrison's Negro, the housekeeper of the Harris's, was called for; he confirmed that a band of blacks had taken Judy away that night. I told the court everything I learned from Napoleon; I was listened to attentively. The evasive testimony of Miss Swingley, the arrogance of Mrs Harris sowed doubt in the minds of the judges; the declared crime was too great and the evidence too insignificant.

"Lies! It's a lie! Mrs Harris screamed.

"Did the black man say all this before he died?" Garfield asked me, warning her with a raised hand.

"I didn't hear his confession, but he'd always told the truth, all his life."

"I was with the black in his last moments, General." When I sent for the chaplain yesterday, that he would turn from a confessor into the main witness and the honour of the Volunteers would pass into his hands. He slowly reached out for the leather- bound Bible and did not immediately speak. Conant closed his eyes, the drooping eyelid lighter than his sunburned face, the chaplain's thin, expressive lips moving soundlessly. Even Joseph Scott, a man of devout piety who would not allow the cashier give any of the officers their money before the chaplain received his monthly hundred dollars pay, even he now looked upon the chaplain as an enemy. But the citizens of Athens fearlessly waited for the testimony of Conant: he was received in the house of the local priest and in other houses, whose walls were hot from heated speeches in which Lincoln, every evening, burned like a heretic at the stake of the church.

"I was talking about immorality in the regiment," the chaplain said quietly. "The wild, rebellious habits of the commander, the cruelty of the men called to kill, even in the name of a higher goal: bad grass sprouts in such a field ..." He seemed unable to collect his thoughts. - We expect good from the wicked, but do they gather grapes from thorns or figs from burdock? I spend sleepless nights, crying out to God and finding no answer. "Ships," it is written in this book," he raised his hand and lowered it again on the Bible, "however large they are, they are driven by a tiny helm. What is this helm? If not faith and conscience, what else can be the helm of the ship of humanity? Yes, faith, reason and conscience, where there is only mind and no faith, vice rises to the surface....

After all the filth and mercantilism of the past two days, the judges listened to the man who turned his thoughts to God,

"Yesterday I returned to the camp despondent. Greed, lies, blood on the square – it lay like a stone on the soul; I despair that neither my rank nor work made the soldier better. And I sat down to write a letter, General, a letter to Don Carlos Buell, I wanted to ask for a company, to keep my dignity, but to try myself there, where, if one believes John Turchin, there is no room for mercy ..."

The inhabitants of Athens stared at me with resurrected hatred. I am the devil, because of me the shepherd needed to take up the sword.

"I had written a few lines, when the soldiers called me to the dying man: I went to the thief and met a kind man. Yes, Madam,

Negro Bingham is an honest and kind man, you did well when you forgave the runaway."

Little Mrs Harris stood in front of the court, stepping back a little, and looked at the chaplain with gratitude.

"He said to me: I mourn for you, Father, you stand with the Bible between hatred and hatred, between white and white, between brothers. Yesterday God chose a black to restore me to reason, to say: Here a shepherd of souls is needed! A lit candle is placed not under a bushel, but in a candlestick so that it shines for everyone.

"Didn't he talk about his stepdaughter, about Judy?"

"Bingham died sorely troubled about the widow, about Judy and the children."

"The Harris's will take care of them, Your Honour!" the mayor said.

"The court awaits, Chaplain," Garfield inquired.

"I have nothing to say, General. The dying man revealed to me the same thing as to Colonel Turchin. He did not have different words for the world and for the confessor.

If Helm's cavalrymen had burst into the square and Jack Harris's head had appeared in the court's window, the sensation would not have been so great. Garfield calmed the room with difficulty; But now there was a noise outside the door, there was a sound of feet on the stairs, shouts and altercations.

"Are you saying, Chaplain, that there was no violence against Judy, but there was coercion of the masters and intimidation of the black girl?"

"So, that was what the Negro told me."

- Wasn't it a feverish delirium?"

"He died with a clear mind, General."

The doors were flung open, to show people crowded behind the threshold: an officer of the guard, Captains Ruffen and James Guthrie, soldiers, Abraham and other Negroes - barefoot, in tattered trousers and shirts.

"Colonel!" shouted Abraham from the doorway: the Negroes were being crowded. "Judy's here, we found her on Elk River.

"Let everyone come in!" Garfield ordered.

The Negroes stepped onto the dark parquet, their bare feet feeling unsteady on the polished wood, they took care escorting a small black girl with an expression of fear and madness in the slanted eyes of the mulatto. She moved strangely, as though against her will, protecting her still childish breasts and face with her hands. Judy did not even see Scott and me on a separate bench, nor the judges: for her there were only the Athenian patricians, the people who were in the house of her masters,

and Mrs Harris herself. Before us stood an unfortunate girl with lips bitten into blood - a young, hunted little animal.

"Come to me, Judy," said Mrs Harris with gentle authority, and Judy sauntered over to her.

Mrs Harris put her arm around the negro woman's shoulders, and she trembled as if preparing for a ritual dance. "Dear Lord in heaven ,what have they done to you, Judy!"

Mrs Harris cried, stroking the girl and looking at her from her dirty feet to her black, litter-stained hair. "Did you force her to come here?" Garfield asked the Negroes.

"We asked her," Abraham answered, "and she come."

"She will follow everyone, she no longer has the will."

"Who you are? Jack Harris people?"

"We are his enemies, General!" Abraham said proudly. "We'll never be Harris' Negroes, we're John Turchin's Negroes!!"

"These are my soldiers," I came to his rescue.

"Blacks can't be soldiers," Garfield protested. "And you know that, Turchin. Take them away!

Abraham rushed to me, the soldiers grabbed his arms.

"They did as they threatened... They did it... believe Abraham. He was being pushed out; afraid that I would not hear, he almost shouted, They promised that they would do this to her every night ..."

When the door closed behind the Negroes, everyone saw Nadine next to Judy. But Judy backed away from her outstretched hands.

"General, look at this girl," said Nadine, "the violence happened today, yesterday, not two months ago."

But Mrs Harris did not remain in debt:

"Those blacks who stole her could have played with Judy all night!" she called. "Judy, tell me, who was it who hurt you?"

Judy was silent, her long arms gathered her dress under her knees; a hopeless look tore at the soul, her face was heartrending..

"Say, Judy, and then we'll leave. Who did this to you?"

"Yankee…" Judy breathed.

"Soldiers?"

She nodded.

"A long time ago?

"Yes, ma'am… It was a very long time ago.

Conant approached her, but he too frightened her with the severity of his eyes and black clothing.

"Yes, ma'am, the Yankees… The Yankees did it… a long time ago…" Judy kept repeating and nothing more would she say.

INTER-CHAPTER FOURTH

"Considering everything that we heard about Turchin and his actions, we expected to see an uncouth peasant, to see a typical product of royal power. However, he behaved like a man of a deeply noble soul and thus won my heart … Still, in a few days he will probably be discharged from the army."

From a letter from James A. Garfield, the future 20th President of the United States, to a friend   Summer 1862

Chapter Twenty Seven

Even a seasoned soldier would not have guessed what kind of strange cavalcade is moving along the road from Athens to Huntsville, among cotton and maize fields, in the shade of oak forests, cedar groves, past lonely, heat-weary pines and sycamore trees, along the line of cypress trees, those dark sentry of estate parks. In front of them were several Ohio cavalrymen, followed by five young colonels, more confident in the saddle than at the judging table, Joseph Scott, Nadine and I, our faithful friend Medill, and with him old Silverman, a photographer who hardly ever parted with his white coat, glasses and a straw hat. And behind us are horsemen - a judicial lieutenant and three cavalrymen.

My hands and Scott's hands did not pull the shackles, we were free to drop the reins and fold our arms on our chests. This was not an arrest, but the liberties we had in Athens were curtailed;

I was not allowed to take with me even those company officers whose testimony was required at the trial in Huntsville.

Garfield rode away from Athens on the evening of the day black Abraham brought Judy into court. The road to Huntsville ran past our camp, and I happened to spot Garfield with his orderly as he held the stallion and hesitated whether to turn back to camp or go on his way, straight ahead.

The hazel bushes at the tomb of Napoleon hid me, but it was not for nothing that Garfield spent his youth on the border land; his ear caught a rustle, he sensed someone's presence there, and boldly rode into a hazel grove.

"A dispatch has arrived," Garfield said as he jumped to the ground and removed his hat.

"The Court's transferred from Athens to Huntsville."

"Athens is an uncomfortable place; talk about plunder, but the town is intact."

"Was he a decent person? Garfield looked at the grave."

"I mourn him like a brother."

With that word "brother" the old spring thunder rolled over the sky of Alabama; I suddenly saw again something that did not come to mind for years: the pile bridge in Novocherkassk, the top of the carriage raised by Sergey's quick hand, his reddish eyes on his long face and his prophetic words: try it, try it, see how they bake bread there, with a hard crust or only crumb, and what the servants are flogged, plain cubs called batogs, as God ordained,, or are they weaving whips from a crocodile? How did I renounce you, brother, how did I call the holy name of another, a black-faced person? Was he my brother, or I had learned to use idle words?

"I never knew a kinder person than him."

"I don't think we'll meet in Huntsville." Garfield held out his hand to me. "I will resign as a judge from the case."

"And Turchin will regret this, Mr. Garfield," I said. "But a man of honour cannot be a judge in this tribunal."

He paused, listening to the sad, plaintive trill of buntings, and his gaze followed the flock of red birds.

"We're still a young nation, Turchin."

"The nation is only just taking shape."

"The Confederates, too—they consider themselves as a nation, and regard us as a rabble."

"The nation will be completed by the war, and it depends on whether we fight courageously and honestly depends on whether a healthy nation is born."

"She will be healthy!" Garfield exclaimed with the ardour not of a general, but of a young politician-senator."

"You saw Judy and you knew that our soldiers didn't touch her. But Judy will die before telling us the truth. This is our defeat, and it can't be redeemed by the capture of towns.

He jumped into the saddle, serious and gloomy.

"Farewell! However, I haven't yet received official consent to leave the tribunal." He chuckled with mirthless slyness. "My best hope is the enemy, that he'll force Buell to shift me from the courtroom into battle. Business has turned to failure in Alabama."

A smart and conscientious man who galloped away without thinking what pain he was causing me. Garfield had never had a day's training as an officer. It was bravery and successes that opened the way for him to the general's stars at the age of thirty. I was in my forties, I understood the war, from the Alleghenies to the Mississippi, and I could have changed a lot with divisions, but my brigade was taken away from me, and then the regiment. With a heavy heart I had left Athens. What the famine did not do, the Missourian ambushes, the swamps near Cairo, the broken wagons on Beaver Creek had failed to do, had been achieved by Buell's velvet glove – velvet but poisonous as sumac leaf.

The regiment did not know desertions - now we've heard about them. Individuals left, one by one, fled to Chicago, looking for places in active regiments; the late Howard's company lost four when Tadeusz Drama was thrown into custody for being challenged to a chaplain's duel. Some of our Negroes have disappeared; anyone could be killed and hunted down in the vicinity of Athens, and there is no way to find out where the black bodies crumbled to dust.

We left for Huntsville. The beating of hooves raised flocks of brownish buntings into the air, their sonorous trills and a short sad cry awoke in me the image of a distant steppe, sultry, but not stuffy, with free winds from the Caspian and the Black Sea. I closed my eyes and saw the steppe blue from the feather grass, the noiseless hawk above my head and inhaled the smells of childhood. Sometimes I found Nadine far from the road and our fellow companions: Over the years, we increasingly thought about the same thing, saw the same thing, in an uncontrollable imagination. Nadine's heart was imbued with maternal tenderness for me; the war taught her to share kindness between me and the soldiers of the regiment, but the hour came for them to wave their caps and hats in farewell to us, escorted us to the road, and all the forces of her soul converged on me alone.

From time to time we came across horse patrols, sentry soldiers, lagging baggage carts, wagons pulled by mules. Around noon, we heard the quick blows of the hammer on anvil and turned off the road. In a cedar grove we found an iron forge-on-wheels, of the 4th Ohio Cavalry Regiment, which, as part of my brigade, first to charge into Athens. The Ohio Volunteers surrounded us: people from another state, given to me to take Huntsville and Athens - Ohio lumberjacks, farmers, rail and sleeper layers - how happy they were to see soldiers from Illinois. You should have seen how they singled out Scott, me and Nadine from the whole cavalcade; how out of six colonels in saddles only one *savage Cossack* was recognized as close, how our horses were cared for, whether the horseshoes were worn out, wasn't there at least one nail missing? The sun burned unbearably, and Nadine and I could easily breathe the resinous air of the grove, among cheerful fellows in tunics or even undershirts, among the clanging of iron spurs and hammer blows on an anvil placed in a barrel of sand.

What could I answer them about the court case, being two steps away from my judges? Empty excuses, pleas in the manner of our ancient Russian wisdom: "God wont desert me, the pig will not eat me", or "the devil isn't as terrible as he's painted." I myself could not guess anything true about the movement of the court, although I did not expect good things, seeing that the judges were putting into business sheer nonsense, even that old complaint against me, kept by Buell, a complaint of the shipowners, the owners of the scows, borrowed back in mid-February for crossing the river and taking Bowling Green, from which the successful campaign of 1862 and our advance to the South began.

"I took these scows, but only in order to take Bowling Green with the smallest losses," I answered the court. "What have you done with these ships?" I didn't hear the question. "I will take other people's scows again, I will take it wherever there is a need for it, I will take it on every American river." - "I asked you, where did you put the ships?" "We didn't have time to look for a pier; we poked into the shore and immediately into battle. That's right, the owners had to look for their junk downstream. Let this be their offering on the altar of war."

... Napoleon was buried hastily, at dawn, at this time in Alabama it is impossible to delay the funeral of the poor; there is no ice or rare potions for him. The negroes came, they sang their psalms, and Conant sang the black one, and, looking at the chaplain, I hoped that a saving ferment was going on in his embittered soul. Napoleon's widow and younger children arrived, all but Judy; she took the side of the masters, and now, along with the mortal body of their stepfather, the Negroes buried the ghost of their stepdaughter. At that hour, from the regimental negroes, I also heard for the first time a military song, which many subsequently heard:

We looks like men a-marching on
We looks like men at war!

Could Abraham compose this song? Back in Elizabethtown, I had found him at night at the Zouave Gazette typefaces, he was startled, but I calmed him down, told him to finish the job and give me a proof. The verses were inept, with errors in everything; but by that time, I knew what miracles the musical nature of Negro talent could work with unpolished words. And the refrain in his verses was glorious: *the black soldier will prove that a slave has become a man!*

The Negroes who brought Judy from Elk River were pushed out of the courtroom, they fell into the hands of the military police, then they were handed over to the authorities of Athens. I went under guard to Huntsville, not knowing the secret of Abraham's disappearance or death, to be fair, the secrets of his death.

As a prisoner, I entered the town, taken by my brigade at the end of April. Then, I burst into the cobbled streets of Huntsville with the Ohio cavalry; then Mitchell ordered a fanfare, now Buell and Mitchell's officers escorted me through the streets of the town to a secluded apartment.

The oppressive days of Huntsville dragged on; days and nights in a house abandoned by the owners, a life of contentment, in a shady garden, among neglected flower beds, dependent on the staff kitchen, under the supervision of an elderly Scotsman, our gloomy attendant. Such was our prison and the punishment of loneliness after a year of regimental brotherhood. Often in my mind I saddled my horse, pulling my hat down on my forehead, galloping off to Stevenson or Decater to get on the train and going to Chicago to open the eyes of Governor Yeats to how unfair Buell was to the Illinois regiment. "Buell wants you to flee," Nadine said, reading my mood. "He's waiting for your reckless move, not knowing how to handle a new trial." In the evenings we didn't light the fire: when dusk fell, we tried to play music - Nadine sat down at Carpenter's untuned harmonium, I took the violin - but no, it was no use, my hands would not play in tune, the familiar music did not fit the occasion, the light violin in my hand seemed like a blasphemous trifle. It's not the music's fault; we were vain, offended, perhaps conceited.

Regimental duties in the 19th Illinois fell to Captain Raffen; he invented excuses for sending officers to Huntsville, and they found their way to our gate. Tadeusz Drum came galloping self-proclaimed, in the middle of the night, bypassing the guard posts, and sat on a garden bench waiting for my awakening. He stayed with us all day, told us about the Athenian guardhouse, about the chaplain who sought his release and met Tadeusz at the exit from the prison, and not in the robe of a priest, but in the uniform of a lieutenant. Conant told the Pole that he now understood the Turchins better than before; but would like not to see such people in the army; those who did not share the past of the country, does not know its future - they were knights of devastation. Drum started talking about Jadwiga; the memory tyrannized him, as if everything was not over yet and a miracle was possible, one day she would stand next to him, tender and fragile, and Beaver Creek would turn out to be a bad dream. Nadine agreed with him that if a person still lives for you, then she is real and fills your life.

I listened to them and hoped that partly Nadine was talking about us, about what I would be for her or she for me, if the worst happened. I believed, but held my breath, not knowing whether we were as strong in our love as Tadeusz Dram, and whether I could hold on to Nadine's memory. She — yes, she would live in me, with every feature, the twilight light of her gray eyes, her voice, her earthly flesh, would have survived three lives, but fate will spare us that—anything but that—I will be the first to go. And, hardly breathing, I thought jealously only of how long I would live in her memory: the absurd, clumsy destroyer of our life, the destroyer without remorse.

Twice Buell came to Huntsville; the Rebels under General Bragg and John Morgan pressed Buell's army, preparing his complete collapse in Alabama and Tennessee. Buell did not call me, and I did not seek a meeting; I didn't have words for him. And the day before the start of a new, closed trial, we had an unexpected guest, General Ormsby Mitchell. He rode at a loud gallop, threw the reins on the railing, and walked towards the veranda with the determined step of a soldier. He was a courageous man, rather ugly, if you take his large, frog-like mouth, and deep wrinkles from the wings of the nose, and sharp, stiff eyebrows, and the severity of his shaven face, apart from the ambitious strength seething in him, from the mind, noticeable in the look of large dark eyes, from a large, noble moulded forehead. But one could not separate them: he was all a receptacle of energy, everything, from dusty expensive boots to a dark mop of hair that covered his head more securely than a cap. Mitchell's slender figure, his heavy, elbow-length, knightly gauntlets, the winged black cape brought the smell of battles into our solitude, and I stood on the veranda, at home, ejected from the life of action. In the twilight he looked like a black man, the whites of his eyes sparkled, a high white collar separated his face from a black cloak.

"I'm not asking you to sit down," I told Mitchell. "Prison inmates are not masters of their cells.

During the spring, your brigade was the best in the division ..."

"We soon rotted under the Alabama sun!" I interrupted him. "It can be seen that it is bad for the northern soldier."

"You could have avoided the trial; I won the right to execute marauders, but you didn't even want to shave their heads."

"Get the death penalty for our executioners, general!" Even a soldier who is forced to put on a rebel uniform risks his life, and the well-meaning murderer, the poisoner of wells, the garret shooter are all protected by our gentleness. I did not find marauders in the brigade.

"Were there really none at all?" he exclaimed in disbelief.

"This is true grief; even you do not believe that your soldier is good and honest! Who will believe him?

Mitchell annoyedly raised his hand to the silver clasp of his cape, but refrained from opening his uniform with a full set of stars obtained for him by volunteers.

"Colonel Turchin," he now appealed to my sense of fairness, "but there really have been excesses in Athens. Stores were looted...

"The soldier did not steal; I will not take a piece of ham from his hungry mouth."

"But where is the line between this piece of ham and a smokehouse broken into and pillaged?! He kept in mind a sheet of the court record.

"If this is Jack Harris' smokehouse, it must be emptied or we'll never crush the rebellion."

"We are people of the same blood, the country is split here. The wealth of the South is the wealth of all the States.

"You still think nothing but blood!" But the abyss between social estates was dug not by blood, not by race, but by corruption and property. By guarding the enemy's property in the South, we are giving away our victory.

"In Alabama it was so close!" burst from Mitchell with the deathly bitterness of a loser. "If only Buell had not hesitated, I would have taken Chattanooga and the route south to the Gulf of Mexico is open. We would've ended the war!

Suffering opened up in a sudden frankness with me, in a deaf and furious irritation with the command.

"You can't end the war without starting it; we press the enemy, but we have to defeat him."

"I want to help you, Turchin. He pulled the gauntlet off his left hand and slapped his knee with it. "Damn it all, can't I help an honest, stubborn man?"

"There was a time when you could," Nadine intervened. - Joseph Scott wanted to help, too, and found himself on the defendants bench."

"So, James Garfield isn't coming back, I asked.

- You know already? So much the better. Buell's appointed a closed tribunal, without newspapers and third-party witnesses." Mitchell was waiting for questions, for indignation, but we were silent. "There's an order not to allow Mrs Turchin to the tribunal"; don't tempt fate, madam, this time you'll be treated to military rudeness."

"The Colonel will appear in court without me," Nadine said calmly. "I wouldn't have come at Buell's request either; I need to go to Chicago."

Mitchell thanked her with a deep nod; as for me, I ignored her, assuming that she was protecting her pride. I could not imagine then what Nadine was preparing for me, and Mitchell, and Buell and his tribunal.

"I suppose Buell isn't demanding execution, or is he? Or Penal servitude" Nadine asked.

"No, only a dishonourable discharge from the army."

"And you find dishonour easier than execution!" I exclaimed.

"Life is lighter than death," Mitchell chuckled sourly." It's easier for anyone else, except for you and me!"

"You appreciated me too late, General."

"I have always understood you, Turchin. But your fate is not my only concern," he went on with familiar harshness. "I don't want the verdict to be a stain on the honour of the army."

"And to your honour!" I answered directly. "That's what's troubling you, Mitchell. With what did you come to me for?

"To give you some friendly advice. Don't call them to revenge. Even now a compromise can be found."

"Compromise - It's more likely that a wild bull will accept a red rag than I will that damned word!" No, Mitchell, if it's a fight, then a fight to the finish!

I never had the opportunity to find out what motivated Mitchell: remorse, an awakened conscience, or a sly request from that fox Buell.

383.   A Group of " Contrabands."
[FOR DESCRIPTION OF THIS VIEW SEE THE OTHER SIDE OF THIS CARD.]

Negroes who fought for the Union Army, Huntsville, May 1862.

Chapter Twenty Eight

Huntsville didn't care about me when I rode to the courthouse in the company of a puny-looking major, two soldiers, and a dour Scot. The major did everything to make us appear to passers-by not as part of a judicial machine, but as free army horsemen. He challenged me to talk, he himself spoke intelligently and accurately, complained that the Rebels turned out to be unexpectedly strong, relied on Mitchell's military genius (not a gift, but a genius; maybe this is a property of the language or an American habit, but we have the word "genius " you hear more often than in Europe; having a glorious but short history, we are in a hurry to announce our geniuses to the world), and he called me sir: "Yes, sir" ... "Just think, sir!" ... "You yourself know this, sir "... "Huntsville is a nasty little town, sir!"...

We left the indignant crowd of reporters outside the door - Medill and Dawson could barely push through to shake my hand - and entered the gloomy headquarters room, where two dozen uniforms were already awaiting trial. In the depths of the hall there was semi-darkness, but I thought there was a too familiar figure, I closed my eyes, accustoming my eyes to the twilight, and, opening my eyelids, found in the place where I had seen Saburov, another familiar face: a Missouri major, from whom I had taken the horse under Lieutenant Maddison . My accusers wasted no time; Buell's sleuths infiltrated everywhere—to Fuller in Chicago, to General Hurlbut Missouri, , to Cairo, to St. Louis; but I felt neither shame nor repentance for the picture of my war, as I had fought it, even drawn with a brush, dipped only in the dark and poisonous colours from the palette.

The new court was rapid. The Court did not return to the complaints of the Athenian storekeepers, although that dirt was not all left behind the threshold: there were mentions of the Athenian "excesses", "ruins", "pillages of stores" were also commemorated, as a regrettable reality, already condemned by morality and the holy name of property. But in general, the tribunal held on to army affairs, to insubordination and to failure to follow orders from senior officers. It was as if Scott had been forgotten - it was not about events being dealt with now, but, if you like, about ideas and morality. Scott felt confident as long as the inquiry kept to the facts, to actions for which he was ready to bear responsibility. The general moral ideas seemed to him an abstraction; Scott is not one to rush history. And the staff colonel, now the chief judge, had got my range in Athens. I marched straight into his hands - angry, impatient, inexperienced. I was surrounded by enemies, and what remains for a surrounded soldier if not a sword!

"Did you keep blacks in the regiment?" the judge asked

"They are with us still."

"You trusted them with weapons?"

"Negroes are our best scouts. However, when they go on reconnaissance, the weapons remain with the regiment.

"Did you know that the president banned the arming of blacks?

"I heard about it."

"Did you also hear that Lincoln turned down General Hunter's offer to create Negro regiments?"

"We'll create black regiments if we don't lose the war.

"But if a Negro can serve in a white regiment," one of the judges, who sympathized with me more than the others, was surprised, "should he be given the opportunity for promotion?"

"At least a black officer will earn his uniform by bloody labour, and not buy it for money.

"You do not honour your own uniform, Turchin!"

"Among us, army ranks are bought, like the Church positions was in the Middle Ages; but how can you respect that which is for sale!

"You are insulting the court and each of us, Colonel!"

"No honest officer will be offended by the truth. Some traders who have bought a regiment with dollars can be got rid of by promoting them to generals. And where do these generals go then? To the High Command!

I did not name Buell, but his name is now on everyone's lips, and not with loud adjectives, but with a revealed misery of his worthlessness. Silence fell in the room; the judges were silent, and I was in no hurry, I waited to see if they were smart enough to avoid the pitfall.

"How did you feed the blacks?"

"At company kitchens, like other soldiers."

"They don't show up in the accounts, and our clerks say no pay was provided for them."

"I pay it myself; circumstances force me to a shameful division: in battle together, but payment separately."

"Isn't that why you're seek requisitions?" Because you are paying them out of food savings?"

"This economy is the income of company officers. I pay with my own money, and if the colonel's wages were twice as high, we would have twice as many black soldiers. I shall walk out of the war a beggar if Congress doesn't change its attitude to the blacks."

I could not have looked so far ahead, my hours in the army were already numbered.

"How much do you pay Negroes out of your own pocket?"

My frankness also created inconvenience for the judges; it was no criminal pleading "not guilty" who stood before the tribunal; by yielding on formal charges, I won in opinion, in the eyes of even people hostile to me.

"Did they tell you that a woman has no place in the regiment?"

"Yes, a rare senior officer refrained from inappropriate remarks," I answered defiantly.

"But you? Do you think neither the law nor the charter is written for you?"

"Nadine Turchin is a worker in the regiment, and not one of the last.

"But Mrs Turchin also interfered in the military affairs of the regiment."

"Only twice a year; Augustus Conant forgot to warn you." I imagined Nadine in an empty house, her impatient glances from the gate to the road, whether I was driving, whether the horseman was jumping to call her to court, to listen to the judges' apologies.

"Once near Larkinsville, in the valley of the Paint Rock River; when volleys hit two of our companies from the mountain and the Volunteers came under fire from the hills, and there was some confusion, Mrs Turchin really drove up to the artillery pieces and said to the battery commander: "Push your battery forward!"

And again the other time was here, right here, near Huntsville. We took the town and almost a hundred and fifty miles of railroad line, our forces were very stretched out thinly. Part of the supplies train was uncoupled, and the Rebels tried to seize the cars.

Then Mrs Turchin gathered a detachment from the train brigade and army stragglers and saved the train. If she hadn't been the wife of the brigade commander, he'd have recommended her for honourable mention. That woman will stay in the army as long as I stay! I finished abruptly.

"You inspired the soldiers to disrespect for the church."

"I am a stranger to any religion, but from the first days the chaplain was given a large tent with a cross and a platform."

And so it is in everything: to a question that offends honour or the moral idea of life, I answered boldly; - take me as I am, I am not looking for indulgence or loopholes. They opened an account of my contributions in Missouri or Tennessee, hoping that I, like Shylock, would start a bargain; and I did not dispute the exaggerated accounts. They accused me of cruelty, claiming that a ruined cellar would doom the family of a hostile planter to starvation, and I answered them with examples of the true inhumanity by the Rebels, not inferior in cruelty to the Madrid auto-da-fe or the Sivash massacre of Tamerlane.

The chief judge laid out intricate skittles on the green baize of the table, I grabbed a bat in my hand, as I once did in Novocherkassk on the parade ground, and knocked out the skittles with one blow. I almost egged on the court, urging it from a trot to a furious gallop, if only to leave the hall as soon as possible, where there was no one close, except for the downcast Joseph Scott. But I spoke the last word slowly, hoping to remember my own words; Medill was waiting for them outside the courtroom door, they would appear in the Chicago paper, and typographical America will begin to cut and reshape them in its own way.

I was talking about the strength of the Confederacy, which has neither a capital nor fortresses, but has raised a large and desperately fighting army. The entire strength of the Confederacy is in its army, and only in its army.

And all our strength must be directed solely against this army: to fight it with a large mass. Many people believe that the war will quickly come to an end - the Rebels will not have enough supplies and war materials. Naive hopes! When all the energy of the people is given to the army, such an army cannot lack anything. If the planters house is burned down, the Southerner will live in a hut; if the hut is demolished, he will make a shelter of branches, but his generous land will again give birth to wheat and corn, and hundreds of thousands of bullocks and pigs will be preserved among swamps and forests. A resolute people can endure any need for a long time.

In 1812, the Russians were seized with passionate patriotism, reaching almost to fanaticism: wherever the French went, every house, every hut and haystack fifty miles wide was burned: a huge, terrible fire engulfed Moscow, and only the Kremlin, with its stone cathedrals and palaces, survived. Its high walls and towers stood gloomily in the middle of the burning capital...

The epaulette crowd, with cries of "Shame!" stopped their angry muttering, they began to listen to my words, gloomily, but attentively.

The Confederacy would not strike their flag because of any stranglehold on supplies, I assured them. The British, as before, will give her weapons, shells and bullets, shoes and blankets. More than once I returned to the war in northern Alabama, spoke of the inevitable abandonment of Huntsville, that in the court where an officer of the Republic is now being tried, the Rebels will still praise the Lord who sent them such indecisive generals. The courtroom was waiting to hear Buell's name called, but I deceived their hopes by speaking not about

Buell, but about the fact that at the decisive hour we see the same dirtiness and baseness among politicians.

"Generals with a tarnished reputation are placed at the head of armies, people who do not know the difference between a column and a deployed front are appointed major generals. Among the junior officers and enlisted volunteers there are people who were born soldiers: they show miracles of courage and presence of mind under enemy fire, but none of these people have a chance of promotion due to the absurd system of seniority production.

"Our politicians make appointments and promotions in the army so insulting to military honour that an honest officer loses all respect for military rank and, filled with disgust, dreams of leaving the army as soon as possible."

"At the hour when we cede to the Rebels what was gained by the blood of soldiers, I am inactive and stand before the tribunal, and that is bitter," I said in conclusion. "But if I alone were accused, who knows whether I would voluntarily appear in court ?! However, together with me you judge my soldiers, yield his honour to the enemy, and here I am in front of you, and I stand here as a defender of the volunteers.

"If our hierarchy has not yet earned any praise, then our volunteer is the true hero of the day. That's who a knight without fear and reproach! As soon as the alarm sounded, he rushed to arms, asking for no other reward than the right to go into battle against the rebels. He fights, despised by the South and not appreciated enough by his generals; he is a bargaining chip in the game of politicians, to whom the interests of Athenian shopkeepers are dearer than the honour of their own soldiers. He is smart and soon becomes a good soldier; he is a brainy technician and capable of any job; he happily combined in himself both the French élan - this impulse that allows him to master any enemy fortifications - and the stubbornness of the Englishman to stand to the end. All the battles won by the Nazis were won by a volunteer, and if the volunteers were defeated and died, this often happened due to the intervention of the command. And I bow my head to the volunteer and say: no praises, no laurels can be enough to reward him for his steadfastness, for selfless courage, for his intelligence and devotion to the country! .."

A defeated regiment was moving through Huntsville: it would have been better if it had skirted the town, but Buell's headquarters gave more passion to the tribunal than to command of the troops. The Volunteers wandered in despondent groups in tattered shoddy clothes, with blood frozen on their lips and resentment, and looked at us unkindly, smart horsemen riding towards them. The ambulances brought the cloying smells of carbolic acid and pus, and the shrapnel-riddled canvas tops did not hold back the groans of the wounded. The major who addressed me as "sir" rode beside me; when we were squeezed by the regimental stream, his sharp knee touched mine, and he immediately apologized. The major knew everything - the number of the retreating regiment, the names of senior officers, the circumstances of the defeat - and this knowledge was bitter, angry, devastating.

"Please don't bother yourself, major, go on with your business," I told him. - I will not run away, I'll wait for the sentence."

"My name's Quentin, sir. Quentin Connolly. I listened to you, sir, and could repeat every word of it."

"If you do that, my friends Medill and Dawson will make you famous in their paper."

"I'll help them. I want people to read your speech."

"Excuse me, I thought badly of you this morning, I don't like convoys who are too polite."

"You're right, sir. I have two shortcomings which prevent me from making a military career:

an independent mind and politeness. Look, they bring beams, iron brackets, sandbags – to fortify Huntsville. And the Rebels will bypass it, it will be given to them for free, passers-by and they understand that; but the commander ordered the town to be fortified - and they strengthen it, and they removed the regiment from the front, with no sense and mediocre ... to digging up the ground ...

"Major! I cried. "You would be an excellent chief of staff in my brigade."

He smiled, a sad smile.

"I always get tempting offers from senior officers on the eve of their disgrace or death. At our gate he said to me, "You're making one mistake, sir. You want to be a citizen in a country which prefers subjects."

I appreciated his words; my mind clung with him, but my heart was in a hurry to leave him, to hand him over to the journalists, to shake hands with Joseph Scott, and go as quickly as I could hurry to Nadine, who did not come out to meet me.

Silence on the shady veranda, silence in the house, the soundless stirring of the white lace curtains at the window; the late afternoon calls of birds in the garden. I called Nadine, she didn't answer. He called louder, already at the door, the voice echoed in all the nooks and crannies of the house, and again silence. The house is untouched, without traces of violence or hasty packing. Nadine's army uniform hung in the closet in the bedroom. But her travel bag, the two dresses, an overcoat, and her brown, high-laced boots were gone.

On the table, I found a note.

"My soul, Vanya! I have gone away to help you. I did not tell in advance, as I know you would not allow it, although what I am doing is not bad, but good and will not damage our honour. It is hard for me to disobey your wishes, but it's even harder to see how they torment you. Don't worry about me, I am in no danger, I do not want you to be under a cloud; we live one life, you and I, one thought and one breath as one "

I was not the man in those days to humbly accept her departure; a heavy rush of blood darkened my thoughts, I felt I stood in an ungrateful world. A devastated, hollow world void of all sense, with the meaning taken out, in strange streets, among strange soldiers, officers roaming between lies and truth, between career and honesty... Drop everything - someone else's house, the dishonest trial and the inevitable verdict - drop everything and hurry to Chicago! That's where she went, that's for sure and wherever the Illinois regiment is, we were vassals of Illinois.

Medill informed me that I had missed the afternoon train and had to wait for the morning. But in the morning the major came with soldiers; the pronouncement of the verdict demanded ceremony. Quentin

Connolly marvelled at my appearance after a sleepless night, my blood-shot eyes, dishevelled beard, and revealed the verdict to me. The court had decided that I was guilty of conduct unbecoming an officer, but did nothing that would not befit a gentleman. It also found me guilty of acts that were prejudicial to good order and military discipline. The court sentenced Lieutenant Colonel Scott and me to discharge from military service.

"Six members of the court recommended that you be pardoned, John Turchin," Connolly said. "They found, and it is written in the verdict, that the offenses were committed in a strained atmosphere and were more an omission than a crime. However, Don Carlos Buell ordered that the sentence be carried out.

"So, I am free I can go to Chicago?"

"You must hear the verdict, and then - Chicago, or wherever you want".

"Except the South, Major!" Except for the Rebel South," I said, not without pride. "It seems that it's only there they still know my worth.

I again crossed the threshold of the headquarters, to listen, standing, to the verdict already known to me.

INTER-CHAPTER FIFTH

Headquarters of Company K, 19th Illinois Regiment.
First bridge south of Reynolds station.

Dear Mother,

Your letter was received, therefore, it was not intercepted by the troops of John Morgan, who for the last 4 or 5 days occupied Gallatin and cut off the connection between Louisville and Nashville for this time....

There is no need to ask "why" Morgan will not be silenced and "when" they will finally do it. General Buell, with his kid glove policy, is using all his influ-ence to dismiss the energetic Colonel Turchin out of service. You can imagine our feelings when we learned that the sentence of Turchin to be dismissed from the army. Yesterday, he left for Chicago, and from Company A, seven and a half miles from Huntsville, to where our company stands, the guys ran out when they saw him, and waved their hands after him until he was out of sight. Where the train stopped long enough, Turchin shook hands with the soldiers. He often, struggled to hold back tears, seeing such affection for him by his former subordinates, which only increased, not decreased after such grief happened to him. It means that he was truly loved if, having won such affection for himself from all whom he commanded, he retained it to this day ...

Huntsville is being fortified, but only, in my opinion, to give work to Negroes and pay their masters. Here everyone is of the opinion that we cannot stay in this area. The Rebels probably want to get their hands on the Negroes of Tennessee and Kentucky, I don't know if they can get it. When I went to Nashville, I met Mr. and Mrs. Temple by chance and was introduced. She's in Huntsville now.

General Buell has ordered Captain Guthrie to return to his company and wait there for a replacement.

There has been no other regiment, that has had as many detailed as the 19th, so many people were assigned to other places. To whom only he did not supply quartermasters, commissaries [20], and clerks for everybody. And also they took officers from him to other regiments. The reason, I think, is that since our regiment was one of the first to be formed, more intelligent men enlisted in it, without waiting for commissions than in other regiments.

Now all officers and privates are ordered to return to the regiment, and the ranks are larger compared to what it was at one time. True, an exception must be made for certain companies, where men, without thinking of another way to express their displeasure with the existing order, are deserting and goig to Chicago.

Give love to everyone at home from me, your loving son

J. S. Johnston.

Chapter Twenty Nine

J. Medill. For John B. Turchin.
Chicago Daily Tribune
Chicago

"...There are people who are fearless in the face of the enemy do not have the courage to raise the voice of a citizen. Better if I wish Ormsby Mitchell hadn't come to us in Huntsville, I would have left the image of a bold general and a player of military fortune; I now know that he is also a shrewd man seeking departmental advance. When I heard that I would not be admitted to the tribunal, I said that I would leave and named Chicago. Do not blame me for my secret departure: plans; you gave me the conviction of equality, so let me act with the directness of a republican. I know that you will be furiously muttering terrible Don Cossack curses, but remember how hard it is all for me.

In my mind, I was already driving up to Michigan, knocking on the door of the Chicago house of Governor Yeats with a hammer, but a scrap of stiff paper fell into my hands, which had lain for three years in our trunk. It was the visiting card of Abraham Lincoln, Illinois Central lawyer! In Mattoon, he jokingly offered his legal services if we were again suspected of counterfeiting banknotes. Chicago is forgotten, I'm now going to Washington, to the President; you gained for the Republic the pure gold of victory, Buell has declared them forgeries, let the lawyer-now-become President protect us.

And now here I am in Washington, alone, in a boarding-house not far from the Arsenal. The area around Washington and the capital itself is a military camp. Everywhere there are company campfires, the blazing of torches—my train arrived in the dark—the night bugle calls, pitted earth, embankments, the black silhouettes of batteries, moving figures visible even in the twilight and stationary masses of wagons. In the light of day you can get lost among the army Babylon; soldiers, cavalry detachments, baggage trains, open warehouses, all in feverish action, as on river wharves, brilliant officers and elegant ladies excited by the smell of powder in the air. From the town you can see the army on the banks of the Potomac, the white spines of tents - thousands of them! — the smoke of campfires going beyond the horizon. It seems that the ploughmen had abandoned the land, left their ploughs and peaceful oxen, and yielded everything to the troops. Shut your eyes, take a step in any direction and stretch out your hand, you will certainly touch a bandolier or a sabre, an epaulettes, a company mule, a bridle or uniform buttons.

And what is unusual for the eye of yesterday's European is the two-story brick White House; not every court dignitary on the Neva would consider this dwelling worthy of himself. But simplicity reigns, morals are simple; a soldier with a bayonet did not prevent me from entering the porch. I was told that the President was with McClellan's troops, and he was expecting him by noon. I left Lincoln's visiting card with the official and asked him to say that I was asking for an audience.

You can guess where my legs next took me: I rushed to the hospitals, of which I had heard so much. The two best metropolitan hospitals of the capital are the crown of the generosity of patrons, evidence of what feats a money man is capable of, if only he was allowed to hold in his hands not a gun, but a cheque book, not to shed his own blood, but to pay for someone else's. One hospital is a real palace, in the building of the Patent Office Museum: the air is clean, the marble floor is covered with thick rugs, the sisters of mercy - Catholic nuns and Protestant nuns of the local monastery of Thanksgiving - appear inaudibly before the wounded, like guardian angels. There is enough space,

the linen, is better than you would see in many hotels; a suffering soldier holds on, poor fellow, with the last of his strength, so as not to groan, not to howl, to relieve his pain, not to seem ungrateful in front of the ladies-patronesses who rustle their crinolines, crossing the halls, like sailboats crossing Michigan. Another hospital was a former stable, but it was even better for the soldiers. The richest banker, Corcoran, owner of a mansion on the Rue Lafayette, equipped the chambers in a long row of barns, cowsheds and stables. The buildings are partitioned off, the wood is dry and washed, fragrant, the floors are watered - earthen and plank, - there is not a speck of dirt anywhere, the bandages are clean, as if they are changed every day, and nearby warehouses, endless warehouses with mountains of linen, medicines, the medical instruments, according to which more than once we shed tears, not knowing how to relieve the suffering of soldiers. Here everything is closer to the soldier, who himself did not come from marble floors; he feels at home here. In my presence, the squire approached the bedside of a wounded soldier, who was more in need of a confessor than buns. She brought him her tray of stale brown buns, and the soldier, squinting his eyes, whispered: "Hey ... old woman! Are your buns sewn together or hammered together with nails?" "Those were his words, you can't say it better.

In the marble hall of the Patent Office I was caught in a storm; doctors and fellow physicians bustled about, nuns rushed about, carried away bed-pans, sheets and blankets tucked in. Then some majestic ladies entered, and among them the most majestic, aging, round-faced woman, restless, with an expression of vain worldly worries and labours on her tired face. She is wearing a bulky, motley hat, bearing all the fruits and flowers of America. Every instant the lady played a role, subordinating to it her steps, her meek, pious voice, and the regal movements of the cook's full hands covered in dark gloves.

I was standing behind a marble column. One of the ladies handed out a pamphlet to a soldier; he glanced at it, and burst out laughing.

"Why are you laughing my son!?!" said the ageing lady.

"But ma'am, it's really damned funny. This lady gave me a treatise against dancing, and both my legs are off right up to the backside."

He tossed the brochure *"The Sin of Dancing"*[21] up in the air and caught it, showing what kind of dancing he was destined to do from now on. The men laughed on the neighbouring beds, and my own face cannot express enough holiness, and the lady stared at me:

"I see that you are having fun too...my dear." Who you are? Why are you here"?

Tenacious, round eyes examined me, finding no agreement between a modest dress and an independent look.

"I'm a regimental paramedic in the Illinois Volunteers."

The impression is that I called myself a general or a captain of the legendary "Monitor". The lady held a flimsy handkerchief to her face and spoke quickly and capriciously through the fabric:

"What is it, Miss Dicks! Oh, how stupid, stupid!"

It was Mary Lincoln herself, the pious wife of the President, an unfortunate mother who buried her beloved son in the winter. Of course, it must happen when I have only one hope for the President! Desperate, I wandered through the crowded capital with people and wagons. The Capitol on the hill is not yet completed: the figure of Liberty lies on the ground, broken into bronze blocks, as if thrown from above. But it has only just been cast and will be lifted up piece by piece, making a whole, and this will be done faster than another whole, for which the war was being fought. The Union is destroyed, the formation of the Republic and of the nation itself are incomplete. These thoughts sobered me; I hired a cab and told the driver to hurry to the White House, I kept urging him on until, like Mary Lincoln, seemed to doubt whether I was out of my wits.

Lincoln immediately came out to meet me with the words:

"I've been expecting you. I'm glad, very glad, Mrs. Turchin!

Think of the man at our table in Mattoon and imagine that he is still emaciated, his cheeks and eyes are sunken, his lower lip protrudes with an expression of mockery familiar to you, and his eyes are the same - intelligent and not letting go. He led me into the study, ceremoniously raising my hand. How difficult his life must have been, how badly he had gone to McClellan's troops if he was warmed by the memory of Mattoon!

"I'm have come here on business, Mr. Lincoln."

"Nobody comes for anything else these days." He sighed. "Even beautiful ladies are not without worries. I remember everything: I remember you, your insufferable husband, the brave boy...

"He's dead!" — burst from me. — "Vilely shot in Alabama."

The President lowered his eyes, regretting that I was telling him about deaths when he had generals for that.

"His name was... Thomas!" See, I remember! I soon forget only those who manage to beg me for a lucrative position."

"Then I risk nothing: I'm not here about any position."

"Now I will surprise you. He opened the bureau, took out a wad of cloth, and unrolled it in his palm: it was a stone! - Did not recognize? It's a Mattoon stone, it broke your window. I then put it in my pocket: after all, the first attempt on my life".

"Why, have there been any others?"

"I haven't noticed, but my guards say that I don't notice due to absent-mindedness. He swaddled the stone again and put it back in his bureau. "If anyone wants to kill me, he'll do it, even if I wear chain mail armour. There are a thousand ways to achieve this. Now sit down and talk about your business."

"Don't you know from the newspapers about the trial of Turchin?"

" Newspapers tell lies, I prefer to forget what I read in the morning in the evening. Yes, and the place must be freed for tomorrow's lies."

I did not like his evasiveness.

"In moments of despair, I would tell myself: tomorrow we will hear the voice of the President, he will stop the trial ..."

"He who has the right to pardon, does not dare to hurry: otherwise what will remain of the administration of justice."

I told him about the Missouri war, the needs of the Volunteers, and the disaster at Beaver Creek.

"But you talk like that," he interrupted me, "as if you were a witness to everything?"

"I don't leave the regimen;: I am a regimental paramedic."

"In the saddle?"

"Not in a carriage, Mr. Lincoln. And not in a wagon. In the saddle."

"How is that possible?" He got up from his chair, and I felt I had to get up too. "Isn't that a bad example?"

"Bad examples are always willingly followed: but after all, not a single lady has mounted a saddle!"

"You are an insufferable couple! Is this Russian blood?"

"We have Republican blood. Today, I took the right to fight for the Republic, and tomorrow I shall be with those who demand another right for women - to vote in elections."

"By the time I'm retired, you'll be driving your banderillas at another bull!" He pulled out the same huge handkerchief in which the stone was wrapped, and wiped the sweat from his brow. "Couldn't it have been done without a trial?"

"When a brave general is defeated, he thinks about how to strike again; mediocrity - looking for the guilty."

He rose and sat down again, absent-mindedly, he took off his jacket without asking permission; he went to the window, running his fingers through his hair; when he heard the carriage approaching, he said quietly: "Mary has returned ..."; he covered his chest with his hands, so that the fingers almost converged behind his back; several times, putting a bony finger to his lips, he stopped the secretary who appeared at the door; and listened, listened, when I fell silent he began pacing the study; and suddenly stopped in front of me:

"You've just passed through Alabama, Tennessee and Kentucky - what do the crops look like there?"

"It's all coming up... the corn, maize." I was ready to be angry. - The grain looks well, they're sowing a lot of wheat now."

"What a pity; Richmond will take it all. And we need it too!"

"Don't worry about the grain there, Mr. President," I said angrily. "Even without backing down, Buell wouldn't let the hungry soldiers take even a bushel of grain."

"Are there enough men there? I mean, soldiers."

"There are plenty of soldiers; it's the determination to fight that's lacking."

"Well, Washington, how did you find it?"

"This is my first time in the capital..."

"And what is your impression?

"So many soldiers and officers, I only hope that they would not shoot each other at the first

alarm!"

He raised his hands and laughed, looking around and as if regretting that no one but him could hear me.

"If I gave McClellan as many soldiers as he needs every day, your cab wouldn't be able to get to the White House. Soldiers would have to sleep standing up! The Swede Erickson, alone, has done more visible good than all our generals combined. He built an iron "*Monitor*", and at least I know what it is: "*Monitor*" floats, moves, shoots! McClellan is a great engineer, but he has a special talent for making motionless engines.

"Mobile generals have a hard time," I said bluntly. "The baggage trains cannot even keep up with the infantry, and if the southern grain is forbidden, the soldier is forced to retreat again.

"Everyone keeps rushing me!" he said with a bitter smile. "I dismiss some generals from service, I bring others to justice, I myself fear the third, well, and the fourth, perhaps the most energetic, I defend against the local packs. But you also have to fight, Madam regimental paramedic.

"In the regiment they call me Madam!"

"I assure you, Madame Turchin, that if we had our own Napoleon, we would not appreciate even him." He approached me and looked into my eyes. "Now you have tasted everything, berries and thorns, tell me: don't you regret that you crossed the ocean?"

"We were not looking for everything ready-made, in shops and storehouses."

He listened to the footsteps on the stairs.

"I'll introduce you to Mary ...my wife. She is a dear, homely woman and is also burdened by all this ... He spoke hastily and raised his eyes to the ceiling, indicating that he meant the White House.

"I remember Turchin and won't leave his fate to the whim of chance ... We must wait for the sentence, Don Carlos Buell!" he exclaimed with a grin. "Can a man with a name like that be a bad general?!

Mary Lincoln entered. In her at-home dress, with an open, gray-haired head, she seemed kinder than in the hospital ward. I stood huddled against the window so that Mary would not recognize me at once; but Lincoln took me by my hand:

"Mary, this is Nadine Turchin. She can tell you things that no Washington gossip knows, not even Mrs. Doubleday. She's contraband, an illegal army paramedic. When I was still unknown, Nadine and John Turchin ventured to feed me dinner; in retaliation, I invited Mrs. Turchina to dine with us today.

Mary Lincoln nodded to me with cold casual nod, and the President stared at us in surprise. So, in tension, lunch passed in the same strained atmosphere. I cannot describe the change of dishes at the table, silent glances, meaningless words.

For that matter, I had no time for a meal; all my thoughts, then as now, were of the verdict awaiting in Washington, and of how far the independence of the President can extend..."

I left Huntsville and received this letter in Chicago, Nadine correctly calculated where to look for me. She came next; the President's decision was delayed.

Meanwhile Chicago, offended by the Ohio judges, decided to hold a public reception in my honour at Bryan Hall, where the city fathers decided to give me a new sword with a gilded hilt. Illinois did not yield its officers to the Ohio general; there was justice in this, and also the jealous rivalry of the states, and the offended honour of Chicago, who had given the best sons to the front. The crowd saw Nadine with me on stage; I led her to the stage to the friendly shouts of the people. A year ago, ironic to Turchina's uniform, they now wanted to see me and Madame side by side and in our union - the stubbornness of Illinois, the challenge of the military hierarchy, the persecutors of liberty. The crowd blessed us, and I did not let go of my tender, timid hand.

Seven years ago we escaped the lectern, and here is another wedding, in disgrace and unexpected joy, in front of thousands of people, in the old Brien Hall, and kindness in the eyes of others, and the right to life approved by the crowd, when love is not locked in an alcove when she, too, is freedom and fulfillment.

A group of citizens, led by Medill, appeared in the hall, they were suddenly joined by a major. Approaching the stage, he confidently seized the sword from the hands of the editor Medill, and then I recognized the senator's son, a young businessman, one of the directors of Illinois Central. He put on his uniform before me, but he made short trips to the army, to headquarters, to food depots and returned to Chicago for a new promotion, so one could also wonder that he was not yet a General. I was ready to receive a sword even from the hands of a street sweeper, but not from an alien organizer of my own life.

And he was already walking up the steps, with a sword in his outstretched hands; he is wearing the uniform of one of the Chicago regiments, intercepted at the waist, fresh from the tailor, but the regiment - I knew this - was in trouble, in a difficult retreat from Corinth, and the handsome man was curled, pomaded, fragrant.

I asked. "Who are you and why not at the front?"

He looked up at me with roguish eyes and gave me a well-known name in Chicago.

"A big name does not exempt you from service, on the contrary, it requires a feat. Are you from the hospital? Or sent on a mission?

The major did not dare to lie, and I turned to the governor:

"It would be my duty as an officer to arrest this military man; he voluntarily left the army. But he represents Chicago, and I will have to deprive myself of that pleasure. However, I would like to get the sword from other hands.

When Medill took the sword from the major, and gave it to me, and hugged me like a brother, the ovation showed that I had done the right thing. But the triumph did not end there, fate decided to reward us at once for insults; an officer from Washington came out of breath into Bryan Hall, he opened the package, and in it was Lincoln's decision - the general's patent in the name of John Basil Turchin.

So I had to speak a word to Chicago, to people of all classes and trades. As soon as I started, someone from the crowd waved a light handkerchief, and once, and twice, and looking closer, I saw Saburov. He quickly put the handkerchief to his eyes, and took it away, and again put it on, showing that he was touched, subdued, that he forgives me all suspicions, because he is kind to me, generous and happy, happy, as for his own brother. This scoundrel had an amazing physiognomy! It was as if he wanted to explain to me that he was with me both in joy and in sorrow, that only the two of us could understand each other in this rude crowd, two nobles in a strange world, among an alien mob.

An amazing physiognomy: decades of wandering, the need to be cunning and crawling, to stay in tutors and intermediaries - nothing changed the self-confident master in him. These thoughts stopped me for a moment, interrupted my sentence, alerting the crowd in Bryan Hall.

And then I put it out of my head; the people in the hall were closer to me, I gave an account to my fellow citizens.

Saburov did not look for me when I left Bryan Hall. Decades passed before I, once more in my life, saw him: a white old man and all the same gentleman.

INTER-CHAPTER SIXTH

"... after the trial, Turchin returned to Chicago, where he was given almost royal honours: the population of the city was outraged by the decision of the military tribunal. A personal friend of the Turchins, a surviving veteran of the 19th Illinois, subsequently told the author of these lines about a public reception that was given in their honour at the former Bryan Hall in Chicago, where the opera is now. It was Captain John Young, now a prominent figure in the electrical industry. He was on leave at the time. He said that such an ovation had never been given to anyone in Chicago, whose population was well aware of the merits of the 19th Illinois. There was such a crowd that the building could not accommodate all the people. In the midst of the festivities, a United States Army officer walked down the aisle and handed Colonel Turchin an order for promotion to brigadier general, which, at the request of President Lincoln, was sent to him directly to Chicago;

it was one of many popular decisions that President Lincoln made to assuage popular anger at the military bureaucracy. But Lincoln believed that Grant himself would have died, perhaps "without receiving any honours, unmourned and unsung."

And when the huge audience gathered in Bryan Hall learned what kind of order this military messenger had delivered to their beloved Chicago soldier, there was such a storm of applause that in many ways rewarded General Turchin and his wife for the worthless humiliation to which they had been subjected.

The Turchins immediately returned to the front, more popular than ever, both in the army and among the people in general."

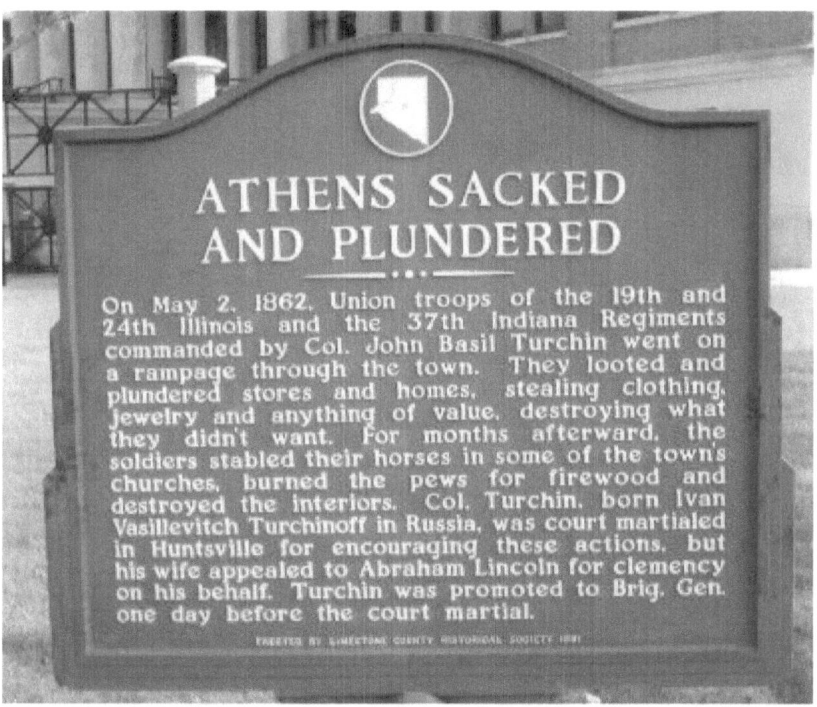

ATHENS SACKED
AND PLUNDERED

On May 2, 1862, Union troops of the 19th and 24th Illinois and the 37th Indiana Regiments commanded by Col. John Basil Turchin went on a rampage through the town. They looted and plundered stores and homes, stealing clothing, jewelry and anything of value, destroying what they didn't want. For months afterward, the soldiers stabled their horses in some of the town's churches, burned the pews for firewood and destroyed the interiors. Col. Turchin, born Ivan Vasillevitch Turchinoff in Russia, was court martialed in Huntsville for encouraging these actions, but his wife appealed to Abraham Lincoln for clemency on his behalf. Turchin was promoted to Brig. Gen. one day before the court martial.

BOOK THREE

Union soldiers outside Fredericksburg, Spring 1863.

Chapter Thirty

And Washington, having elevated me, forgot about the newly minted, uncomfortable general. And really, what else does a warrior need: after all, they saved honour, pulled him out of the mud by the ears, made him a general, and with that they showed the way to positions and the pension fund. Lincoln presented me with a general's patent, as the French monarch gives a buckle to a loyal subject, as our unforgettable - orders, as Lord Palmerston - a parliamentary smile. *The President forgot about me among many worries, and, offended, I took off my new uniform and put on the old one, the colonel's, scolded. Motivated by the requests of my friends, Richard Yeats reminded Lincoln of me, and the President wrote a resolution on his letter, showing much kindness and little determination: , if possible, from units of his own choice, and will be sent to where we now face combat missions, in Kentucky.*

The Minister and the Commander-in-Chief must have had objections or had no time at all to think about a general ordained not by the military bureaucracy but by the whim of the president. For half a year I was left to Chicago hospitality, pamphlet brooding and fury at the same Henry W. Halick who had succeeded McClellan. And he was from the scribes, he graduated with honours from West Point, he studied the French strategist Jomini - I don't know what he was like in the department, but it would be better for him to avoid the war and live a century as a military genius in peacetime. What stars of military art would have risen in our firmament, if, unfortunately, war had not broken out. But the war began, soldiers were dying like never before in American wars; McClellan was stuck in the swamps of Chickahominy, the noble Fremont was chased off by the compromisers, Buell fled pursued by Bragg and Morgan, and Halick achieved a semblance of success at Corinth and became commander-in-chief. He posed as a genius, an arbiter of the military destinies of the nation, pretty much soaked his throat with Madeira and Tokai during Lucullus feasts in honour of failed victories and climbed on strategic stilts from which he no longer saw the sinful earth and the armies rushing about on it.

I was waiting. Dear Chicago has become a gilded cage for me; I made up my mind to start engineering, and on the same day I dismissed it, in the hope that in the morning a military courier would demand me. Chicago paid me a colonel's salary, and I could go on a spree: many craved this agility from me - pistol firing into the night sky, rampage, feats worthy of the worst rumours about my Russian war, but I was sober, and gave my salary to city orphanages, announcing in the Chicago Tribune that he did not intend to use free salaries. Chicago then numbered a lot of uniforms with two rows of shiny buttons and a fair salary - how they avenged me for the dollars given to the orphans!

I replayed in my mind other people's battles in the autumn of 1862, and then the coming winter, found mistakes in them, and suffered, and understood that in the distance it is easier to see miscalculations in orders and the movement of regiments, that in the field I would not have avoided failures. I would have to accept that I did not come to court with my war; perhaps Fremont, the Jacobin of America, if we fought a little longer side by side, would want to put a hard saddle on me and put an iron bit in my mouth. Where there to reconcile! Even Nadine's story about dinner with the President and their remarkable conversation at the very end, when pie and peach compote were served, did not completely wash my eyes. Lincoln ate quickly, casually, as if performing a burdensome duty, and asked about me. "Explain to me, old, stupid man, could your Hannibal change a little? Shorten, change your uniform to fit the army he serves so well? "It's not a problem to shorten it," Nadine objected. "They demand of him that he finish it up to a chaplain's coat or monastic vestments!" I'll tell you, as they say in Russia: what is in the cradle, such is in the grave! .. "So America, through the mouth of the President, awarded me another nickname: Hannibal.

Society's sympathy has narrow boundaries. Today you are a hero, a man of the day, some zealot will throw a new top hat in your honour, so much so that it falls under the heels of the crowd. But weeks will pass, they will cool down and ardent and, sighing, with a lean face, they will sign a petition in your favour or a money sheet. And oblivion is not far off. How unlimited and bitterly was revealed to me here, far from the battles, the philosophy of the loud Robespierre's kitchen and deliberately unarmed Carbonari!

In those months, letters became our joy. Alexander Raffin often wrote: from Nashville, where the 19th was stationed, given to the corps of General Negley, when Buell briskly ran north to ignominy and retirement; wrote after the battle of Stone River, where our regiment immortalized itself with a bayonet charge, and where the glorious Joseph Scott - convicted and then returned to the regiment - found his mortal wound. "What a blessing," wrote Ruffin, "that Scott had Athens before his death and the opportunity to exalt himself in the eyes of the regiment. He had done his job well before, but after the trial and sentence, Scott was looking for bullets, and not blindly, but in a well-waged battle, always ahead of the game, in the most uncomfortable and dangerous places. So it was on Stone River: Scott led the regiment in a bayonet charge, won it and died. Will I be worthy, in the duties assigned to me as a regimental officer, will I be at least partially worthy of my teachers? I am sure that if Captain Howard had not died on Beaver Creek, he would now be at the head of the regiment. Letters came from Fenton and Johnston, shy, with details of regimental life; they also wrote about Balashov, hastening to please me with news of the courage of the Russian volunteer, of the almost reckless courage of this kind, silent man. Tadeusz Drum informed us that in Nashville the regiment received back the Negro Abraham; he miraculously escaped from Alabama, was taken with injuries to one of the Nashville infirmaries and, looking out of the window of the ward, saw Barney O'Mullen.

The nineteenth remembered me; and I seemed to myself a mossy stone, under which not only the spring of life would not burst, but even the wind, the breath of this life, was too lazy to bend over the stone. But

Nadine did not know rest: a paramedical service in a military hospital; new writings, and in them Russia - as if America had not been for six years - Russia, provincial ladies, nobles, landowners- zhuirs and those drunk from the circle, county people and serf peasants. She went to the infirmary, I saw her off for a long time, standing at the window, and began to read new pages of her manuscript. If she dared to write in Russian, even the pale shadow of George Sand would fly off Nadine's pen; but no, it was written in French, there was still a hope that they would print it and not disappear. She wrote more simply now, the engraver cut the plate deeply, scattering dry dust with a crash. Even then, I cowardly drove away the fear that Nadine's writing would not fly off the table to people, decay in obscurity and Russia would not bow to her talent and soul. Is there a crueller punishment than this, when a person is born to give a lot to people, but they do not take from her, and she is in obscurity, then not heard of? Her fate is more bitter for me than anything else; I became what I wanted; - as a citizen, a worker, that robbed Nadia of life, did not realise even a hundredth of her price.

The year 1863 came and I became unbearable; after a sleepless night, I used to look like an old man, I put a cup of coffee on the table with precautions so as not to break it - my hand trembled from suppressed, driven inside strength. I would rejoice: it happened! The great thing happened - the President announced the abolition of slavery; at last the mighty horses made their way to the pass, where the best sons of America were waiting for Lincoln. I've been waiting for this hour as a believer in the second coming - and now the word is spoken, the war will move to another stage, the South will become rabid from Lincoln's proclamation; the war of retreats, Buell's war will become impossible, the hour of my war has come - and I am inactive. I was hobbled, lost in the correspondence between Chicago and Washington, squeezed into a general's uniform cleaner than a straitjacket. Black resentment closed my eyes, the world was darkened, and when I had already decided to accept the invitation of the railway company and looked after an apartment on the outskirts of Chicago - Kenwood, a telegram from Washington returned me to the line. This happened in March 1863.

General Rosecrans, who replaced Buell, assigned me to the old brigade, but not for long. Garfield happened to be nearby, he understood better than anyone how exhausted I was from sitting in Chicago. At Garfield's urging, I was put in command of the 2nd Cavalry Division: I mounted the 39th Infantry Regiment as a regular Dragoon Regiment, and the Chicago Board of Trade Battery into a Horse Artillery Battery, the only horse artillery in the West in those days.

"I trust you and your wife, General?" asked Garfield.

"With me; where is she, an orphan, to be."

"When she was sitting in the hall, we, the judges, had a hard time: there are such faces, they have truth and purity."

"Here they are in Huntsville and they didn't let her in court!"

"I know; That's when Buell and I parted ways."

"I did not keep up the conversation, I put Buell in the account and him."

"He's doing badly," Garfield continued. - They talk about the reserve. Will be waiting for a new appointment."

From Halik? Or from General Lee? "A brave man shouldn't be merciless," Garfield grumbled. "Buell won't serve the South."

"When they skimp on the fate of the republic, I do not want to be spare. However, Halik will find a place for him, they are of the same clay."

"God! The divisions have not yet been accepted, but slander the commander-in-chief. When will they learn to live?"

I did not overcome this science until the end of my days, but I accepted the division, went through the Talahum campaign with it, then they changed me with General Crook, he got the cavalry, and I had a brigade of five regiments in the division of General Baird, and I was at the head of this brigade in the memorable Chattanooga campaign, which I led in the attack at Chickamauga, you can read about this attack in books. 'Chickamoga' is a native, Indian word, it is a river of death, as if the natives looked centuries ahead and foresaw the autumn of 1863, when the blood of their white enemies turned the water of the river red.

American history will not forget the last Wednesday of November 1863, when Grant stood on the little hill of Orchard Knob, near Chattanooga, Tennessee, and watched the battle, a mile away, on the bluffs of Missionary Ridge. Parts of the 14th Army Corps of the Cumberland Army were advancing there, and among them was my brigade, and hundreds of nice guys from the 19th Illinois. On the Missionary Ridge, nature also came out against us: steep slopes, a heap of granite rocks, ravines, fallen trees, bushes beyond which one could not see a person even from two fathoms, trenches of rebellious shooters, row after row, at full height, and guns of all calibres. General Bragg found his position impregnable, he calculated everything on this cold, piercing day, everything except our holy thirst for revenge for the blood of Chickamauga. History does not often gather so many first actors on the stage at once; Ulysses Grant commanded here, General Sherman attacked Missionary Ridge from the north without success, Sheridan, Wood, Johnson, Thomas, Baird and other brave officers led the regiments into battle. We went to storm the hill, called "the north-western corner of hell", crawled up steep cliffs, seizing rows of lodgements, climbed rocks, made our way along ravines and stone crevices, walked under the muzzles of cannons, leaving the dead, wounded and those who fell into lists of "missing", or, simply put, torn to pieces by a direct hit. Mission Hill was taken, many guns captured, among the first to raise the colours on the summit were the 19th Illinois and my brigade of Baird's division. We drove the rebels to the thick woods beyond the battlefield of Chickamauga.

Grant did not give either Thomas or Sheridan the order to take the hill, he calculated that by capturing the lower ranks of the lodgements, the soldiers would stop, waiting for orders. Unperturbed, Grant, seeing that the attack had not stopped at the appointed place, turned to Thomas: "Who ordered these people to storm the ridge?" "I don't know," the general replied. "I didn't order." "Perhaps you ordered, Granger?" Grant asked Gordon Granger sternly. "It seems that they went on the assault without any order. These fellows have only to start, then the devil himself will not stop them. Grant immediately urged the messenger - Captain Avery - to Sheridan, and the captain delivered the same answer: "I did not order to storm, but we will take this ridge." Sixty banners led our soldiers forward; when a dozen standard-bearers fell, a dozen others picked up the banners, pierced by bullets and darkened with smoke. An hour later, Grant stopped at the boulder under which Balashov, wounded in the shoulder, had taken refuge from the November wind, and, looking at the big man, asked why, without receiving an order, they rushed to the steepness of Missionary Ridge. Balashov recognized the commander, but did not rise from the ground, he lost a lot of blood. "Turning back was like falling into the underworld for us," said the slow Balashov. "Standing where we stood was no better; we crawled ... In the end, we ended up on top, crossed the mountain and - down ... "

Our victory at Missionary Ridge demoralized Bragg's army, opened the way for Sherman to Atlanta, to Georgia, to the ocean - it is considered the turning point in our great war. My name was not among the promotions: not then, not later, after Buzzard Rust, after the battles at Resaka, Chattahoochee, Mount Kinisaw, or in Georgia, where I moved with Sherman. But what is the need; Nadine is with me, and there are no more tutors to instruct us, black regiments are nearby, fighting bravely, in increasing numbers; Buell was expelled from the army, and our entire raid with Sherman was the triumph of my Russian, as the Athenian judges called it, war. I got mine in full - the filial love of the volunteers, the friendship of the company and regiment, the tenderness of Nadine, the look of her gray eyes, capable of lifting me from the ground even when wounded. And I became her sick soldier, the infirmary sufferer: with two wounds, with a sunstroke in Georgia and with a fever in Chattahoochee.

Both times I was wounded happily, right through, by rifle bullets, and I did not lie down for long. Although the war was slowly tilting in our favour, the enemy was once again at Washington. General Sherman caused considerable damage to the Confederacy during our march through Georgia to Savannah, on the Atlantic coast, but if he had destroyed General Hood's army, the damage would have been catastrophic for the South. The South responded to Lincoln's proclamation with a tightening of slavery, a massacre at Fort Pillow, where hundreds of Negroes were executed - seeing the spectre of defeat, the South took out its fury on the slaves, preferred to give them to heaven, but not to the Yankees. And at that time the North again entered the potholes of compromises; the armies of the North gave the politicians the first liberated states, and the politicians turned their teams and rolled back. The soldier conquered Arkansas and Louisiana, every mile was paid with unprecedented blood, and politicians saw in these states erring lambs, repentant sons, pranksters who don't even need to kneel - enough is a simple apology, said indistinctly, to the ground, so that it is not clear whether this is an apology or a new curse. The princes of rebellion, enemies of freedom to the black mould of the family crypt, were forgiven, if only they would verbally recognize the abolition of slavery and half-heartedly, with a careless nod, bowed to the Union. We've seen enough of these Jesuits back in Missouri; yielding to strength, they mutter words of loyalty, and their eyes are filled with bile, rebellious blood is knocking at the temples, and it seems that outsiders can hear how she strikes a short word: revenge! The people of the liberated state are again pushed aside by a noble hand; the Negro and the white poor are not allowed to the ballot box, the elections are the same, hundreds of thousands of soldiers' lives have bought not freedom, but new slavery. I walked along this land together with a soldier, we moved as if through a swamp, laid a bloody path, made a vain road, behind our backs the swamp closed, and now it is yesterday, bottomless, treacherous. Having recognized the President's proclamation, the slave owner did not give free rein to the blacks, only those who rebelled, risking their lives, and joined the regiments of the North, became free. The South also corrupted our officers, seduced them with a share of economic robbery, cotton speculation, the possibility of owning land, the patrician caste of the South.

A thought came to me more and more often that when the war was over and we straightened our backs, looking back, we would see that we had been deceived and that the world, hacked by our military ploughs, ploughed all over from Texas to Virginia, was the same, as before, on a lie. and privileges of the blood. Is my war over? Has it not come time for me to strike my guerrilla strike, not where Thomas is fighting Hood or Grant is fighting Lee, but in a different Theatre? That's when I started writing for Medill and Ray's paper; I wanted to sound the alarm, but, apparently, my voice was not yet strong. Then the sun-stroke hit me, the fever at Chattahoochee threw me back into bed, and I took a thirty-day vacation. While we were driving from Georgia to Chicago, the horizon was covered with clouds; the dodgy Robert E.Lee got the better of Grant, the rebels threatened Washington. But a light appeared in the clouds: Congress passed the bill, it did not allow Louisiana and Tennessee to return to the bosom of the Union with a mask of innocence - the bill closed the doors to the demagogues of other slave states, ready for a false oath, for a crafty oath of loyalty.

At the beginning of July 1864, Congress adjourned the session, and the supporters of freedom expected that one of the first Lincoln would take this bill of Congress from the table and put his signature on it, making it law.

INTER-CHAPTER SEVENTH

"I am sending you a small photograph of Mrs. Turchina. This afternoon I will have to visit Madame General's wife. She sent me a note that she wanted me to come and tell her in detail about this "mutiny" in Charleston, pc. Illinois. She is a great politician with us, always aware of all the events of the war, she has been inseparable from the general for almost three years now, accompanying him on all campaigns, and with all her heart she hates the "copper-headed snakes". [23] She looks beautiful, smart, and everything about her is femininity itself.

Major James Connolly of the 123rd Illinois
to his wife in Charleston, pc. Illinois
Dispatched from Ringold, pc. Georgia
April 10, 1864

Chapter Thirty one

On a stuffy July night, the window of a room in a Chicago boarding house was left open, and I was awakened by street sounds at the intersection of State Street and Randolph.

On the first floor, in the twilight of the drawing room, still curtained off from State Street, a teenager was dozing in an armchair under a portrait of Washington; he heard the creaking of the wooden steps.

"My name is Charles, sir!" he rushed to me. "Your name on the mailing list; I took it myself at night.

In front of me stood a young typographer, thin and polite.

"The list is printed in the newspaper, but I thought you'd better be the first to know about it. Oh no sir! he was offended, noticing that I put my hand in my pocket for a tip cent.

I extended my hand to him, and he shook it with dry, hot fingers.

"I don't think you're rich, sir." - Yellowish, fox-like, with a narrow chin and lively, searching eyes, Charles's face turned red with embarrassment. "This is a cheap boarding house. Mr. Medill wrote in the paper that you spend your money on Negro salaries? But there are so many blacks ..."

"Negroes are joining their black regiments now."

"Is it good, sir?" he asked seriously.

"It just so happened: you don't have a black friend, do you?"

"I have not yet met such a young man: neither black nor white."

"You didn't look for him among the blacks."

"I will think over your words, sir."

We passed the brick building of the concert hall and turned onto Randolph Street. Charles silently kept up with me; I remembered him from the first meeting: sharp knees, narrow shoes with knocked down toes, a frequent step and a sunken chest, behind which a kind, hasty heart beat.

"I'd be a good soldier," he said after a pause. "But I can't sleep on damp earth, and a soldier can't do without it."

"It is necessary: both on damp earth, and in a swamp, and in snow."

"And if in the cavalry?"

Night falls - and the same cavalryman is on the ground: he wraps himself in a blanket and sleeps.

"I can't even wear a blanket if the fire goes out," he said carefully so as not to get into trouble. – "Can fires be lit?"

"Sometimes it's impossible, it's dangerous."

"You see, sir! And I don't drink wine, I can't get warm."

"Are you a Quaker, Charles?"

"We are Catholics, but my father abused wine. And I, sir... I want to be a senator! First as a lawyer, like Lincoln ..."

I put my hand on his thin shoulder, he was tense, tremulous, like a stupid deer.

"Sly! Do you want to climb into the Presidential chair?!"

"That would be too much, sir. Although I was also born in Kentucky, but I live in Illinois!"

I burst out laughing; he distracted me from my anxious thoughts. The Atlantic has not brought letters for a long time. Nadine's father at first threatened to be excommunicated, then he became unhappy in moaning and entreaties, and convinced that we could not be returned, he withdrew and fell silent. Neither he nor the adjutant wing wrote any more, and with his departure for Missouri, there was nothing to wait for letters. My father did not hope to break his son's stubbornness. Our family has not yet got rid of the Cossack, wayward root. Anything happened to us: accidental losses, the obscurity of Turkish captivity, secret death, or the return of a son already buried in the church to his father's doorstep. Two letters came from my father: one to Long Island, and in it the land councils - it found us on the day when we said goodbye to Rowland, the second - to New York. He did not write tenderness, but the old, Catherine's time sheets of writing paper shone, next to the watermarks, his love and hope for the best. Soon he died, Sergey wrote to me about the death of his father. His nature is light and impulsive, and he wrote as he spoke - briefly, unceremoniously, with mockery. Such letters are good

for frequent intercourse, but sending them across the ocean, not knowing when the answer will come, is as difficult as it is difficult to hide a smile in a postal envelope.

We wrote to three overseas addresses about the change of life, leaving the Long camp, to Prince Lvov, to me in Novocherkassk, and to Herzen in London. And what, like a letter from Herzen, and not a letter, but a package and in it - a book printed in Europe, Nadine's book, a thought born by her, and faces, faces, living faces, and she is a mother, a happy woman in labour, the light of typographic children! Alas, it was not a package that lay on the palm; a thin, one-leaf letter from St. Petersburg, news of the death of a retired colonel, Prince Lvov. The aide-de-camp wrote, reproachfully reporting between the cold lines about the death and funeral service, about the modest funeral of a veteran forgotten by everyone and about the money left by the deceased, about an amount close to twenty thousand Rubles, which he is ready to bother for Nadine from a noble, generous, able to forgive the government, from the emperor himself, despite the insult inflicted on the court by Colonel Turchaninov. He also informed us of his close trip to Berlin and Vienna, no longer with concerts "which Europe has always been waiting for" ("he has become old, his hand is not firm, however, firm, the public would accept, but the artist himself knows his limit ..."), but for meetings with the enlightened people of Europe, "in order to see for yourself the admiration of the world for today's free Russia." Nadine and I were given a reproach, to the government and the crown-bearer - the most perfect devotion and kneeling, so that the postal censor would not even have a fleeting thought that the adjutant wing appoints us a rendezvous in Berlin or Vienna.

I sat down on a bench against the wall. The eyelids drooped by themselves, they were burned from the inside, but no tears were shed. Everything floated under me—the stone floor, the heavy bench with a carved back. Charles's voice, "You dropped the letter, sir," came from afar, like the anxious voices of soldiers as a sunstroke threw me to the ground in Georgia. They were then amazed, not hearing the shot, seeing that everyone was safe, and the general fell on the ground cracked from the heat, lying with his leg stuck in the stirrup, and his darkened temple nearby. with a raised hoof of a horse.

"Unfortunate, Charles. Mrs. Turchina's father died."

"How bitter, sir! I was weary of my father while he lived, and when he died, I wept."

"Prince Lvov was a glorious man and a kind father."

"Did you say prince? I'm a republican, sir." - Charles faithfully stretched out in front of me. - But when the prince dies, the flags must be lowered ... Forgive me for causing you grief. But the letter has come, you would have received it, perhaps in a more difficult moment.

I looked at him in surprise: he had ripened like a diseased apple, turning yellow early, in a blush and wrinkles at the same time.

"Now you're going to the funeral..."

"He was buried a long time ago, Charles, and no one lowered the flags: there are more princes in Russia than we have senators."

"I have not yet seen the funeral of the senator, but the general was buried with me; there was a lot of music and people were crying."

The little republican lived with all the prejudices of class society, trembled, burned with them, and I did not dissuade him; telling him about Russia, about Nadine's dead father, I belatedly blamed old Lvov and mourned the dead. In those moments I blessed our poverty, our simple clothes, our entire difficult existence; there was also a grain of our truth in it, not bookish words of rightness, but a living grain of justification before all those close to us and before Russia itself.

Every step along Randolph Street brought me closer to the boarding house. Here are the stone walls of the concert hall, and the corner of the boarding house - no matter how long it takes, the wide steps, made of oak blocks, will approach, I will climb up, cross the threshold and say:

"Father died..."

"Charles, shut up about this for now. I held up the letter in my hand before slipping it into the inside pocket of my uniform.

"I know how to keep secrets, but my mother guesses everything in the eyes. She is sick, my younger brother and I support her. Mr. Medill was so kind and hired him. My brother would very much liked to meet you.

"Good. Wait for me, Charles, I will introduce you to Mrs. Turchin."

A new face, participation in someone else's half-childish fate could soften Nadine's shock. I went upstairs, stood touching the doorknob, listened to the splash of water, then soft, barefoot steps on the floor-boards, and could not, did not dare to enter with the black news at a moment when Nadine was especially happy, after a long blissful sleep. I stood at the door, guessing Nadine's movements, the rustle of skirts pulled over her head, the rustle of boots pushed out from under the bed with a bare foot, the barely audible voice of a comb in long hair - I ran down the stairs to Charles and asked the hostess to tell madam that I had left on urgent business to the publisher.

"We'll go to George Fergus, Charles."

"But I owe a lot to Mr. Medill, is it convenient for me to visit another printer owner?"

They are friends, there is nothing to talk about.

I didn't know Fergus before the war. It was said that the passion of the abolitionist and the Presidential campaign of 1860 made him a publisher; Fergus then bought a printing press, typefaces and paper, typed and typed the speeches of the Republican idol and records of his political duels with Douglas himself, sold the prints for a penny, or even handed them out for nothing. A university scribe, a young husband and father, an impractical man, he blew his wife's cash and dowry, and things were heading for bankruptcy when Fergus's father's friend, the editor of the Chicago Democrat newspaper John Wentworth, provided him with a loan of money and advised him to publish several books that would come in time. Fergus reprinted Beecher Stowe's recently sensational novel, Hinton Helper's Threat of the Crisis in the South, and several other serious books, which, however, found readers. The owners of newspapers liked the young unmercenary, learned publisher, capable of translating another essay from French or German, standing behind a type-setting cash desk; they were ready to see Fergus as a Chicago polymath. He was patronized by Wentworth, and Medill and Ray allowed my pieces from the Chicago Daily Tribune to be published in monthly pamphlets; I have been sending them from the army to Chicago since the spring of 1864. Fergus's letters to me in Georgia attracted me with a determined mind, an interest not in battles and troop movements, but in the organization of life in the states taken from the rebellion. George Fergus did not agree to recognize the Confederacy as mischievous states.

"The weapon knocked out of their hands," he wrote to me, "they will skillfully replace with another: demagogy, political, secret rebellion, a conspiracy that will help them return what we got at a terrible price." When we met, we got close: he turned out to be younger than thirty, a thin, careless man in clothes, a heavy head slightly thrown back to his back, and his eyes look out from under lowered eyelids, guessing your thought in advance. Long-armed, with a funny, simian loose upper lip under a dry nose - he might have seemed unpleasant, but as soon as Fergus smiled, spoke, flashed his eyes - and you were drawn to him, drawn to listen, to look into the original, lively face. And next to him is Horace, the perfection of proportion and tone; fair, pink, blue-eyed, with a black crown of hair, she did not take her loving eyes off George.

Charles was amazed at Fergus: he expected to see anyone, but not so young a man dressed modestly, so that if their fate had been reversed, Fergus would have come to the printing workbench, and Charles would not have lost his face in the dirt in the publisher's chair.

"For God's sake, take off your uniform, buy a good suit, and I will clean and mend this monster," Horatia told me, pouring coffee for the three of us and putting out a cherry pie.

"Nadine won't let another woman iron my uniform."

"I'm not afraid of her," she played her game. "I'm afraid of General Sherman and my George, but you and Nadine - not at all ..."

Fergus this time was deaf to his wife's game.

"Than to darn fine uniforms," he exclaimed, "wouldn't it be better to sell them to a junk dealer!" Did you know that Lincoln pocketed the Tennessee and Louisiana Bill? When power usurps the scarecrow and tramples on laws passed by Congress, war becomes nonsense.

"But, Mr. Fergus, Lincoln is a great man!" - Poor Charles got up, intending to leave the house where his idol is blasphemed. The greatest since George Washington!

"You have forgotten Richard of England and Napoleon Bonaparte!"

"Yesterday you knew no equal to him," I reproached Fergus.

"We nearly broke our necks watching his prowling ship, hoping he wasn't deceiving us, but the enemy. Now even the blind will see whom he leads by the nose. He waited until the session of Congress closed, the sheep went on vacation and suddenly noticed that they were shaved bald and Abie was knitting warm underwear for the Confederates from their wool ...

"But, Mr. Fergus, the President didn't veto the bill; I read the night telegrams!

"He acted more cunningly: no veto, but no bill either; and the scoundrels who have declared themselves the government of Louisiana, Arkansas and Tennessee remain in power.

"I'm leaving, sir," Charles told me, appalled at Fergus' injustice. "I thought you would protect the President, you, his general!"

"I'm not his general, I'm an officer of the Republic."

"You know, Mr. Fergus, that mankind is ungrateful," Charles attacked, "and a people, even such a great one as ours, is in a hurry to trample on their prophets."

Fergus did not know that Charles was preparing to be lawyers and senators, and grumbled under his breath:

"Good placement of commas, boy, many believe that all difficulties can be solved by murder ..."

"I hope this does not apply to our war?" I asked.

"Alas, sir! This applies to every life taken," he said with adult sorrow.

"Georgie!" exclaimed Horatia. "How well he speaks!"

"I think that Mr. Lincoln, too, mourns everyone who is killed…"

"Did you decide that because he has huge handkerchiefs?"

"No, because the President has a good heart and dreams of peace."

"Nice occupation in the middle of a war started by rascals!"

"Difficult, but that's why God found him in a law office in Spring-field and put him over the whole country. He drove out the bad generals and appointed Grant, he freed the Negroes…"

"Every time he does good with a delay of two years, and if by one, he considers himself a great sufferer.

Why shouldn't even the president live by his own wits?

"Well, you brought the rogue to our house!" Fergus exclaimed cheerfully. "He will make a career if he wishes."

Charles prepared to continue the dispute, but the appearance of Nadine prevented him. She brought with her a short gentleman, in whom I did not immediately recognize - the Mattoon stationmaster Hansom. The tobacco beard is shaved off, the sideburns are shortened; grasping, bright, cucumber eyes looked out from under the violet cylinder, winking in an amazing way at the interlocutor, and boring him. Gone was the once belly and provincial bagginess: a light, active gentleman entered the room. But even if I were blind, I would recognize his voice, deep, singing when Hansom wanted it, an organ voice, with only one flaw - he had no connection with the soul.

At first, Hansom's appearance seemed to me ghostly, I saw only Nadine, even the Ferguses and Charles disappeared somewhere, the Petersburg letter burned through the lining of my chest, and Nadine approached joyfully, slyly, with her eyes asked to send Hansom away, she was guilty that she had brought him here.

"God sent you to me!" Hansom exclaimed.

"I'm afraid it's not him, but the bullets and the fever." I introduced him: "Mr. Hansom is the stationmaster at Mattoon."

"Uh, you got it all mixed up, John," said Hansom cheerfully. - They said all sorts of things in Mattoon when they found out that you were given a regiment: you know this audience from the Democratic Party - in their eyes any Irishman is a robber, a German is a hoarder …"

"You know, Hansom, why the States need immigrants."

"We are clean, Turchin, no matter what they say about us. People come to us so as not to die of hunger, and we give them shelter, work, and they live. Tomorrow blacks will come for free work, do you think we will push them away? No! In each pair of hands - a shovel, a pickaxe, a hammer, a saw, a trowel, a crowbar ..."

"What do you want from me?" I interrupted unkindly. – "I am a soldier, the time for drawings is not ripe."

"The war is in our pocket, Turchin," he regretted my blindness. "And don't extort new money for guns, the orphans need the money..."

"Just think what pockets they have!" Fergus' face hardened.

"Hansom, I will not discuss military matters with you," I said.

"Excellent: war is behind you, Illinois Central is behind me. You have a rare gift, Turchin, you get along with the workers, and this is the main thing now: the war has corrupted people. When I read about the trial in Athens, I said to myself: damn it, this is so similar to our Turchin - to smash the rebellious nest to smithereens. They should have been thrown to their knees, they deserved it, Turchin. Illinois Central has invested everything in the war: generals, volunteers, dollars, now we want to receive bills, we want to build. Why did we fight, Turchin, if you want to leave the good of the South to yesterday's masters? "The passion, the enthusiasm of the businessman, the holy faith ruled him, he expounded the great plan of Illinois Central. We must spread our arms wider: the right one through Quincy or St. Louis, west, beyond Kansas City, beyond Colorado and Utah, to California, the left one east, towards the Atlantic, through Pennsylvania and Massachusetts, and the legs should stand firmly, reaching the Gulf of Mexico and Florida. The road to the South will pass through four states: Kentucky, Tennessee, Mississippi and Louisiana, people who know these regions are needed, they will lead the construction artels as bravely as they led the regiments ...

Nadine marvelled at my patience, and I suddenly remembered our departure from Georgia to Chicago, farewell to the brigade, as if we were parting forever, and not for a month, which my ailments demanded. It was getting dark, things were packed, the violin was packed, Nadine suddenly got up from her bunk and removed from the post a saddle worthy of the London arena - a gift from the volunteers of the 19th Illinois. "What do you want a saddle for? I asked. - Leave him on the batman, - "We will not return here ..." - "Why?" I was surprised. "Don't you feel like we're not coming back here? Sherman will go to the ocean without us ... Let me take the saddle. This happens with Nadia: causeless sadness, detachment, a look that seems to be falling into a black void. I took the saddle from her, saying that I was ready to carry it on my shoulder, not to take it off even at the president's dinner table, if he would honour me; I clasped her hands, pressed them against my hairy face—how happy I would be if I could make her completely happy too. We were in a tent, with one of Sherman's black regiments a hundred yards away, and we heard the most ingenuous and kindest of all Negro songs: "Ten dollar a month! Three of dat for clothin! Go to Washington. Fight for Linkam's darters!"[24]

Nadine's saddle ended up in Chicago, mine was moving with a batman to the Atlantic, and Hansom was waiting for the black soldiers to let them drop not cotton seeds or corn kernels into the ground, but lay sleepers on it and lead mounds through malaria swamps.

"Why are you trying so hard, Hansom?" I asked.

"But I'm now one of the directors of Illinois Central; we gather people who are not alien to the interests of the company.

"I don't care about them. I also don't want the rebels to keep their ancestral lairs and their safes...

"You see! Hans was delighted.

"But not for you to get it all."

"Mr Fergus!" Hans begged. :Explain to this stubborn one that the money, since it is printed or minted, must belong to someone."

"In a republic, money should belong to the people," I said. "He needs to be in charge of them."

"Oh, yes, people!" A wild thought in the mouth of an educated man revolted Hansom. "Our people are noble, they will not take free money.

What is the way out? And there is only one way out - the people will earn this money and they will move into his pocket."

More than once I have observed how outwardly the officers changed when fortune smiled upon them, but this could not be compared with the change in Hansom, with the activity awakened in him, with nobility and gloss. Only playing with the top hat betrayed him as a nouveau riche: he found grace and aristocratism in manipulating an expensive top hat, either smoothly moving it, as he must have seen in the Theatre, then putting it on edge in his hand, then clapping, then drumming distinctly with his fingers, then skidding behind his back - an impatient huckster, a man of the crowd, in the worst sense of the word.

"We have found you a place not just an engineer, Turchin: you will be the boss on the construction of the road through the whole state."

"Turchin once occupied Huntsville, and with it a hundred and forty miles of railroad," said Fergus. – "In one morning!"

"Foreign roads are easily taken and even more easily lost: Turchin took, General Buell gave, - he showed his awareness. - Richmond valued Turchin highly, at fifty thousand dollars, now calculate what we value his talents for: five hundred miles of road at forty thousand dollars a mile! That's what he will turn!"

Nadine and I looked at each other, I gloomy, from the dreary desire to drive Hansom away, Nadine with a silent request to take him as he is.

"Mr. Hansom," she said, "our Thomas is dead."

Hansom mixed up, he knew us childless in Mattoon.

"Remember, messenger Thomas? He joined our regiment."

"It's a pity!" Hansom remembered, but his heart did not respond. "It's a pity that young people are dying, otherwise it's impossible, not for old people..." He raised his top hat, as if aiming from a gun. "War is for the young."

Someone drove up to the house: the springs creaked, the sound of footsteps was heard on the stairs.

"He was shot in Athens, Alabama."

"We will build a road there, it will be a monument to our volunteers. The road will pass Athens, then to Birmingham and southwest - Tuscaloosa, Jackson, New Orleans."

Poor Thomas had already dissolved, his dust a grain of sand, a handful of dust lay at the base of Hansom's railway embankment.

Joseph Medill and his son, Jerome, entered the room; when the war began, he studied at Philadelphia College and now chose not the army, but the engineering field, all in the same Illinois Central. Father and son rejoiced at Hansom, and it pricked me painfully, they seemed to me members of a caste forever closed from me, more existing than the past of the country and our war. Medill also picked us up on State Street to invite us to a Sunday picnic in Michigan. I ungraciously refused, and Hansom, deaf to the hearts of people, again began to persuade me, until I stepped towards him and firmly grasped the lapels of my coat.

"Victory is not yet in your pocket," I said. "General Lee is still closer to Washington than you and I are to Richmond. If I leave the army, it will be to prevent the usurers from losing in an unclean game what we won with blood."

Hansom bowed, saying that he would find me again, Medill, as he left, winked at us - he was an easy person, he loved me, but treated me like an outlandish specimen of a mammal.

"Here, Charles," I said to the young man, "the general can afford to talk to Hansom like that, but the future senator cannot."

Now the best people in the army, but they will return.

Medill called me from the street. I leaned over the window sill and put my finger to my lips.

"There's a letter waiting for you in the mail, John! From Petersburg."

I nodded. Medill found my mystique amusing.

"And what about the notice of inheritance? Will you leave us?"

Medill's son wiggled the reins, the horses moved.

"Even if they offer me the crown, I will not betray the republic."

"Take the crown!" he shouted, twisting around in the carriage towards me. "Hansom and I will come and establish a republic for you."

Medill waved his hand at me, not realizing how close to the truth his left about inheritance is - how easily we pronounce this word, forgetting that only death opens the way to inheritance.

The day was windless, the stuffiness was advancing on Chicago from all four sides, converging on the dirty streets so densely that the lungs could hardly take in air. Nadine was not bothered by the heat, Georgia tanned her face, she was still full of holy awe, holding the Fergus' daughter to her chest, and she moved smoothly, easily, in a pleated skirt of green foulard, with a wide belt, in a white cambric blouse, in open shoes, which gave her a feeling of festive, almost forbidden freedom after heavy military boots. We heard the sounds of the band from the Chicago arena, where the Volunteer depot was still located. On this Sunday morning, we hardly came across any active military, or veterans in rags, or even cripples of the war, which Chicago had pretty much. And suddenly - an orchestra, and, audibly, the step of soldiers along the pavement, and then the song of Ellsworth's fiery Zouaves burst out in the cramped streets.

A cab hot in the sun dragged along the street;

"Hey!" I shouted at the driver's back, and he pulled on the reins. I helped Nadine into the cab, sat down next to me, and said, "Camp Long!"

He did not immediately realize where I needed to go.

"To a military camp. I will show you."

"It seems to me, General, that there is no camp there now."

Behind a gentle hill is an avenue of elms, the leaves dry and dusty from the long heat; people didn't drive here often now, grass broke out between the old ruts; elms were left behind, copses, maples, carved foliage of young oaks went. Nadine didn't ask why I was going to Camp Long, as if she needed it too, languished in Chicago, but she didn't guess what her soul was asking for, but I guessed. All the way she was silent, finding my hand, squeezing it so that the blows of blood converged into one; if I asked her what she sees in the blue and green gap between the cab and the back of the driver, where the trunks of birch trees floated, the rough bark of elms, as if torn by the heat, and light clouds - if I asked her about this, and she would answer with words that were and in my heart: I see Russia. We have often been struck by the mysterious resemblance of grass and trees within the bounds of the Earth, the voices of birds, a feather dropped in the field by a swift or a lark, an acorn knocked out of a hat by a fall, the first drops of rain, the slow step of oxen, the chestnut kind eyes of a working horse, the ringing of grasshoppers. This resemblance gave us not only the bitterness of memories, but it also contained a symbol of our faith; can't people also recognize each other, over random differences, skin colour or blood prejudices! If nature has so generously ploughed the land for general sowing, why do we not dare to trustfully throw our grains into the ploughing? Why do we brag about a special click of the tongue, light and even hair in front of curly and dark or milky, pink skin, forgetting that even a Negro has it pink in the palm of his hand ...

"Pay him to keep him satisfied," Nadine said in Russian. — "Let him go with a good heart."

We found Camp Long at the crossroads of time: no longer a soldier's home, but not a wild land either; sun-baked paths, un-grassed ruts, dark circles where Chicago loam burned for months under a company boiler, bark knocked over by a wagon, a broken undergrowth, a hook driven into a trunk, or a piece of hemp rope on a birch. A flattened cardboard cartridge caught my eye, then a collapsed boot of a volunteer, a tattered cap, a cigarette butt.

Here I stood with Grant, here I saw the restless Chicago companies, poor Howard, the haughty face of Tadeusz Drum, and the black figure of Augustus Conant. Only a thought can instantly embrace years and spaces: the Missouri and our ardent inexperience of newcomers among hostile estates, the brown Mississippi, old steamers, barges, gunboats, quays full of soldiers; Howard's night alone with his big brother, and another night by the wreckage of the Beaver Creek bridge, and reached the South, Tennessee and Alabama, the blow of the northern blade, penetrating the South deeper than others, and a march through Georgia that I could only dream of two years ago . The war was still

going on, but like a misfortune and liberation, I felt that my gunpowder, cannon war was over.

We stopped at a camp dump overgrown with weeds; crumbling
wagon wheels, grandfather's carriage made of oak beams, a broken
boiler, broken rifle butts, decayed harness, rusty stirrups, a leaky knapsack, and on top, as if still dreaming of saddling a living ridge, is the
skeleton of a hard Indian saddle, broken, with eaten skin and rags.

"You said we weren't going back to Sherman... We weren't going
back to the army."

"Did Hansom throw you off balance?"

"No. We have one life, and every day of dishonour is irreparable. A
loved one must die so that, with grief, this simple thought is affirmed."
I took out a letter from my pocket. "Our father died: I owe him everything, he gave life to you, and prevented me from dying ..."

In vain I continued to speak: Nadine's grief was stronger than my
words and compassion. She did not notice the outstretched hand with
the letter, only one word lived in it - short, inevitable, like an eyelid
that finally trembled, like an instant spasm - she, who had accepted so
many deaths, could not see his lonely death.

"Dead!" she repeated, as if it were not me, but she telling me the
bitter news. – "He died, Vanya ... I left him!"

"Remember: your father was old, he was seventy, and the long life
of a widower, wars, wounds from the Caucasus..."

"You won't understand this either, what it means - the soul flies away.
Not to God, without a choice where to fly: she leaves us, she no longer
needs us, neither memory, nor thought is ours, nor repentance. Give!"

She left with the letter, hiding behind birch and hazel bushes,
rushing about with her misfortune, circling in the copse and, approaching me, left again. I did not disturb her, she lost her whole family in
her father, leaving no living soul on the ground. In the most difficult
days of the war, her father appeared in front of her in an unbuttoned
uniform or in a dressing gown with a cord, touchy and condescending;
he stamped his foot in anger, as if it were a small one, clutched his
bristly white chin with his palm and, without thinking through the

punishment to the end, forgave, he told everyone that his children were honest and noble, and that they were stubborn and extravagant, this was such a disastrous, vain time ... I didn't move until I saw Nadine running towards me.

"You are alive, you are alive!" Nadine muttered, stepping back, touching my face and shoulders with her hand. – "How I need you to live! .." At such moments it is easier for believers, but I do not regret that I do not believe; to each his own. "Today you will submit your resignation letter, and I will write to Petersburg, let the money be given to orphanages, and your father will be the donor; not his runaway daughter, but his father—the uncle will arrange it.

INTER-CHAPTER EIGHTH

Radom, Illinois, May 9, 1879
From a letter to A. D. Kiger, Esq.,
Librarian of the Chicago Historical Society

Dear Sir.

I don't have a single copy of my 'War Reflections'. I would be obliged to you if you would ask Mr. George F. Fergus to rummage through his office and find some of the pamphlets, and if he did not find them, let him go to John Walsh of the Western News Company, who had them on sale. I myself would like to have several copies, because I am asked from other places to send copies; it is obvious that these pamphlets will be found and come back to life after they have been buried somewhere for so long and everyone has forgotten about them.

As for my colonization of "Polish lands", I must say that my plan was quite successful - for the benefit of the Illinois Central Railroad company, but not for my own, because I am now as poor as I have always been and must work by farming to live somehow...

When the painful time is over, I will write an article about this colonization, even though I myself have nothing to do with it now, and send it to you.

Yours sincerely
J. B. Turchin

Radom, Ill., December 10, 1888
Mrs. George G. Fergus

Dear Madam,
The notes of the part for violin and piano "Memories of Mozart" by Alar,
which I received by mail from Lyon and Healy, I am sending you by mail, as
you wished. I'm afraid I'm giving you too much trouble, but it's all your kind-
ness. We both thank you from the bottom of our hearts for your wishes for a
joyful Christmas and wish you and your family the same.
With friendly feelings, your
J. B. Turchin

Radom, Illinois, April 10, 1889
Mrs. George G. Fergus

Dear Madam,
Thank you so much for taking the trouble to get violin strings, and
I'm just amazed that if I just want to get something and ask one of your
family members for it, it's always up to you and you always do it...
Hello everyone, I remain your devoted servant
J. B. Turchin

Radom, Illinois, January 21, 1890
To John Moses, Esq., Librarian

Dear Sir,

Your letter of 17 January was given to me by Mr. Fergus. Unfortunately, I still do not have a detailed presentation of all the facts about which you ask me to report. Many of these can be found in Van Dorn's History of the Cumberland Army. The most accurate idea of everything can only be obtained from the documents of the War Office, and I do not have them yet, and I do not know when I will have them. I temporarily suspended my work on the history of the Battle of Missionary Ridge.

Yours sincerely

J. B. Turchin.

On April 14, 1865, Abraham Lincoln, the 16th president of the United States, was shot by John Wilkes Booth while attending the play Our American Cousin at Ford's Theatre in Washington, D.C.

Chapter Thirty Two

These are the first papers that came to Vladimirov not from Turchin. Horace Fergus knew John Moses, librarian of the Chicago Historical Society, who succeeded A. D. Kiger in this position, visited the heirs of Moses, they managed to find several letters. The general's letter to Kiger was found to be a draft; Turchin whitewashed it in Chicago, at the Fergus' home, a few hours before his hasty departure for Washington. In the spring of 1879, Turchin was promised admission to the military archives to study documents about the Battle of Chickamauga. Colonel Robert N. Scott, head of the War Records Office, was finishing up the preparations for publishing a huge mass of documents from the North and South and wrote to Turchin that he was ready to provide him, at his choice, with copies of the reports of federal and rebel officers. Turchin rushed to the capital, and - in vain: the door of the archive this time, despite the location of Robert N. Scott, was closed in front of him. Another five years elapsed before Turchin was able to obtain lists of even his own reports to Baird and Grant.

Other papers passed to Vladimirov from the general. The old man took them out of a leather chest with a miser's hand, and from there came the smell of paper caught in the fire; it seems that not only the "History of Lieutenant T." was burning. Will Turchin also give him "La Chólera" - a Pskov, peasant story, in which, even from the torn out phrases, one feels more strength than in "The History of Lieutenant T."? The past century was generous, it gave Russia geniuses, but it is bitter to lose a chronicler. Vladimirov did not recognize himself: how did his interest in old papers become stronger? Isn't the clue that Virgie is next to him and she hands him every piece of Turchin's paper as if she were holding at least the first draft of the Declaration of Independence.

In the last week of February 1901, Vladimirov twice went to Turchin - the old man's story was interrupted, he had to open new, non-military doors, and he hesitated, unlocked, suddenly remembered with resentment that he had asked for the St. he writes like a slave in a galley, but the need came to read someone else's, they don't give him, they don't hear his hunger ...

Turchin's desk was covered with piles of papers; crumpled sheets lay on the floor, the work did not stop even at night - in the snow-drifts of papers, two brass candlesticks with cinders, in influxes of stearin (*Stearin candles are a sustainable alternative to conventional paraffin candles.*), unopened candles were thrown on the bed, next to the violin; the old man must have dozed at night in a semi-armchair. "What are you working on now?" Vladimirov asked cautiously; their soul ties were broken. "Everything is over the same, over the same," Turchin persistently blocked the guest from the table with his gaze. "I have to say that it didn't work. I have no assistants, dear doctor!.." Vladimirov rose off his chair in an offended manner. "If I skimped on visits in the last two weeks, it was not without reason: Horace's daughter Fergus was seriously ill ..." Turchin stared at him mercilessly, as if he had not known before: meanwhile, Vladimirov was dressed as usual, and Turchin looked strange, homeliness had disappeared - a blue velvet jacket and soft shoes, he is wearing an old frock coat of coffee colour, his waistcoat is buttoned up, the collar of his shirt is pulled together under his beard with a black bow, on his feet are heavy, loud shoes in this room. Maybe he was waiting for someone or changed his habit and began to go for walks? "Humanity hurts! As long as we take care of individual sick people, it languishes…" This alienation from the house of Fergus was unkind. "This is our destiny, General, to help society by treating individual sick people, but if we kill, then also in units, and not in thousands, like generals ..." - "Do not call me a general; I've been wanting to ask you for a long time." "Why can't I, but Fergus can, and Johnston, and Foraker?" "The general happened to me, you know, it happened, but it might not have happened." "But you call me a doctor!" "Because this is your business, and here is my business," he pointed to the manuscript. "This is my business, and a long, long time ago... Did you say the daughter of Mistress Fergus?" he turned back. "Wasn't Virgie sick? .. For Horace, her whole life: the Fergus buried their youngest daughter. It happened a long time ago, in 1889. George then became unbearable to live ... "

Imperceptibly Turchin reminisced, he stood at the window, staring at the gray, slightly snowy and cut by a deserted road plain. The old man had changed, the profile roughly sculpted a stern, steep brow, a sunken cheek with a fold from the nose to a mouth closed by an uncut mustache and a stiff, matted beard; the face more sharply opened its structure, lengthened.

Yes, it was 1889. 'Chickamauga' just came out. George Fergus wrote a preface to it that was flattering to me and intransigent to Buell. Today what? - Buell is dead, the noble Garfield died even before Buell, died, like Lincoln, President and also at the hands of a scoundrel. Buell was soon thrown to the side-lines by the war, but he already signed the battles like no other, sitting in the chair of the president of the railway company. His pen did not aim at me alone; It turns out that it was not he, Buell, who lost Tennessee and northern Alabama, but Mitchell and I—Secretary of War Stanton gave Mitchell the right to execute his soldiers, but he did not execute him and thereby ruined the campaign. The dead general—and Ormsby Mitchell died while the war was still going on—couldn't answer the slander, his son did, publishing 'Courage and Suffering' after the war, but I picked up the gauntlet myself.

A quarter of a century has passed since Buell retired to the reserve, to the maple cloth of the mansions of Indianapolis - at such times the nations forget about the last war and start new ones, but everything is alive with us: Buell and Turchin, Turchin and Buell, vile Athens and Major Harris in the robbery saddle, everything is alive and will be alive until the equality of blood comes. He paused, curbing his own passion. The book is out: Joy. Let them read hundreds - it's still a joy. On the road of lies I rolled out a big rock, a sort of boulder from Missionary Ridge, he and Buell's descendants can't budge. Joy! - he said for the third time with the inertia of memory, the old man's eyes no longer responded with joy. - Life puts grief next to joy; the little daughter of the Ferguses died, and in the spring the compositor of "Chickamauga" was gone, Charles died without becoming a senator. He was buried by the printing workers of Chicago, he was already their leader, prophet and conscience, I am proud that the lead letters of "Chickamauga" were fitted by his hand. And soon the turn of George Fergus came ... I played at his funeral, I knew that the indifferent would not understand, they would twist their mouths: a bald old man, like a beggar, with a violin in a cemetery ... Plays, plays, instead of screaming, breaking the violin on grave granite.

He fell silent, as if he had discovered everything worthy of a story; nothing could bring him out of this silence. On the next visit, Turchin did not stand up to Vladimirov, like an old man, moving, turned to the door, with a pen in his hand, complained that time was wasted on trifles, on food, on sleep, and he wants to be free in his actions, to bring to the end business ... And Vladimirov decided to try his luck for the last time.

A book from my father arrived from St. Petersburg and a letter to Turchin, respectful, with questions about the town of Radom, although the town did not turn out, even a supernumerary one, there was a village with a decrepit church, a post office, a tavern and two shops. The book was Vladimirov's undoubted trump card, and the other trump card was Virginia Fergus. If, together with Virgie, he does not soften the general, then this is hardly possible at all. From Turchin's friends, Vladimirov learned a lot about him, and the news hurried him. The past winter in Radom had a hard time for the Turchins. They were left without hay and sold the cow, saving her from starvation; Nadine's hands took so little milk from her that the cow was no longer a nurse for them, but a familiar living soul. The stoves were rarely heated, passing by, the Radomites had not heard the general's violin and the echoing sounds of the piano since last autumn; cold, Turchin's fingers did not hold the bow well. Leto Turchin wrote as it was impossible for a young man - from dawn to night roosters, even writing a letter seemed to him a sin, but there was no peace - everything extraneous, everything that was not a line, not a thought, seemed like a criminal waste of time. - Turchin became stingy with words, he heard, heard his Nadine with a slightly drooping old man's ear, his back bent over the table, heard and loved, suffered her involuntary rejection, repented with a cry of the soul, a plea - to understand and forgive - but he did not find enough words for her either. He caught himself when Nadine was burning papers, he saved a lot, took them out of the fire, rushed to write down what was lost from memory; asked her to sit next to him, to be with him. The Turchins entered autumn and the early cold, like birds suddenly grown old without feathers, without strong wings for flight, without fat under the skin ...

The fields went out the window of the carriage, deserted; the ploughman has not yet brought out the working horses. With a long cry of alarm, the train burst into the bare March forests, throwing wisps of smoke into the tangle of black branches and dark needles; the train is short, and the locomotive, with its screams and the smell of burning, is nearby.

It was the working hour, the General did not hear the knock. Virgie knocked harder, but no one answered.

"Ivan Vasilyevich!" Vladimirov called softly.

Silence, silence, and Virgie pushed the door.

The desert of the table, the desert of the plank floor, with crumpled pieces of paper on it, the chair and semi-armchair pushed back against the wall, the rolled-up mattress and the grinning iron bed net. In the empty room, the window seemed unfamiliar, and the table huge. The emptiness of the room was so sudden that they were silent for a long time, looking around. Then Virgie picked up the papers from the floor and spread them out on the table; nothing connected to them was read. With one Turchin removed the dirt from the pen, on three lines the crossed-out lines were barely readable: "... *to which Frederick the Great and Napoleon laid. They had that ... from above, the talent that they were able to combine ... ",*" *... just like General Seidlitz ... in history, as an exemplary cavalry general ... ",*" *... mediocrities led ... turned out to be the most bloody ... long, and when geniuses then they...*" Vladimirov took out his father's book from his coat pocket and put it on the table, as if, despite every-thing that had happened, the book belonged to the person who worked at that table.

"Do not grieve, Nick, I beg you!"

The abandoned room did not have complete control over Virgie's feelings: she connected the Turchins with other rooms and other windows. They hurried to Turchin's, but found themselves face to face in a strange room - the bare walls, the jug on its side in the washba-sin, the iron mesh of the bed, the bare window - everything shouted to them: hold on to each other in this desert, only in you two - living warmth , restless blood.

"We'll find him, Nick," she said, not recognizing her own voice. — "Nick! Nick!" she called, as if she couldn't breathe in his face. – "You only pretend to be an old, strict doctor, and you are a boy ... Nick ..."

Let's not judge Virginia Fergus, the daughter of a Chicago publisher. She was a faithful wife to Vladimirov in their long Russian life. In St. Petersburg, and later in Saratov, she sometimes yearned for Chicago, but worries about Nick and her three sons left her little time.

But not Virjie and not Vladimirov are the heroes of our story ...

There was a hurried knock on the door, and without waiting for an answer, a tall, thin old man entered, exclaiming "OK, General!" and slammed the door, as if he were afraid to chill the room. With a quick movement, he pulled out his pince-nez from his coat pocket.

"He didn't get there, damn it! You'll have to excuse me."

He lifted his soft hat over his gray head.

"Mr Christian!" Virgie called out to him.

"Quite right, Miss Fergus," said the old man. He might have added that it was difficult to recognize her; Christian could see everything when he put on his pince-nez, and he understood everything, even though he was single.

"Please meet him, he is Nick Vladimirov, a Russian.

"Very glad," the Postmaster held out his hand to Vladimirov; the old man was somewhat reminiscent of Lincoln, perhaps Christian lacked the power and originality of Abraham - the same beard, long wrinkled face, big nose, eyes under thick eyebrows, but also a thin breed, as if he was a born aristocrat, and not a printing worker who became a postmaster.

"The General wrote to me about you once."

He asked not to be angry with Turchin, the general is busy with a huge job, and only one who has lived Turchin's life, knew Europe, parliaments and courts, old wars and the whole war of freedom against slavery, can do it; the old man can no longer rummage through the archives, and several old soldiers help him. And now Christian took a short vacation, found the necessary documents in the Chicago military archive.

"I haven't written for a long time," said Christian with bashful pride. "The general predicted the future for me and was wrong.

"Father regretted that you stopped writing."

"One day, when I woke up, I realized," he put aside his joking tone, "everything I could and wanted to say, others speak better."

"What is the general writing now?" Vladimirov asked.

Christian looked at him warily.

"Only he can tell you."

"Are you going to Radom?"

"He wasn't going there before warm weather," Christian replied evasively.

Vladimirov decided to leave Turchin alone. But even in active Chicago, next to Virgie, he did not find peace in his soul, he could not neglect his moral duty, which is higher than insults and worldly inconveniences. At the end of March, he and Virgie arrived in Radom. A wooden church, gray houses under wood chips and dark iron roofs, a street driven into dozens of ruts, dilapidated fences, a stone mill, dead to new grain, rural desolation and longing. During the night the last snow fell and melted, adding dampness to the dark trunks and roofs and blackness to the ploughed gardens, the sun hid dully behind the motionless clouds. Has Turchin lived here for thirty years? Why didn't he run away from here? What plague stopped Radom; in the third year of his existence, five hundred families bought land here, where are these people and their descendants? Behind the low population of Radom, Vladimirov saw a secret, a collapse that took too much strength from Turchin. Was it not because of this that he fell silent, as soon as he ended the war and saw the Radom wilderness ahead, black forests, heard a wolf howl behind the walls of a farm with an adobe floor, a dull, funeral ringing of bells donated to Radom by the Franciscan fathers? Who devastated this land, quickly, in one generation, took away from it the centuries-old oaks and the voice of the bells on the necks of branded cattle, who measured and arranged it so boringly, dividing it with fences, chipped dams, corrals, revealing gray cattle and gray houses? .. The father's book showed the primitive life of Radom: pagan fires from burning stumps and trunks, the first clearing, the first felling and the first blow of the hammer on the anvil, still without forging, without fire splashes, from daring, to hear the voice of the forge in the virgin wilds and know that they would rush in fear, wild goats and the wolf will add speed, spreading on the run, and close squirrels will fly in all directions. Radom then was born, promised to be, to go around the neighbouring settlements on young legs; now there is nothing left in the village of the ambitious and the proud - early decrepitude, a bitter effort to stand on his feet, to hold out, rattling window panes when the Illinois Central steam locomotive rushes near.

They passed the church and soon found themselves at the Turchins' house.

- They are at home! Virgie exclaimed. Smoke was rising from the brick chimney, woodpiles were stacked at the porch, firewood was piled up, freshly chopped logs retain a weathered whiteness and a delicate, not yet alcoholic smell of wood. Here you will see Nadine.

Virgie pulled on the bronze handle, the door didn't budge. I had to knock: first on the door, then on the window. Nobody answered.

"Is there another entrance here?" Vladimirov asked.

But the kitchen door, too, was locked; they returned to the porch, knocking on all the windows, looking around at the road to see if Turchin might show up.

A steam-horse wagon loaded with unpeeled, dry-leafed corn cobs drove out of the alley, a farmer in high boots, a drooping hat and a waistcoat over a gray shirt, walked alongside the horses. He slowed down, patted the horse's rump, guiding it forward, while he fell behind and took off his hat.

"Hello!" Virginia replied.

"What doesn't open?" he chuckled.

"He went out somewhere," Vladimirov looked back at the house, the iron-carved crown above the chimney sharply appeared in the sky.

"Before they had it openly, now they are constipated. He lived here with him alone, chopped firewood."

"What is he doing there, closed?"

"And he writes everything! Burns papers and writes new ones."

"How do you know that he burns papers?"

"Through the smoke." And then the boys look out the window. He suddenly stiffened his face and tore off his hat again. "Good afternoon, Mr. Turchin, they are waiting for you here."

Turchin stood at the door. He nodded to the farmer and did not take his eyes off him until he went on his way. Turchin in a gray over-coat with a narrow velvet collar, in boots, with a hat in his hands, but without a tie; a wrinkled shirt collar peeked out from under his beard.

Virgie ran to Turchin, he slammed the door and propped it up with his old man's sloping back. She kissed him on his unshaven cheeks, smelled—with squeamish pity—the alien, hostile smell of old age.

"This is Novak, the son of Jan Novak, the first settle there," Turchin said after the farmer. "To be honest, I didn't immediately hear your knock," Turchin lied awkwardly. He thought about something in his mind: he was looking for how to save his loneliness.

"Thank you, we've borrowed the old man," Turchin said in Russian.

"I am not one of the benefactors of mankind," Vladimirov said coldly, "I have my own interest, my own self-interest. You asked for the father's book, he sent it at once, but the way is not close."

Turchin abruptly accepted the book, grabbed it with his old man's webbed hand, leafed through, stepping back to see better, moving his lips, marvelling that life had passed, a whole century on another planet, and he was reading the Cyrillic alphabet, and everything was close to him, every word, every line and letter outline. His knees were weakened, his head was spinning, the steppe heat of the Don region breathed on him with native letters, his ears filled with the hubbub of birds flying over the spring Aksai, the calling voice of his mother, the quiet cough of his father, his nostrils grabbed the warmth of a white-washed stove, half a hut.

"Misha struck me as kind at once... as soon as he entered the office on Washington Street..." the old man muttered, catching the lines with his eye. He was particularly interested in something: - The famous ropes from the Kronstadt factory ... well, I remember. Demidov's rails, anchors, chains... a samovar riveted from one sheet: that's how they surprised the world! Mr. Alisov's sheet music printing machine?

"It was in Philadelphia at the exhibition," said Vladimirov. "In the Russian department."

"Vonlyarlyarsky's loader! Won-lyar-lyar-sky. You don't speak right away. And what's that? Marine views of Aivazovsky? Well, well, well; Vasiliev's flax, popular prints, Belyaev's bast, matting, lambs. Well, what about Alisov's music printing machine, what's so special about it?

All this is already old, a quarter of a century has passed, and Russia has already changed. Only marine species of Aivazovsky survived.

"That's right," Turchin handed him the book. Take it.

Why, the book is a gift.

"I can't waste time," Turchin whispered, almost in Vladimirov's ear, "not an hour. I have to finish mine," he asked for understanding and mercy. "Later, my dear, later ... I am now bad, angry, cruel, otherwise it's impossible ... I promised you and I'll tell you ... I'll tell you right now: what a year to say, what an hour, everything is one; whoever is under-standing needs a few words - here is my Pompey," he pointed to the Radom houses, "both Peter and Troy, look!" And suddenly, changing, he vigilantly looked around the yard and the street. Christian arrived.

He brought papers - priceless papers: memory is strong, but papers are needed, you read the old paper, and the memory will immediately get on its right feet. I also have other assistants, and they suffer from me, so I take advantage of what they endure, you have something - you have come and left, you will forget insults...

"Will I still leave?" Vladimirov grinned and took Virgie's hand. "That's who I love."

Virgie threw herself on the old man's chest again. Turchin ran his tangled, stiff beard through her hair, pressed his ear to the top of her head, fussily, clumsily intercepting her shoulders and back with his hands.

"Here you are - you've barely arrived, let alone give him the best girl in Chicago... You'll take her to Russia, Vladimirov," the old man said commandingly. "Let her fly in Russia… It's crowded here, you see, let her breathe with Russia… Once I took away from Russia a priceless treasure… an ornament of the earth," the old man said quietly. "He took it away, gave it to another land, and you will return the light of Russia … Russia is great, and the good is noticeable in it; I was orphaned, took away the beauty, and you will bring … " In a fit of generosity, Turchin exclaimed: " Let's go, I'll show you the city!"

Virgie decided to stay in the house: ahead of Turchin - he wanted to enter his room before Virginia - she opened the door - and chaos appeared to her eyes. A draft drove from the fireplace up into the chimney black, twisted sheets; there were a great many papers in the room: on the round table and on another, called the ladies' table, on the windowsills, on the floor and on the chairs.

"Ivan Vasilyevich!" Vladimirov pointed to the fireplace, where weightless black pieces of paper rustled and spun. - Annoyed with his wife, and now they themselves have begun. Vladimirov knelt on an iron sheet in front of the fireplace, trying to grab the paper dust with his hands. "How are you!" Vladimirov grieved.

Turchin lifted him to his feet; never before had observer Vladimirov seen Turchin's face so bright and kind. He held his guest by the shoulders and, bowing his head, looked sideways with blue eyes from already indelible, heavy folds.

"And you, too, with a heart, with a heart, now I see! That's why this angel fell in love with you! They discover something in a person that we cannot understand with the male mind. This superfluous is burning," he pointed to the ashes, "working props. I'm in charge, and the excess - down with it! Otherwise, it won't be long before you go crazy," he said suddenly, suffering.

They left the porch, crossed the yard; Turchin stopped at the gate, looked around at the house and the outbuildings, as if he expected that life would be found there: the cow would turn and moo, wondering why she was being kept locked up when the first grass broke out on the hillocks.

Turchin said on the street:

"Let's go to the rails; it all started with them."

INTER-CHAPTER NINTH

From the book of M. M. Vladimirov
"Russian among Americans"
(1872–1876). SPb., 1877

Page 77. All my subsequent efforts to find a job did not lead to anything. In some places they said directly: "We will give you a job, but you will have to wait a long time for the money." I went to another place - it's full; in the third, the previous answer. The financial crisis that erupted across the country further worsened things in Chicago. Factories, plants, mills cut their workers by half, and sometimes more; some are completely closed, or they work 3 or 4 days a week. Terrible clouds were approaching from all sides.

Before leaving St. Louis, I obtained the address of a Russian who left Russia as a colonel immediately after the Crimean campaign. Entering one of the houses on Washington Street, I turned to an elderly gentleman with the question: "Are you Mr. Turchin?"

- Are you Vladimirov? he objects.

"How do you know?"

"I got letters from Detroit about you. After all, you have been living here for a whole week and you won't show your eyes.

"I worked and hunted for the boss," I answered and told what had happened to me.

Page 78. A long conversation ensued about the most important events of the last ten years. Turchin did not meet the Russians for a very long time and spoke with visible desire about the affairs of his native land. The life of this gentleman, as he himself told it to me later, was not a peaceful pastime of the current days. During the internecine war, Turchin drew his sword for the northerners and for the difference, which was once written about in Russian newspapers, was awarded a general. Listening to his stories, I was proud that one of my compatriots took an active part in the battle for freedom, for progress. Now Turchin is an agent for the sale of land owned by the Illinois Central Railroad.

On one of the following days, at the invitation of my kind host, I went to see him not at his office, as on the first occasion, but at his apartment, where Mr. Turchin introduced me to his wife. After a few usual questions: how do you like America, Chicago, where were you, what did you see, etc., our conversation turned to Grant.

"Do you know what," I said, "I am struck by these merciless caricatures of the president, such as, for example, a thin, dumbfounded horse, which is dumped into some ditch because of its worthlessness, or a reveller with an eternal cigar in his mouth, etc., extremely caustic . After all, the Americans themselves honoured him with such a high honour and should have some respect for him. What do you think?

"Caricatures," answered Mr. Turchin, "like the ones you mention, are of tremendous importance. Why look at Grant as some kind of inviolable being, even if he was appointed President of the Republic? Americans don't want to have idols; Grant for them is the same creature as everyone else, with the same advantages and disadvantages. Well he performs his duties - he is praised; badly - they will be dumped into a ditch, like a worthless, dead horse."

"But you must admit that I lose respect for the president if I see him in the figure of a dead horse.

Page 79 – It is true, but you will never become a free, independent voter if you pay attention to dignity. Take a person in addition to his external differences, and if he is worthy of the honour of being president, vote for him. Will his moral qualities increase even a degree from the fact that he becomes the ruler of a powerful republic? "Leave aside dignity, in essence, meaning nothing. Here the president, according to your judgment, is good, according to others, he is no good. Ruthless caricatures begin, which, if they irritate you, are not at all because they do not have the slightest respect for dignity, but because the president, as a person, is worthy of respect."

"But you are right," I said.

Page 88. My affairs fell more and more into decline. There was a week when I made $2.10, the next week I made $1, and then not a cent. running in one frock coat, the cold pestered me terribly; there was nothing in my pocket.

Page 89. ... hope for work has almost completely disappeared ... Now I. V. Turchin came to the rescue;

The two of us went to his acquaintances, but here we ourselves have no work, there we no longer need workers, and in the third place the workshop is completely closed. My "bank burst" and Mr. Turchin was kind enough to offer me his means of subsistence. Taking this opportunity, I testify to him my sincere gratitude, since this was the most difficult time for me of all the most difficult ... Day after day goes by, newspapers bring news of the indignation of the workers; clouds are moving over Chicago, prophesying a riot. The workers are almost forcibly expelled from the city; I had one similar experience in the free Jobs Bureau. I was told that some farmer had a place, but before they gave me his address, they wanted to provide me, at my own expense, with a travel ticket so that I could not leave. I refused, despite the difficult circumstances. I come to Mr. Turchin and tell him this.

"Yes, if you want to go to the country, then I can give you a job," Turchin said.

"With great pleasure! Where?"

"In my colony."

"Great, let's go"! I answered and began to get ready to leave.

Chapter Thirty Three

Remember how Hansom called me to lay rails through the forests and swamps of the conquered South? Peaceful generals, unlike Buell, were not afraid of the South, they equipped regiments from the poor, gave them picks and shovels, hammers of masons and anvils, axes and trowels, and delivered sleepers and rails to the Theatre of their war better than the artillery department - guns. Hansom kept up with me for a long time; he desperately wanted to throw a lasso around my neck, and then my neck was strong, forehead wide, my eye was quick - I didn't give in. He left him, for a while, but he left. Soon a peaceful life began, day after day, other passions, other temptations and sins. Courage is already at a different price, not less, but more; it happened that an officer, unshakable in the sight of the enemy, was lost in front of a strange table, in front of a sheet of paper and a searching look,

We ended the war in the autumn of 1864; the Confederate government was cracking, about to collapse, in the rebellious states we had captured, tricksters and rogues were in charge, they publicly cursed Richmond in their Capitols and prayed for him in the middle of the night. With an army brigade in Georgia or South Carolina, all I could do was win new lands for the Tartuffes of the rebellion; pen in hand and Fergus' presses, I had the hope of fighting for black freedom and equality. We settled in Kenwood, a suburb of Chicago, where we lived before the war. That Chicago, old, generous to us, wooden, not yet burned by fire, was not too big; he was inferior even to St. Louis.

We lived in Kenwood; lived the colonel's salary, collected in a year and a half, when the treasury began to pay the Negro soldiers. And the war was still going on, it had its heroes - old and new - Richmond still bit, spraying poisonous saliva, tens of thousands more mothers mourned the soldiers they had born. The North fought successfully, and who among the victories wants to listen to the voice of a sceptic, reminders of battles given in vain or unkind prophecies! My pamphlets did not go briskly. They were appreciated by a few who wanted to look ahead, and therefore were not afraid of the past. The Republic craved authority, I encroached on it: I exalted the volunteer, the armed people, but as soon as the talk turned to generals and politicians, my mouth was full of sharp pins. And now my enemies valued me more than my friends; they responded to issues of War Thoughts with whistles and curses, Richmond scolded me, which was a good sign that I was still a danger to the rebels. I advised the president to uproot the poisonous tree of slavery, to expel the executioners who started fratricide from our country; He called to reinforce the paper freedom of blacks with deeds, to give them the lands of yesterday's masters. The South hated me, and the Chicago layman shook his finger good-naturedly at me; I deceived him, he wanted to see me next to Ulysses Grant, in the close areopagus of the victors, *(NOTE: The Areopagus council was the most respected court in Classical Athens.)* and here is my gratitude: I live in Kenwood, in a poor apartment, go in a frock coat, make friends with free-thinking sceptics and compose the most boring matters myself. Is there a greater sin on earth than to deceive the militant hopes of a man in the street who has never fought!

Lincoln held on for another term in the White House, and Richmond became rabid: the most dangerous of all is a mortally wounded beast: for him there is no morality, no precaution or barrier that would keep him at any other time. Fate itself favours him, he succeeds in the impossible, his last blow with a clawed paw is fatal, so that both the winner and the vanquished fall dead side by side. I have followed European affairs and seen how eager England is to have her honour hurt, and if the honour of England is hurt, the English will not fail to act. And France fell under the yoke of an adventurer, and he was waiting for the Confederation to be on the brink of disaster in order to interfere in our affairs, especially if England would provide ships to transport soldiers to Mexico, where little Napoleon has a second home. And if they really intervene and manage to split our country, then it is easy to imagine how devilishly they will laugh when they see our once powerful republic broken and dismembered.

I collected my military-political plays into one book, gave it the name "Military Reflections", wrote two final chapters: one about the intervention of Europe threatening the republic, the other - addressed to the people and the President; its title is "What needs to be done?" - speaks for itself. The name on the book is not Fergus, but Walsh: my friend was then in trouble, almost went bankrupt on scientific publications and asked Walsh to print. Walsh did not have time to put the book in the shop window, when a miracle happened: everything sold out in one day, I received the book that had survived in the printing house, the only one with which I went to Washington to Lincoln. But the bell on Walsh's door didn't ring very often that day, the buyer bought the book in bulk, twenty copies or more, someone bought everything to burn the book or drown it in the gutter. The strange fate of our books!

I have Chickamauga, but we couldn't find War Thoughts twenty years ago. It's a sin for me to complain, but what about Nadine! To write something rare that no one else could write, and not to see not only a book, but even a typographical raw sheet, a page, a typed line - what an execution! If the machine is not invented by such and such and in such and such a place, then years later it will certainly be invented by another sharp-witted head in another place, and if it were not for Raphael, if he had not been born, he would never have existed.

I didn't have to give Lincoln the book. The President drove off with his wife to the Ford Theatre, well, and how this performance ended, you probably know. No matter how we reproached Lincoln during his lifetime, a heavy longing descended on us: with one shot the South achieved a lot, the gallows did not pay off our misfortune. The brigadier general was not taken away from me: but I made myself a general without a uniform. Returning from Washington to Chicago, I took off this uniform as well, it was my mourning for Lincoln - no one else ever saw a general's uniform on me. We did not take to the streets of Chicago at night when we were transporting the dead president - the living Lincoln, as we knew him, should not have given way to a long zinc coffin, a hearse, a rite. The roar of the crowd, the rustle of tens of thousands of feet, the breath of grief reached the outskirts of Kenwood, the night sky brightened with thousands of torches, until the funeral train left to the southwest in the direction of Springfield. In the morning we wandered through the city as if through ashes: thrown torches everywhere, the smell of burning tar, collapsed boardwalks and bridges over ditches, windows smashed by the onslaught of the crowd, crushed hats, gloves, scraps of paper trampled into the street dirt, withered, rare in this the season, the flowers, the broken chairs that got on the sidewalk for some reason, and the flags, the flags, the flags of the Federation, almost on every house.

We went with the book to Lincoln in Washington poor, and returned beggars. So I had to return to Illinois Central in my old role as a voyager technician. It was then that I travelled the whole state from corner to corner and chose this land, the southern oak forests of Illinois. Do you see, behind the turn of the track - an iron bridge across the ravine? Formerly, a stream ran there, and above it stood a bridge on oak piles. In the winter of 1872, we brought here stone, iron farms and forges, in February the frosts were released, on one warm night the snow melted, and we settled the workers in tents. The forest then approached the road itself, only at the ravine it thinned out: there is a clearing with a grassy mound and a trace of an old felling overgrown with hazel. After the first explosion, a man climbed the hillock: gray-haired, long-haired, bearded and in a military overcoat. He was thin, tall, leaning on a long gun. We tore the ground with gunpowder for a new bridge, assembled it not far from the old one, on which trains were still running, and every day an old man appeared on the hillock. The explosions scared away the animals; the iron roar of the riveting drove them away from the road, and only the old man, like a daring forest scout, went out onto the hillock. All sorts of things were said about him in the tents: as if he had lost his mind, he lives alone, with an Austrian gun, as if he is a sorcerer and considers himself the owner of all the lands between the Ohio and Mississippi rivers.

Once I could not resist and went to the old man.

The work behind me was quiet, everyone wanted to see the meeting of the general with the goblin. The old man stood hunched over, not putting on his hat, and not moving. When I was about twenty paces from him, I heard a calm voice:

"John Turchin! That's the meeting."

If it were not for this voice, and even blue, sad eyes with lowered eyelids, I would never have recognized Tadeusz drama in him. We parted ten years ago in Athens: and I have grown old in these years, and he, in my opinion, has lived a whole life. I rushed to him, he - to me, at the bridge they decided that it was rubbish and we grabbed hand-to-hand. From the crowd they ran to my aid, but they soon noticed that our fraternal embraces.

Under a military overcoat, Tadeusz wore a black frock coat, tight trousers over boots with spurs, a fresh shirt caught by a waistcoat tight as a corset, the silk ends of the tie darkened under a gray through beard; it was still the same dapper Tadeusz, independent and with a proud posture. He led me to him along a forest path; as soon as we entered the forest, Tadeusz whistled, and a horse came out to us from behind the gray nut trunks. We walked, hearing her even, friendly step behind us. I told Tadeusz about the rumours about him.

"That's right," he said. - I live alone, but right now in my house - a friend, you will see him, colonel.

"I was a general, Tadeusz."

"But for us you are forever - a colonel, who was not humiliated even by the general's epaulettes. And Mrs. Turchin - madam?"

"She is alive." I felt that he would like to know about her, but was wary. "And still with the obnoxious colonel."

"I live alone," he repeated. "I try not to kill unnecessarily. I brought Yadwiga's ashes here, I can't live apart from her. I got here, as a veteran of the war, one hundred and sixty acres of land, but I didn't put up a fence, so people don't know where my land ends. Well, what about a sorcerer? Let's wait a bit, maybe you will call me a sorcerer.

On the way to the house, Tadeusz told me about his life. James Garfield did not punish Drum severely for challenging the chaplain to a duel. Tadeusz was taken to the guardhouse, but he soon returned to the regiment. However, Tadeusz Drum and Augustus Conant were not such to forget about the thrown glove. Conant was looking for a bold act to leave for the time being the clothes of the church and put on a uniform; Drum, remembering his wife, cherished in himself hatred for the defiler of her honour. And when the news came from Huntsville that I had been expelled from the army, shots between them became inevitable. They got together at night, silently showed each other the pistols, and dispersed to their places. Conant wounded Tadeusz in the thigh without even getting a scratch himself. The matter came out. Don Carlos Buell took away Drum's company, believing that both were punished on an equal footing, after all, Conant was removed from the post of regimental chaplain. But Conant was given a company in another brigade, and Drum became a disobedient officer, adamant, dangerous in cruelty to the enemy. If Buell had held on longer, Tadeusz would have ended up in jail; the replacement of the commander saved him. But it was not easy for him to serve; he became suspicious, arrogant to senior officers, and in the spring of 1864 he transferred to a Negro regiment. He expected to become a regimental officer, but at the army headquarters he was humiliated, given not a regiment, but a half-killed company. He accepted this, too, without a word of objection; since then, Tadeusz let go of his beard and mustache, finding satisfaction in the early gray hair, as if he wanted to spur his own life, to drive it faster to the place where he would unite with Jadwiga. The black soldiers called the thirty-two-year-old officer "white father" behind his back, he soon became the head of the regiment, but the black ones were difficult for him. He confessed this to me as a sin, he never knew how to destroy the dividing distance and attributed this to lack of time: if he had one more year, they would have reached full brotherhood. Tadeusz deceived himself; He became a brother to Howard in a matter of days, he accepted Thomas at once - his ardent heart did not need years for this. Somewhere lived in him an arrogant breed, the nobility of instinct: between him and the Negroes lay nature, physiology, and honest Tadeusz was horrified to discover this vice in himself, was executed and could not break himself. I remember him from Missouri and Alabama, how he hurried to kindly receive a black man, to accept him as an equal, to put his hand on his shoulder, and in this very haste the lesson was visible. Regimental Tadeusz ended the war, as he entered it - as a captain. For almost five years the war carried in a bloody stream Tadeusz Drama, born the son of a Kansas pioneer and a Polish immigrant. A young teacher from Overland near Kansas City, Tadeusz Drum, showed himself to be a born officer and a one-of-a-kind marksman; I do not believe that Augustus Conant was saved in a duel by chance, if Drum wanted it, he would have killed him, and would have killed him in such and such a place as he would have liked. The war was over, and Tadeusz had to decide for himself where to go. He didn't go to Kansas; his mother died, he broke with his father before the war, when he mounted a horse to hunt abolitionists with other farmers. A familiar Irish officer from the Milligan brigade advised Tadeusz to take gratuitous 160 acres of land from the treasury, and they settled in the southern corner of Illinois. Tadeusz dug up the ashes of his wife from the land of Indiana, went down on a longboat along the Ohio and buried the ashes on his site.

Tadeusz's house suddenly appeared in front of us, so impenetrable was even the winter forest. Khoromina from ridges, which you will not immediately find in this forest; A plaque is nailed over the door with the words: "Come, you who suffer!"

"This house took all my money," said Tadeusz.

"Do people often wander into your wilds?" I doubted, looking at the board with the inscription.

The current settlers are afraid of the forest, they rush to open lands, but people live here too.

The house is not divided into rooms, it is all one multi-windowed room, with a stove in the middle and a wide bench on three sides of the stove; wooden bunks along the wall, a cupboard and shelves for homework, books and school appliances, which are used in gymnasiums to furnish natural science lessons.

"Are you waiting for your whole company to come to you!"

— You are the first brother-soldier met in our forest. There are aboriginal Yankees, Irish, several German families, but far away, they don't hear a shot in the rain. And that's why they called me a sorcerer.

He pointed to a corner, there was dark as if a coffin, placed upright, a gray box the height of a man.

"Look!"

In the case was a life-size anatomical model, with exposed muscle structure.

"When I was transporting it from Du Boys, the box fell and opened. The local farmers decided to seize me at night, roll me in resin and feathers, and drive me out of the local forests." He closed the lid of the case. My Irish friend rode up to me with this, and we warned them: we took the model to the farms, and I told them that I would teach children the sciences.

"Where is your farm? Where is the cattle or the field?"

"I was about to buy cows and let them into the forest." Drum chuckled. "Yes, I did not see it again." Here the owner does not see his cattle for weeks; each has its own brand.

"Why didn't you brand it?"

"A red forge brand," he said thoughtfully. "No, I don't want other people's suffering, even cattle. How many of them were, suffering, torment, but for what? We lost the war, colonel."

I understood what he meant, how close to the truth; and yet so far from it.

"I don't know of another war where so much has been won."

"And you didn't notice that we were betrayed?" Tadeusz was amazed.

"Understood and left the army. Not all betrayed, others lacked the courage to go as far as we wanted. But a nation was formed, Drum, a nation that abolished slavery, although it had grown into it to the very top. The south is shaken forever, it will lie and be hypocritical, but we burned the slave stalls. They burned well, this fire will warm me until the last hour."

He looked with secret sadness, ready to object to me, but suddenly his eyes went to the window and he said: "Here is Michalski, my mother's nephew. He arrived in the States in 1856 with a friend, found us in Overland, but did not stay in Kansas, went to Chicago."

Tadeusz's guest jumped off the horse awkwardly, threw the reins over her back and headed for the house. He was wearing a short, cheap overcoat with strong shoulders, but his shirt, under his reddish face, shone white from afar, and his black tie was tied in a magnificent bow. Tadeusz introduced him, but he did not shake hands with me.

"What happened to you?" Tadeusz got angry.

The Pole was sullenly silent, looking at me like an enemy.

"What happened, the devil take you!"

Yes, I once saw this gentleman, I saw it, it seems, under extraordinary and unfriendly circumstances; but where and when?

"I'm going back to Chicago," the guest said in Polish. "This gentleman probably understands our language, and I don't give a damn."

"How dare you!" shouted Tadeusz. "When I told you about General John Turchin, you assured me that you would be happy to shake his hand."

"Oh mother of God! Are you Turchin?! I took you for another, he sailed on the same packet boat across the ocean."

"That was me! Why did you hate me on deck?"

"Leon said that you were from Warsaw, from Alexander's retinue. You are sailing to America for the wedding, to spend money, and we are in the hold, hungry."

"Where is your merciless friend?"

"Leon is gone," he replied mournfully. "He made a terrible mistake, I'm afraid you won't understand it."

Tadeusz waved his hand in annoyance, went to the stove and pushed several unusually long logs into it.

— Leon's heart has always been with the rebels; if the North rebelled, he would be with the North. But, unfortunately, the southerners became rebels, and he followed them; he would have gone to fight for the devil if he knew that he was in the minority. He is a Pole, pan Turchin! He is a knight himself and searched the world for fallen knights to help them rise.

"He's blind!" - The apostolic gray head of Tadeusz, with a steep, rounded forehead and wise eyes, was all in the glare of fire.

— Tadeusz! Michalski pleaded. "He died, not even his grave was left ... It's a pity for the Poles, pan Turchin, oh, what a pity!"

He sought my response and support.

"I sympathize with the cause of Polish freedom, pan Michalski, but I see more than one such cause in the world. Those for whom someone else's freedom is rubbish will not help Polish freedom either."

"Pity the Poles!" Michalsky lamented. "Now Bismarck too has risen to the whip and the gallows, now the Poles will die from him too."

At dinner, Michalsky did not leave this conversation, reproached Tadeusz that he had betrayed Poland, was sad that the Poles came to the States without English, without acquaintances, without money, sorry, sorry for the Poles ... They drank and argued hotter. Tadeusz did not renounce his mother and her homeland, but explained to Michalsky how good it was for him in the regiment of volunteers, where different people came together and became brothers, bypassing the abyss of blood. I liked Druma's speeches, his small gray head held high on a strong neck, his faith in humanity. He went out onto the porch with a gun and fired into the air, explaining that in the absence of wind two or three of his friends would certainly hear and respond. About an hour passed and neighbours began to drive up to the house - two Yankees, natives of the region, a young Irishman and his wife, all on horseback, and an old German man in a tiny gig. Drum threw firewood into the furnace and over the red-hot stove, an excellent fire lit up inside the house, so that we wept and coughed from the smoke; Tadeusz was happy with his neighbour, Michalsky, he forgot me for long hours, left from inescapable grief.

It was a happy evening for me too. And two days later I was standing in Hansom's office, tempting him with a business project. I laid out my reasons for him and repeated to myself Michalski's favourite proverb, who, when drunk, advised every guest of Tadeusz not to lose heart, but "biorac Boga na pomoc, a djabła w garsc!"[26]. *Take God to help, and the devil into a fist! (Polish)*

Chapter Thirty Four

I appeared at Hansom as a businessman who planned to hit the jackpot. In a coffee frock coat and trousers of gray cloth, with a trimmed beard, Saburov would have liked me too. Everything in the States was being updated, Chicago was being born anew from the ashes, and I had to stand before Hansom with a claim to novelty.

"Has anything happened, Mr. Turchin? " He barely raised himself in his chair, the ladder of success separated us too much. "I'm not engaged in construction, but in the monetary part."

"So I hit the target; I want to get rich in Illinois Central."

He was silent, weary of the conversation, from which he expected nothing.

"Ten years ago you extended your hand to me."

"Nine years, Turchin," Hansom interrupted. "Nearly nine."

I was so fed up with the South then that if you put a piece of gold under each sleeper, I wouldn't bend down after them so that some bastard would not think that I was bowing to him.

Warlike words I spoke in a tone of remorse; Having already struck a match, Hansom did not light a cigar, but spread his hands. It was a gesture of reciprocal regret.

- Now General Turchin decided to take up his mind! "My plan is the simplest, I don't even have a piece of paper with me, everything is so simple. Our company, along with permission to build the road, also receives a land grant - a land concession - a square mile of land on both sides of the entire railway. You and I know what the construction of the road and bridges costs, and we could get this money back."

"You decided to trade the land!" Hans chuckled. "This land, by the road, is bought when the best is dismantled."

"The best land is where it's within easy reach of a wagon to ship grain, bring tools and machines," I said with conviction. "I would like to start from here. There was a map of Illinois behind Hansom, and I pointed to the south of the state. - There are heavy forests here, it is necessary to uproot the earth, but I undertake to populate this region at the age of five.

- How badly do you still know America, Turchin. Nobody will go there. In the West, free land, a lot of steppe, plains.

"But I got to know the immigrants well, Hansom. The West is far away, there are wagons to Kansas City, and then? Further in the wagons, there are Indians, there are arrogant aboriginal squatters, you have to plough the land, not letting go of your gun. And in Chicago now - hungry people, factories cut workers by half ..."

"Did you find a single madman ready to bash his forehead against the oaks there?"

I replied that I knew such people and drew a tempting picture of a colony based on the Poles. After the 1863 uprising and the Franco-Prussian War, many Poles come to the States. They settle in Detroit, Buffalo, Milwaukee and Chicago, but the Pole is a born farmer, he needs land. If you put things right, the Poles, Irish, Germans, and behind them the native Yankees will be drawn to the south of the state. Illinois Central will enter the project with a break-even guarantee - only the one who has contributed the first amount is allowed to land.

"Well, what's your interest, General?" Hansom was worried. "The salary of a land agent is low."

"I'll take my money," I answered with all the swagger I could muster. "I would sell land in parcels of at least forty acres, at seven to ten dollars an acre. Anyone who has bought forty acres of forest is obliged to buy a small plot in the future village to build a house. The colonist pays a quarter of the land, and at the end of the first year, only the interest on the remaining debt, so that the farmer can get back on his feet. The debt is divided into three parts and paid in three years: if the whole amount is paid at once, then Illinois Central concedes a tenth of the price."

"Where is your money?"

Now I will surprise you. I laughed with relief. "I take a commission from the company: a dollar for each acre sold, not a dollar right away, but as the land is paid. I will have expenses, I will open an office in Washington. Need advertising."

Hansom nodded: "I made modest terms."

"Illinois Central is giving me land for a village, one hundred and twenty acres; I divide this land into lots of two acres and sell each for fifty dollars for my exclusive use."

"That will give you three thousand and more than one year," Hansom calculated instantly.

"Illinois Central is giving me more land for a park, and one lot for a cemetery. When the first six hundred acres are sold, the company will build a stop there for trains to stop. And nearby - a barracks for eight rooms, so that the colonists had a place to lay their heads before the construction of houses. And the last thing: the company allows free round-trip travel to those who go to see the land."

"And all this alone, without assistants ?! Try to at least measure and divide the land under the forest into sections."

"I also have an assistant, Nikolai Michalsky, a Pole."

Hansom took three days to respond. At the door he stopped me:

"And you will settle there, general?"

"When business there outweighs business in Chicago, I will settle down. And while I'm driving, let the colonists see me as the Chicago boss."

The messenger came to me the next day; Hansom asked without delay to see the office space on Washington Street. However, he stipulated that the barracks would be built close to the station and subsequently moved to Illinois Central for storage, storage and other services.

If only he knew how much hope I attached to this building, as if I was going to erect not a barracks, but an abode of justice. I did not allow the house to be cut down from the damp forest of Du Boys, I brought dry logs of fir from Wisconsin and looked after that it was built to glory, log to log, and the floors were assembled so that even a sharp eye could not immediately find where the two boards closed. In addition to eight large rooms, I built a hall with high windows, and a kitchen with a huge stove and several boilers. I wanted to gather workers, unwitting sufferers of our rich republic, and give a new direction to their lives. The experiments of Fourier and Owen leaned towards the factory—their time had passed; another factory rumbled with iron sheets, clattered with machines, smoked, sucking hundreds of thousands of workers under its roofs. This factory was rushing along the American rails with an iron locomotive, there was no point in launching an Owen or Fourier steam-horse carriage against it. I also read Thoreau and was grateful to him for revealing Walden to me, but even that experience belonged to the past; in 1873 I planned a hostel of equals, and the first class in my school of life was to be a large log house. Eight families will settle in it; I will arrange everything so that they want to be together; in line to carry out duty, in line to cook food, with nothing to hide from each other. The majority will wish for their own home—I understood this too—but they will come to their home as different people. We will call the blacks here and we will receive them not as servants, we will accept them as brothers. The virgin forests around will give us silence and time to form in the right direction, and we will meet the new age without fear.

Two days later, Michalsky and I were sitting in an office on Washington Street: there were so many sufferers from Austria and Poland, from Ireland and Prussia, that I could transfer here the inscription from the house of Tadeusz Drum - "Come, suffering!". Your father, Misha Vladimirov, also knocked here - he knocked in extremes, thin, hungry, having spent the last cent and not daring to ask for a loan. How easy it is to ask for a loan when you have money in your pocket, and how difficult it is to lend a beggarly hand! In April 1873, only Poles set out with us for the first bridesmaid and the first timber market in the oak forests of southern Illinois. They admired the age-old forest, the noble walnut, the green meadows, the flight of frightened roe deer over the ravines and fallen trunks - the land of Du Bois will enchant anyone. Returning to Chicago, the hungry people spoke so well about the Illinois oak forests that some Chicagoans scratched their heads: if they should change their meagre housing for the paradise of Du Bois. The cashiers at Hansom accepted the first hundred dollars. The future settlement began to be printed in the newspapers, and when it came to paid advertising, Michalsky was more willing to put our dollars on the columns of the out-of-town Catholic Gazette than the Chicago Herald Tribune, which was within easy reach. The colony took shape like a Polish colony, and I was not sad about this: the glorious people, whose language and soul are so close to me, will lay the foundation of the building, we will find other material on the spot.

Fergus and Charles greeted me with sarcasm: they were the lost inhabitants of the city. Charles was looking for truth in the society of Chicago workers, and Fergus followed the same path, albeit bookish, in social theory. He foretold that I would bring forth new farmers, poor or prosperous, resourceful or worthless, just like all the other farmers in the whole world. I scolded him as a business Yankee, a machine-worshipper, and we dispersed, dissatisfied with each other ...

Having taken the land, the colonists expected to spend summer and winter in Chicago, buy tools by the new spring and say goodbye to the city. But Chicago met them harshly - panic seized the city. Factory workers were left without jobs, there was no salvation even on the outside, letters came to Chicago from St. Louis, from Toledo, Philadelphia and Detroit, can't you earn at least a dollar a week here? In the summer, the future colonists tossed grief in Chicago, waiting for a full day and night without children crying for a piece of bread, and when autumn approached, they stretched south, to their land, uproot the forest for arable land, and come what may. A wild berry, a nut, even an acorn, on which the wild black pigs of the oak forests ate, encouraged people, and whoever could borrow a gun and a dozen cartridges until spring saw himself as an eater of a steak or a turkey roasted at the stake. What was their surprise when the train stopped at the new station, still without a name, and they saw the barracks and the smoke from the brick chimney! Here they found a house, a supply of flour, corned beef and molasses - Hansom became generous with the food loan, seeing that it was only ten percent of the money already received for the land. Everything was arguing, the barracks were quickly built by hungry Chicago carpenters, the late summer was pleasantly warm, the wolves retreated, frightened off by the noise of work.

Any sack of flour, a barrel of corned beef or a box of biscuits had to be transported from Chicago: the natives of the region had no grain for sale. Like Tadeusz Drum, they led the life of hunters; roe deer, fanged boars, pigs, foxes and wild turkeys abounded in the forest. If they raised a little arable land, then under potatoes and corn, to feed the family. Yesterday's volunteer, elevated by the civil war to the consciousness of his own strength, but also corrupted by it, was both generous, and cruel, and quick to punish. He willingly accepted an invitation to visit, even if by agreed shots in the air, but he could live alone in a forest lair for a year. The few wealthy farmers who uprooted oak forests and collected granaries of grain for the winter fattening of cattle were shunned by the veterans, hating their encroachment on the forest and the animals living in it. It happened that on the eve of the harvest the fields of wheat and corn were engulfed in flames; earthly passions gurgled on the earth, they, and not lightning, gave rise to fire, shots, murders, about which the police and confessors never found out the truth. And they met my colonists not with bread and salt. The poor man, caught in the forest area, at the first logs piled into the wall, did not have time to put down the axe and pick up a gun, and only the need written on it, only the bitter seal of poverty saved him from reprisal.

When there were no free rooms left in the barracks, Drum took our carpenter Maciej Dudzik to him - they partitioned Tadeusz's mansion into three rooms, in two of them Dudzik's families and his henchman Vojtech Malinowski settled for the winter. Drum bought a second horse and gave a pair to the colony, for the transport of logs and autumn field ploughing. Drum spent his last savings, and we did not lag behind him; before the New Year, I bought provisions and heavy ploughs, brought four working horses from Nashville, and suddenly I had to pay for the barracks with Illinois Central. Hansom apologized that the company decided not to delay the construction of the station premises, and the barracks would have to be carried at my expense, but I can profitably get out of the difficulty by taking payment from the residents. What kind of payment could we talk about: the poor came with bare hands, without a second axe! Often I laid out from myself two or three tens of dollars to help the new settler to grab onto the ground, to enter into the forest, saving penal servitude. I turned over hundreds of dollars, in the eyes of the colonist I looked like a wealthy man, but I was his equal, only a free first-class ticket allowed me to get off the Chicago train like a gentleman.

They called the village by the Polish name - Radom, without deviating from the American habit of populating this land with twins of European cities. Most of the first settlers were from the Kielce Voivodeship, from the ancient city of Radom, founded by Casimir the Great in the middle of the 14th century, and as soon as the name Radom was said out loud, everyone agreed that you couldn't imagine a better one. Illinois Central announced a new station in the newspapers, we began to cut the main street, Varshavskaya, - it was then that your father applied his young strength, from this very place he began: the forest stood untouched, we marked the doomed oaks with a light stress, and for them – a mortal wound.

Tadeusz fell in love with his half-blooded Poles: memory, reasonable, fair, again company commander in the forest war, and all the care is about the farmer soldier, so that he is fed, dressed, and having started the battle, to win it. An old gun fed him, old-fashioned frock coats and jackets hung in the closet, bought when he settled in Du Bois, his shirts were washed on a nearby German farm, Drum trimmed his beard in front of a mirror - and that was the end of his household needs. With each week, he cooled towards Michalsky. I attributed this to mutual jealousy: Michalski, the company's second land agent, often traveled with me to Chicago, and when we met in Radom, we retired for a long conversation. However, the reason turned out to be deeper: looking around, Drum and Michalsky noticed that they were following different paths: Michalsky was in a hurry to build a church, Drum was looking for a place for a school. Michalsky came from the boarding house of the German Catholic Jan Bauer, where the missionary

Karol Klotzke sent services, and kept talking about the temple, that the true Pole, even without a roof over his head, would first take care of the shrine, where he could praise God ... ". Drum objected that it was necessary to start from school, there were children of the Irish, and Yankees, and Germans in the district, and he undertook to teach everyone, bypassing the difficulties of the language.

The dispute was not between faith and unbelief, Tadeusz did not reject God. All the colonists shared the delusions of the century: the poor, who lose their husbands and brothers in the uprisings, uprooted by their homeland, thrown across the ocean by tyrants, would rather find solace in faith than in science. Drum, together with me, thought about the brotherhood of people, surpassing the difference between birth and faith, Michalsky felt sorry for the Poles, only he saw them on the Noah's ark of mankind, only he stretched out his hands, believing that for others there will be other hands, and if not, then is this the fault? The colonist pulled out the veins, wielding one axe in the forest, recalling how his ancestor knitted from the logs of the hut; kneading the liquid clay, covering the cracks with it, ramming the earthen floor with an oak block, and Radom's fame grew. The advertisement was silent about bleeding hands, in the midst of general unemployment and poverty, it screamed about the cheap land of Radom, about future crops and free pastures.

The Church of St. Michael the Archangel at Radom, Illinois was built in 1874.

Chapter Thirty-Five

Before Christmas, a money buyer came to the Washington Street office. There was little Polish in him, except for the language that betrayed the Kashub (*Note: The Kashubians are a Slavic minority group who have historically occupied the area in Northern Poland known today as Pomerania, bordering the Baltic Sea.*) or the Silesian native; artistic, as if no longer masculine, interest in the styling and shine of dyed hair, in the complete symmetry of the mustache - fair, also made and with a curl at the lowered ends. He groomed his face, the soft skin of his cheeks and chin; fading, veined, blue eyes lost out from the sharp, borrowed blackness of the eyebrows, and the flattened nose gave the face an expression of secret arrogance. He called himself Josef Krefta, asked for the topographic plans of Du Bois, and I spread the sheets in front of him. The land is divided into rectangles, we hatched the sold plots; no matter how successful things were, the land was still empty, it could still receive many thousands of colonists. Watching Krefta leaning over the blueprint, I wondered what had brought him to us: his hands, with quick, nervous fingers, were not for an English axe or a heavy plow. Maybe the money would do the work for him, but Radom didn't know the labourers yet—does the brotherhood need a buyer for someone else's labour?

"The best sections, it's true, have already been dismantled," Krefta remarked.

" You can't see it on the map."

"People will not take the worst land as long as there is better land."

I wanted to object that there would be enough good land for half the inhabitants of Chicago, but I did not say it, neglecting the duty of the commission agent.

"How much land do you need?" I inquired.

"Watch how you do things."

"If you pay in full, you get a ten percent discount."

"That is if I buy forty acres. How about eight hundred?"

No one had ever bought more than 80 acres - and there were two such merchants: me and Mindak, a decrepit Chicago tavern keeper, who was building a house on Warsaw Street, not far from the station. Eight hundred acres for cash! Such a merchant immediately corrected our affairs; even I would have received more than seven hundred dollars on my account, hanging at the last cents. But in the scope of Krefta, danger lurked.

"Even if you buy land for seven dollars, the purchase will cost more than five thousand. With this amount, you can live in Chicago.

"If both I and the colonist on forty acres have the same benefit, I will take the instalment plan. But that's not how it works in the States."

"In Radom you will find something that we are not used to."

Krefta was perplexed, looked at me, as they say, with all eyes, from under heavy, bluish eyelids.

"What are you going to do with this land?"

"Maybe I'll sell the forest, get cattle, or start sowing wheat."

"I have to be honest with you: Illinois forest, even walnut and oak, is not worth it, everyone has plenty of their own tree. And it won't take long for wheat to sell: it's cheaper to bring flour from Cairo or St. Louis."

"Why are you inviting colonists to Radom?"

"We save the poor from dire need; no one promised them quick wealth. How long have you been in the States, Pan Krefta?"

"Twenty years".

For a Polish fugitive in 1831, Krefta is not old enough, and he did not wait for 1863 in his homeland. The Pole chuckled.

"I fought here, pan Turchin, but not for long." He leaned over and lifted his trousers, and I heard a click on the hollow leather leg. - On the march, a cannon fell on him and crushed his leg.

I invited him to come to Radom, promised a horse under the saddle and a good guide, meaning Tadeusz Drum. It turned out that the colonists knew Krefta, bowed to him reluctantly, forcedly, and Tadeusz turned his back on him. I was told that Josef Krefta was a usurer, lending money at high interest rates, and running his own insurance business. That's right, he wants to wait until the price of the plots rises and resell them at a profit.

The danger is great, although the thought was comforting that it was not only hopeless poor who came to believe in Radom. I won't sell it, Michalski will, and how not to sell if an American citizen has come with money and wants to buy what is for sale! I delayed the sale, the unforeseen inflamed Krefta, and when it became impossible to postpone the completion of the deed, I turned to Hansom.

"And sell the land to the Pole if he wants it," Hansom decided without listening to me. "Let him take at least eight thousand acres.

"If we start selling thousands of acres in one hand, the land will have two owners, Mr. Hansom. The owner of a thousand acres will turn his back on us, he will probably demand a fee from us for the passage of trains through his land."

"Draw up drawings and bills of sale in such a way, Turchin, that the interests of the company do not suffer. The land must be sold, sold, before it occurred to Washington to impose a high tax on the sale. It was hard to recognize in this man a Mattoon Hansom, a thieving fat man: uncertainty was kept, perhaps, only in his fingers overgrown with red hair, he touched with them the objects scattered on the table, his shaved cheek, chin, key rings and black braid on the sides of his coat. Sell the land, Turchin! Drive it for miles, the company needs money today; trains leave without passengers, people do not even have a penny for a ticket to die away from home and be buried at public expense."

"Krefta insists on a twenty percent discount on purchases of more than four hundred acres," I said from the other side.

But he's investing cash!

"Krefta will divide the land by twenty and ten acres and sell it to make a profit."

"And let it profit! The more money people, the better."

A few days later he informed me that a double benefit was possible, but all cases involving the sale for cash of two hundred or more acres of land passed to him, with me retaining half the commission in these cases. ("You're right, Turchin, the company should know the large land-owners by sight: who are they, why are they climbing into the Illinois wilderness? ..") Tadeusz and I still had a lot of fun with Krefta, and he was afraid to take it for cash, bought 600 acres in instalments and did not close to Radom. Krefta paid the land on time - in 1874 and 1875, he did not touch it with an axe or a hoe, we already began to forget about him, and in vain ...

In November, the daughter of Vincent Kowalski died. Father knocked together a narrow coffin: Dudzik and Malinovsky stood nearby, with carpentry tools, but Kovalsky did everything himself and, when the hour came, he lifted the coffin on his shoulder. He walked without a hat, high, as if in ploughing, raising his legs, hugging the coffin with his arm and pressing his head to the unpainted board, as if listening to see if he would move, if he would not call their father and mother Stefa. Gray-haired from a young age, a patient rural worker, he was the first to cut down a hut, the first to bring it under a plank roof, the first to take out belongings from the barracks, and the first to become related to the holy black kinship with this land. We chose a secluded place in the forest, a small hill - I'll show it to you, there is now Henryk Ludwig's garden - and my father dug a grave on a hill.

Why did I take a violin to the funeral? I could not come to George Fergus without a violin: nature did not give me tears, and the sobs of the heart, not coming out, tore it; with a bow in my hand, I could cry without showing tears, I would like to speak at the grave of Fergus, but it was the right of others, his comrades in the workers' union; I went behind the briar bushes and played without thinking whether they were listening to me. And what brought me with a violin to Stefa's grave? I don't know, there was no other music in Radom yet, the grief of the Kowalskis struck me, a guilty feeling entered my heart that I showed people a way out of hungry Chicago, a way out through extremes, through tormenting work, and I led them, saying that I know where and why: I led in pride that I see the goal, and this goal is not far off, within the limits of human life, and, trusting me, people suffer disasters. Shoemaker Jan Kozelek, a one-eyed Radom nugget and a master of all trades, read a prayer over the coffin, and the violin did not attempt to console the Kowalski's, it reminded them of the sorrow of mankind. Perhaps it was the violin that turned the views of the colonists from domestic concerns to the highest interest; when the carved cross made by Dudzik fell into place, the colonists surrounded me. Maciej Dudzik was the first to speak:

"God punished us, Pan Turchin: he took Stefa for our sins."

"If we are sinners, why punish an innocent girl?"

Dudzik put the axe and the bag with the tool on the ground: he did not dare to argue about God and his inscrutable ways.

"People's law punishes the sinner," Kozelek said, "but God enlightens the people. God has no prisons and gallows; God takes an angel to heaven and tells us: you are not worthy that he should live among you."

Jan Kozelek fled Chicago in order to get rid of two chronic ailments: drunkenness and shoemaking. He would go out in the morning to his plot, a skinny figure in a long coat and a red hat with a drooping brim, hiding in the twilight forest at dawn, a hemp rope thrown over his shoulder dangling behind his back, as if urging the gallows in search of a secluded place. During the day, the neighbours heard the lazy clatter of his axe, occasionally the distant rumble and crackle of the fallen giant, and everyone left work, waiting to see if Kozelek's axe would knock again, if the tree, along with the undergrowth, had broken, and the shoemaker, who was leaving for the forest, as if on a lost war. Kozelek's soul was looking for worldly recognition - he read any prayer without hesitation and the first days of Radom self-appointed served a church service, until the missionary Karol Klotzke, a sharp-nosed, strict mentor who emerged from the depths of the forest, appeared in Jan Bauer's house. Shortly after the war, when the first Catholics, Germans and Irish, appeared in the district, the Bishop of Altona - Baltes - appointed Karol Klotzke as a missionary with a permanent residence in Du-Queen. Before our arrival, Klotzke did not favor the local Catholics: the twenty miles separating Bauer's house from Du-Queen, the lazy shepherd overcame only once a year. When the axes rattled and the Illinois Central steam locomotives began to let off steam at the Radom station, Karol Klotzke began to frequent and, sparing himself with the purchase of a plot, hinted to his flock that he would accept land and a log house as a gift from the colony. As a sin, Klotzke tightly closed his eye, the left eye, which was not on Kozelek's face at all - he lost it in the uprising of 1863 - and in the worst moments it seemed to the shoemaker that the German was making fun of him. Kozelek buried his wife and son upon arrival in the States - they never set foot on free land - they were removed from the packet boat with dysentery and buried at Castle Garden. Kozelek wanted to return to Poland without leaving the emigrant depot, but there was no money for the journey - only a shoe tool was returned to him from the quarantine cell.

Following Dudzik, many laid the instrument at my feet, as if renouncing Radom and our cause. The colonists came to the funeral from the forest, the night frosts hurried, only Kozelek came from the barracks: a week ago he sold an axe, drank money and in the morning sat at the window of his room, looking at the knocked-down heels of Frantisek Gajewski's boots.

"What is our sin, Maciej?" I turned to the carpenter. "That you didn't let hunger suffocate you? Or is it a sin to leave Poland?"

Poland is always in our heart! exclaimed Kozelek, in a proud delusion that huge, strong Russia cannot be kept in the soul with the same love and torment as little Poland.

"They have no guilt before their homeland," Klotzke said prudently. "They're outcasts. Their wounds are holy."

Klotzke bowed before their homeland - even Kozelek's face had disliked him - spoiled by a leaky eye, it kept the stamp of sublimity and beauty - framed by a curly blond beard, with a firmly fashioned sensual mouth, crowned with a preacher's forehead. That is why I did not reach the most hidden depths of their hearts because I did not agree to give a special price to the Polish emigration over any other. In disputes, they did not hold out against me for long, but even when they were silent, they kept in their hearts a prayer without words, like the sound of a shell raised from the bottom of the sea, like the sigh of an organ in an empty church - a prayer that nourished their pride in humiliation and poverty.

"That's not where we started, Pan Turchin," said Dudzik in a low voice.

All our pride! - the shoemaker supported him. - Everyone thinks about his own house, but the house of the Lord is forgotten! ..

The Kowalskis were tormented that they had taken the girl from the dry barracks to the house from the damp forest, on the dirt floor— glasses ordered in St. what price would he have paid if, unfortunately, he had brought his house under the roof before Kowalski?

"Damn me," Dudzik said, "if I put at least one log in my wall before the church!"

A noise arose: everyone praised the carpenter, marvelled at their blindness, asked God for forgiveness and mercy.

"Why don't you build a church?" I asked.

"When people who have despised faith are at the head," the missionary answered for them, "it is difficult to start a conversation about God's house.

"This man is here, Klotzke, and does not hide his eyes; I have no guilt before God, - I turned to the settlers: - Haven't we agreed that here everyone is equal to everyone?"

"We were afraid to offend you, Pan Turchin!"

"The church is a matter for the community, and you decide."

I removed the stone from their souls.

"Pan Tadeusz," began Dudzik, apologetically, "we'll have to put off school...

Dudzik went with Tadeusz to the land of the village, choosing a place for a school. Now he spread to those who wanted to build a church before the school, and was afraid to look into the eyes of his favourite and benefactor.

"Let it be as everyone wants," Tadeusz replied with unexpected calmness. "The general and I have forgotten many prayers, but good people will pray for us too."

Dudzik threw the old Confederate woman on the ground out of place, fell to his knees, as if the vaults of the temple were above him, and made a prayer together with all the colonists. It was a moment of paganism: a closed forest all around, a wild hill with a grave black by night, the close cry of a bittern and the dry rustle of autumn oaks.

"We came here on St. Michael's Day," the carpenter said. "Let's build the Church of St. Michael the Archangel for the glory of God."

They chose a place not far from the station, this is the place, they began to uproot the forest, cut through the street to the railway. White pine trunks brought from Chicago, heavy boards sawn by city saws, brick and stone were dragged on themselves. Collected in Chicago - and not only there - money for the church and its decoration. The blacksmith's forge did not cool down even at night, the brothers Gaevsky, Frantisek and Adolf, handled iron and furs, repeating the native proverb they loved: "Where the blacksmith settles, the city will grow." They built God's house simply: four high walls, a steeply sloping roof, narrow, elongated windows - then there was still neither a bell tower in front of the church, nor this extension behind. Home log cabins stood abandoned; birds flew into the bare rafters, the forest animal crept up to the abandoned walls. The carpenters moved from Tadeusz to the barracks so as not to waste time, he was left alone and rarely showed up in Radom. We have another person who firmly stood with his back to the church: this person is your father. Misha Vladimirov, an ardent overthrower of the church hierarchy, he uprooted oaks on the future Varshavskaya and, as a sin, settled at Kozelek - in the evenings they argued late into the night about God, about the origin of the earth and the universe.

Nadine came over for Christmas. I met her at the train, trying not to miss her eyes when they saw the street cut through the forest. For a moment I saw a mirage, Nadia with her father at the Petersburg railway station, how confused she is, looking for me, wants to see me sooner, to be sure that I am the same and just as confused, but she doesn't see and is frightened if something happened, if they sent me away for some other war; and I'm right here, just reach out your hand. Why is she so excited now, so earnestly waiting, looking for an eye and a hand? If, of all that fate can give a person, I had only this, then I would be lucky.

"What is it, Vanya?" she wondered.

"A Polish church with makeshift benches, with a poor priest, if anyone decides to come to us."

"Not a priest—priest," she corrected me. – "You can't compete with priests."

"I need the colonists not to hide in the forest holes, even the church is good for that."

I listened without anger to her grouchy disagreement: the winter life of Chicago was hard for her. In Radom there is a piece of bread baked in the barracks, cheap coffee with a spoonful of molasses, corned beef boiled in a bowl, in Chicago there is hunger. Here the farmer is sheltered from the wind by oak forests, in Michigan there are fierce winter storms. Everyone is in plain sight, but on the Chicago streets there is poverty behind the shutters, rags, unfortunates who lack the strength to trudge through the city from end to end, listen to refusals and abuse, and the dog screeching of door hinges. And Nadine's day is not enough to run around shelters, women's committees and street kitchens, depots for the hungry, instead of volunteer depots. Every dollar that went into the Turchins' purse was spent either on beef thin soup for an unknown Chicagoan or on my colonist's bread, Nadine prioritized the Chicago trouble over the Radom need.

Missing Tadeusz Drum, I saddled his old horse and called Nadine with me, but she only accompanied me to the Kowalski's hut. Having paid a sacrifice to God, they remained in humble despair on the farm, among makeshift chairs, a table and bunks on an earthen floor. Nadine walked alongside, her hand touching the saddle, imbued with the surrounding silence and the austere grandeur of the winter forest: the mighty branches of oaks, knobby, tight under the rough bark, the knots were like the muscles of centuries fashioned in wood. The path disappeared under the fallen leaf, it rustled under the hooves and tenderly responded to Nadine's step. I thought of another woman, with her son in her arms, at the arena, in the crowded Chicago square, how she glided beside Howard like a barefoot, proud Indian, shuddered at this vision and met Nadine's gaze, shocked by the same ghost of the past. I leaned over in the saddle and with a jerk put her in front of me, my arms were then still strong with the Don, long-standing strength, inconvenient for the enemy. We drove to Vincent's house in silence, thin Kovalskaya, a teenage wife, rushed to Nadine, crouched, wept inconsolably and loudly. Whom did Nadia hold close to her at that moment: the mother who buried the child, or her own unborn daughter? After all, Kowalska was far from thirty.

Drum was lying on a wooden bench, under the brown bear fur, with his hands behind his head. He heard the stomping, the neighing of a horse, my jump on the boards scattered by the porch, but did not rise.

"Hello General," he said.

His face went ghostly white as I wandered around the room looking for a candlestick while I lit the candles.

"Why aren't you glad to see me, Tadeusz?"

"It seemed to me that someone would come today, so I prepared the lamp."

By the burning candles, I could make out the copper lamp of Drum, the glass lay separately.

"Light it up?"

"No need. Sit down, don't think about me, it's better that way."

I realized what kind of boards, thrown by an offended hand, blocked the porch: Tadeusz removed the partitions.

"You weren't expecting me, but Dudzik?"

"They won't come back here, Turchin. He lowered his fur-lined feet to the floor. "I myself don't know how to return to the experience, to the cities once taken, to the parental home, to the rejected friendship. If we had not taken Athens a second time, our life would have been different."

"I do not regret that we took Athens again!" I objected passionately.

416

I argued, but meanwhile the words of Tadeusz had power for me too, and I did not return to the old one. parental home; the classrooms of the cadet corps, the ends of the Nevsky, Warsaw, the Carpathians, Sevastopol, shot with cannonballs and bomb shells, London, the barren arable land near Rowland, the cities taken - everything flew into the past; one day we went back to the old Chicago nest, Kenwood, and weary, looking for a change.

"You will return to Radom, Tadeusz," I insisted.

"Maybe." But how easily they left me, as if they were leaving an inn. And this is Maciej, a nice person. People, people, is this material suitable for your plans, Turchin? Would you like to try training ants?! Everything is ready there, in their big heaps of forests... We will have everything: a school and a church, a smithy and a tavern, a cemetery and a post office, but this is also in other places! There are also slaves there - we will have them too. Slave and usurer - how to do without them."

But you are not a slave! You yourself are not a slave!

"I went to the forest. Now I am a slave to my loneliness."

"You are not a slave! Even if you try it for a tooth, even if you burn it with aqua regia,(Lat."Royal Water" is a highly corrosive, fuming liquid and it's used to dissolve gold and platinum.), the slave suit will not come out. How can we not help others to burn out the slave in ourselves! Man is noble by nature, whoever gave life to the human race, he put in us the beginnings of love, reason and justice.

He sat on an oak bench with his head bowed.

"I was looking for humanity," he said quietly, "and Michalsky offers me county Poland.

"Do you want me to break up with him?"

"No!" Drum responded quickly. - He did not have the fate to become a volunteer of the republic, he has nothing to remember; it is terrible not to have a future, but it is little better to live without a past. Tadeusz got up, cleared the table of some papers, sacks of shot and gunpowder, and a pile of the Nashville Journal, the Washington County newspaper. White German bread, a cold turkey, wild garlic, some fragrant herb, and an open bottle of whiskey appeared on the scraped boards of the table. - Michalsky was too long a shadow of his friend Leon, he once boasted that Leon, fighting in a regiment of southerners, took revenge on St. Petersburg and the tsar. Even now he does not understand that it was possible to take revenge on the monarchy only in the republican regiments.

The memory of the war stirred Drama: the war, in which no nation and nation came together, but the sons of one country, the sons of mankind, divided into two armies, had a magnetic effect on our hearts.

"Some of the old-timers are dissatisfied with the construction of the church," said Drum.

What do they care about the church?

There are many Protestants here. Some run wild in the forest. He chuckled. They consider this land theirs. They were given the forest to the unmeasured, they were allowed to count the acres themselves. They figured - in their minds - from horizon to horizon, from Mississippi to Ohio.

"I went to Nashville on their silly lawsuits."

"Those who complain to Nashville are not dangerous. There are those who themselves are the judges of everything, and the executioners."

I took off my coat from the hanger, Drum looked out the door, into the calm night. A fragile diamond moon shone high in the sky above the motionless mass of the forest.

"Did you decide to go?"

"Nadine is waiting for me."

"Madam doesn't like Radom." We stood with him on the porch. "She rarely comes."

Yes, more often I came alone, alone whiled away the hours at the carriage window, alone considered Radom's needs in my mind.

"Nadine is busy: with the Chicago poor, with the Washington Congress. Our ladies planned to get the right to vote in the elections, so they went to war with Congress, under the command of Mrs. Woodhol.

I'll take you, John. He returned to the house for a hunting jacket and strode over the scattered boards without closing the door. - Madame has not lost a single battle, God grant her good luck now.

I did not allow Tadeusz to go far, said goodbye and jumped into the saddle. I remembered the day Elizabeth Howard came to collect her husband's body; they both, Tadeusz and Elisabeth, wandered blindly in the darkness of their misfortune. Having taken away their only love, fate seemed to circle the two oppressed figures in a common circle. Nadine said more than once that they could be together, their union would not seem like betrayal and betrayal to anyone. But Elizabeth remained Howard's widow and mother of his son, one of the most beautiful women in Chicago, and Tadeusz Drum a forest hermit. From their military mail was born not love, but friendship; Tadeusz wrote to her about Howard's company, about the grateful memory of volunteers, such letters bind a woman more tightly to her dead husband. The thought that he is alive for strangers revives him for a loving heart with the power of reality itself. It is here that the seeds of spiritualism are thrown - in earthly love, in the sorrow of losses, and not in the speculations of charlatans. We love, and the imperishable image lives, appears at our call. Didn't this happen to me too: the mother reached out with a trembling hand to our gate, wanted to open it, go in to her son, but the father did not let him in, stood barefoot on the ground wet after the rain, waited for me to go out to the gate, bow, call into the house, and I wrote something on paper, crumpled it up and filled cartridges with it, rejoicing that I was about to run to the gate, but I still had to finish the word, and when it was written down and I looked up from the paper, there was no one at the fence, and there were no traces on the earth. And the regimental commander, Prince Lvov, did not leave me - he hastily buttoned his uniform, making mistakes with buttons, keeping his eyes on me, filling his pipe with tobacco, and as soon as we approached, he rushed to run, quickly, with his shoulder blades drawn together, as if he was expecting bad things from us, even a shot in the back. First, he ran along the Warsaw parade ground, dragging a fallen spur and hitting the ground with his heels, like a drum, and then I saw him running away Russia, Nevsky, a yellow field with falling grain, a narrow strip of land in the middle of rotten water, behind which Balaklava and a red from the blood of Sevastopol. If he had stopped, we would have explained everything to him, but he did not do us this honour: he left behind him, in traces of flight, banknotes, banknotes, but we did not bend down, did not take, we needed his gray hair, his shoulders bent over the years, exhausted legs, to wash them with warm water, which we splashed from a white English jug while running, to wash and wipe with a clean towel ...

I was pulled out of my thoughts by shots, a wild, forest cry, a fire blazing ahead; it seemed that the station, the barracks and the wood brought from Chicago broke out.

Three horses, crouched in saddles, with black armbands on their faces, were rushing straight at me. My horse reared, covering me, and the riders, recognizing the horse, managed to shout - not to me, but to Drum - that the damned Pole should not go to Radom anymore, otherwise they would burn him, along with his coffins.

I galloped to Radom. The fire went out, the colonists flooded the walls of the church with water. The night guests did not even suspect what a good service they rendered to Radom: the colonist firmly grabbed the forest pioneer's hand with both a gun and an axe. Soon the church stood under the roof, ready to receive parishioners, priests and Mass; the carpenters set to work on the school. Drum roamed the farms, risking a bullet in the chest or in the back of the head - he again became a zealot of Radom.

Since the spring, auctions have gone briskly with rebidding in the office on Washington Street. By autumn, Kovalska had given birth to a daughter - having opened the account for the losses of Radom, they were the first on this land to give birth to a new man.

The first priest, Josef Mushlevich, arrived early and, together with the parishioners, covered the house of God with planks. No money was sent to him from the diocese in Alton, and the colonists lived in need, so that the good Joseph spent his own capital - three hundred dollars, and when they were left only for the road to Detroit, he left Radom.

"Smierč i zona przeznaczona" [27] used to say Josef, and it's time for a priest to say: "Death and paraffin!" The colonists had spent their ardour on the construction of the temple and now indulged in worldly affairs. Father Joseph, although he grumbled that the ringing of a silver dollar sounds louder in the ears of a parishioner than a church bell, did not fall into anger; and a sensitive ear has not yet heard the ringing of dollars in the thinning Radom forest - only the annoying clatter of an axe, the grinding of a shovel, and the cry of the first mules bought cheaply in neighbouring Kentucky.

CHAPTER THIRTY SIX

The train started moving, Vladimirov was dragged inside the car on the move. William Christian introduced him to the people standing and sitting around the coffin, covered with the flag of the United States. Actually, only one old man was sitting; on a white stool borrowed from a Southern Illinois insane asylum with a prosthesis in front. Vladimirov remembered the named name - Barney O'Mullen, it was buried under the ruins of Turchin's unfinished memoirs. But another old man, with a daring little head, he would recognize himself, although he had not met him before - Tadeusz Drum. He is alive, light, slender, imperceptibly smart, in a white shirt and an old-fashioned bow tie under an open hunting jacket.

The coffin is closed and tied with a rope under the flag so that the lid does not move from the shocks. Turchin was dressed in a general's uniform, but through the silk of the flag and the painted boards, Vladimirov still saw the old man in the coarse linen of the orphanage, as he had found him in the brick cellar, on a lonely table. The orderly said that the dead are rare among them, before the end of the sick they are taken away so that they confess in their native walls. "Our patients are no worse than other people: you cannot distinguish them in the crowd, unless it occurs to him to get on all fours and howl. And you don't want to be healthy, or what, howl ?! - He pointed to the body of Turchin: - That's really who was healthy in mind ... "-" Why is he here? Vladimirov asked. "So the time has come. And he was strong in mind; why did he die? - The spring is gone. Are you his son or nephew? Vladimirov shook his head: who is he to Turchin? How long has he, vexed, been looking for the old man, and now he is grieving and executing that he left Turchin and went with Virginia to the wedding, as if the general could wait for years. He knew that the spring was running out, but spring deceived him, happiness blinded him, and he let Turchin go without listening to the end. What happened in Radom in 1875, what explosion shook the unfinished building of justice? "It turns out he is a general. - The orderly straightened the uniform warped on the back of the chair, - I thought that was the nickname; we have both presidents and kings; why not be a general ... "On the table lay not a general, but a short peasant, a blacksmith, a farmer, with heavy hands and a large foot, who had been walking behind a plow for a century on unbroken earth, in the warm ashes of burnt trees. They brought the coffin, Vladimirov hurried upstairs to the doctors, lingered with them and was late for the conductor's whistle.

The train moved slowly, passing by the outskirts of a southern Illinois town, a narrow rivulet, farms, brown stacks of last year, clumps of tall cedars and beeches, meadows ripe for a spit - June had just passed halfway, and the wheat ear began to slope, nodding to the farmer so he didn't miss the deadline. Drum stood at the head of the coffin, listening and waiting for the engineer to speed up so as not to shame Illinois Central in front of the dead brow of her soldier. People huddled and languished in the car: Poles from Radom, a retired officer of the Chicago Club of the 19th Illinois Volunteers, and two young soldiers borrowed from the state military authorities. Barney O'Mullen demanded that they stand at attention at the coffin, and they were drawn to the door, no to tobacco, you can't smoke at the coffin. Vladimirov heard the muffled voices of the Poles, tried to guess if the builder of the church, Dudzik, Kowalski, or Jan Kozelek, was among them. The one-eyed goat, which means he is not in the carriage, is true, late Radomites gathered here ... And again the aching feeling returned: together with the body of Turchin, the quarter of a century that had passed since that time would be buried.

On the half-stations they waited for oncoming trains; stuffiness quickly filled the car, the trills of red buntings sounded loudly.

"Look, they're flooding," said Barney O'Mullen. – "Well, at least they sing. Well, in Russia, is there oatmeal?" Barney asked Vladimirov.

"Why, there is. They look similar, only yours is redder."

"Are they screaming? Are they crying?"

"I'm afraid to make a mistake, I'm a city dweller, and oatmeal is an inconspicuous bird."

"That's how you know our language in your family!" Barney praised; he also considered the Russian doctor as relatives of the general.

"I am not a relative of John Turchin, but the son of his friend."

"It turns out that in Russia everyone speaks our language ?!" the old man cheered up.

Drum was fixing the iron ladder, he wanted to go and quarrel with the engineer, but the engine screamed and dragged the cars towards Mound City.

"They told me at the shelter that you were with the general," Vladimirov turned to Christian.

"He died with me. He died calmly." It seemed that Christian wanted to get rid of the interlocutor, to prevent empty questions. "He left in peace."

"You don't feel like talking?"

"Don't be offended," Christian said. – "Cap Turchin has seen a lot of deaths, but they have not learned to talk about it. When Turchin left Alabama in disgrace, volunteers stood for several miles along the railroad, and everyone wanted to reach out for his hand ... - He hated the green expanse outside the car door, the summer, slumbering, indifferent grace."

"Did he say anything as he died?"

"I did not understand," he said in French: "You should have gone up to Herzen, Nadia ... What a pity that Herzen did not see you ..." And something in Russian.

Vladimirov did not explain. Herzen's name would say nothing to the Stevenville postmaster, but a secret joy came to his heart: Turchin revealed something to him; he knows something, if a few words of a dying man evoked in him a picture of London, a house behind a wall, a gate unlocked by a cook ... "Why is she so unkind to me?" Vladimirov thought sadly. He knew so much about this woman and did not see her. Maybe the general himself didn't want to show her the son of the man they both knew young? But she was able to accept in her heart the Mattoonian Thomas, or Charles, or the daughters of John Fergus. Why is he unworthy of this honour, why does she avoid him? Why was she not in Anna, in the orphanage where Turchin was dying? To live with Turchin all his life, inseparably, so that the cannon fire, and the attacks of the rebels, and the tiredness, and the wrath of the judges - nothing could open their hands and leave him in his last days? Is it possible to understand this?

"I was shown a medical record made at the end of April, when the general was taken to Anna," Vladimirov said to Christian.

"Any formalities?"

"The record does not convince me that Turchin certainly had to be sent to this shelter."

"He had a strong mind," Christian said sternly. "But the light went out. When you are young, you don't realize that this is possible. What is there, in this piece of paper?"

Christian's thin physiognomy, everything from the dry and brown beard, as if made of dyed hemp, to the narrowed eyes on the parchment face, expressed contempt for the world that the general's magnificent brain set out to judge. Even dead, in the cold armour of the skull, this brain was for Christian a more reliable part of the universe than the vain minds of the living.

"He has many friends, was there really no home where Turchin could end his days?"

"You don't know this character, his diabolical pride! Christian was angry. And don't mess with your kindness."

And what, how does he really not know Turchin? He wrote regularly, perched on the corner of the table, and when he returned to Chicago, he whitewashed, rewrote the pages and sent them to St. Petersburg. But will life be revealed behind the pages, or will they crumble, dry, lifeless, like the leaves of a tree whose roots have been cut off?

"You're right, Christian, I hardly understood the general well," he said humbly. "But I would have buried him in Chicago.

"Listen, Mr. What's-your-name!" Barney waited for his moment. "If you are a man with a wallet and connections, put up a bronze monument to the general in Chicago. Right, Tadeusz?"

"Damn it!" Drum said. "They won't let the old man lie down in peace in the ground!"

"Some people think that the richer the cemetery gates, the shorter the road to God!" Barney exclaimed.

"Will you finally tell me what the doctors wrote there?" — Christian was pestered by the orphanage paper.

Vladimirov took a sheet folded four times from his pocket. The postmaster moved away from the coffin, to the very door, and, holding his long arm far away, read the paper. I read it and a second time, surprised by its routine and emptiness, looked back at the coffin, trying in vain in my memory to connect what was written with the dead old man.

"That's what I thought," he said, returning the paper.

"What are you talking about, the fire?" Vladimirov asked.

"We burned the last fire with him near Athens, in Alabama; there was no ash left from those fires. But a lie is written here: he did not complain to anyone about pain and suffering!"

"What did they do there, captain?" Barney rose from the stool, holding on to the coffin.

"You went to him; did he ever complain to you?" Christian asked Vladimirov.

"No."

- And you?" he turned to the Poles. – "Did everyone hear complaints from General Turchin?"

"Mr. Turchin was a proud man!"

"Give me your paper!" Barney moved towards Vladimirov, the train stopped, and the crippled Irishman would have fallen if not for Christian support.

"It's not worth reading, Barney," he said. "And don't call me captain; that's who our captain is - Tadeusz Dram."

"Mr. Drum started the war as a captain."

"I graduated as a captain," answered Tadeusz. "And I'll tell you, William, don't make an idol: the old man was a man. He suffered severely and complained bitterly..."

The train was approaching its destination, slowing in jerks, roaring through the outskirts of Mound City; all eyes converged on Christian, on his disbelieving, unhappy face at that moment.

To whom did he complain?

"To me!"

"In the whole world - only you!" - Fellow soldiers, they became hostile against each other. "Will you believe this, Barney?"

"If the old man was suffering," said Barney conciliatorily, "why shouldn't he complain?"

The train stopped; their carriage is opposite the crowd, a military band, a gun carriage with four horses under mourning blankets.

"The general was denied a pension," Mikhail Florikh said. "I brought the War Department's reply from the post office—you should have listened to how he scolded them!" True, mister, I ran for the general's mail as a boy, they paid me a dollar a month ...

"You were paid a dollar," boomed the postmaster's voice over the crowd, "and you don't have a mind for fifty cents." The general did not ask for a pension; he refused to accept the money!

The gangway was brought, to the sounds of a military band, the Radomites lifted the coffin and carried it away. Vladimirov saw Virgie, - she raised her black-gloved hand, - Horace Fergus and her eldest daughter, unfamiliar women in mourning: all mournful witnesses, but not the widows of these funerals. He recognized Senator Foraker, Johnston, standing apart from the young Howard.

"Pony! Hey Pony Fenton! Mr Fenton! "Barney yelled from the car. "Do you recognize me, James?"

Barney skipped down to the veterans, introduced them to a Russian doctor, a relative of the general, Christian joined them, and the four of them set off for the carriage. James Fenton, a Turchin volunteer, and his comrade Billy Burton, while waiting for the train, managed to refresh themselves in the buffet and chatted about the upcoming funeral. The banner of the 19th Illinois Regiment, which, next to the flag of the country, covered the body, and the gun carriage was brought from Chicago; the veterans would not agree to place the general on the official Mound City cemetery gun carriage—who can be sure that this gun carriage did not carry the secret guardian of the rebellion, or even the rebel himself, to the grave?

The crowd behind the coffin was made up of Chicagoans - Chicagoans occupied the car - Mattoonians, led by the elderly Teddy Dawson, from the inhabitants of Radom and nearby farms. Nadine wanted to bury Turchin in Chicago. She went to the Chicago governor's residence, along with the elders of the 19th Illinois club. Everything was in favour of the Chicago cemetery, but the state governor said to Nadine: "You are a brave woman; you have reached old age, you have given your life to the country and the Union, but you do not have people close by blood here. And I dare to ask you a question, where would you like to be buried when the hour comes? "With Turchin," said madam. "Only with him!" "You see," said the governor, upset. "That's not possible in the Chicago War Cemetery." He named illustrious generals, whom the harsh cemetery charter separated from their dead wives. "But I fought! Madame retorted. "Isn't there also my right to a military cemetery!" "I heard a lot about you, mistress, but you were at the front, violating the charter." "We never regretted it!" Madame interrupted him. "The living sometimes manage to violate the charter with impunity, but only the living. In Chicago, you will not be at home with General Turchin; in Mound City, in the national cemetery, you can lie down next to him. In this I have the guarantee of the War Ministry. That, eyewitness Fenton said, was not the end of the matter. "Now, Mr. Governor," said Madame, rising, "when the affairs of the dead are settled, I want to object to you. You said that we do not have blood relatives in the States? You are mistaken: we have many sisters and brothers here, the best that fate can give, and not an accident of birth. "I meant relatives," the governor said with a hint of arrogance. "Family, clan, family crypts." "We would have had this in abundance in our first homeland; we were looking for another... "- " And found, Mrs. Turchina? "We have seen how a nation is being formed, we have entered into it, by the right of soldiers and workers. We have learned that the difference in blood is nothing next to the commonality of the idea. And they realized that only that people can be considered great, which does not encroach on putting itself above any other people. She held out her hand to the governor. "To see so much in one lifetime is a great reward, Mr. Governor."

Two riders in faded-blue civil war volunteer uniforms held back the step of artillery horses, the carriage moved slowly, Turchin's body under the banner was motionless, but his face seemed to come to life when the cortege entered the alley of elms or sprawling oaks and sharp, changeable glare pulled Turchin's large eyelids out of the shadows , a huge forehead, closed lips and strong cheekbones of the bottom. Horace Fergus and her daughters went ahead; in the gap between the shoulder of Virjie and her mother, Vladimirov saw Turchin's thrown back head, the stubborn and deathly face of a God-fighter.

Where is Nadine? Where is she now that the burial has been decided, the grave has been dug in Mound City, and the veterans from Chicago have arrived at the little station? About a hundred people, counting the onlookers of Mound City, go behind the coffin, hiding their eyes from the unbearable sun, from the musicians splattering with fire, polished pipes - but she is not.

The National Cemetery opened around the corner of the avenue of white cedars; they intertwined the crowns and covered the road with a more reliable shade than oaks and hollow elms. A low stone fence, a latticed gate sunk into the grassy ground, open forever, a light landau with lacquered wings and a pair of raven ones at the gate, and behind them a dense park with rare tombstones; the caretaker's house, the chapel and two shepherds in black robes - an aged prelate, hunched, decrepit, but with lively eyes under the prickly white cornice of his eyebrows, and a young pastor. Vladimirov noticed Barney O'Mullen's confusion at the sight of the old prelate, how the veterans nodded at him as they staggered, and Christian frowned at his top hat, even though they had already entered the cemetery.

"Who is this holy father?" Vladimirov asked.

"Our former regimental chaplain," Fenton whispered. – "At one time he was an officer, and now an important person in the local diocese."

"Conant!" Vladimirov recalled.

He uttered the name so loudly that the gray-haired prelate shuddered, looking around to see who called him, but found no one and signed the procession with the sign of the cross. Vladimirov stopped, he wanted to keep in his memory an unusual figure - Conant left, accompanied by the pastor and disappeared behind the gate. Vladimirov joined the procession in an open meadow framed by broadleaved sycamore trees. Here the soldiers of the civil war rested, lying wide, separated by cut grass, low mounds, white marble slabs with carved names and dates. The love of relatives and the mason's chisel did not practice eloquence here, and purses did not argue with wealth; the soldiers' cemetery, open under the sky, spoke for itself.

At the edge of the clearing, at the border of light and shadow of a sycamore tree, Vladimirov saw two women in mourning clothes, next to the earth thrown out of the grave. Everything flowed past his attention: the motorcade slowing down, the coffin again served for the general, the quiet command, half a platoon of armed soldiers, the movement of the crowd towards the grave - everything flowed without touching him, he saw only Nadine. She stood, leaning her hands on a stick, straight, with her head up, with her face turned to the coffin. The second woman, stout and beautiful, also gray-haired, in an expensive dress, in a mourning cape with sparkling glass beads, put a handkerchief to her eyes, was in sorrow and grief. For Nadine, there was only a coffin, not even a coffin - a random wooden bed - but the one who slowly floated towards her alone, without anyone else's help, simply floated, completely tired, without complaint, hushed, approached her, only her, because all life is their union, only he and she, she and he. And in the first minute and after, recalling the Mound City cemetery, Vladimirov did not see the old woman under the scarf lowered to her eyebrows; the soft line of a raised chin, a parted mouth, eyes in intense expectation, a slightly cut nose, and not lowered, as usual with old women, immobility itself, and in immobility - a stretched string. The crowd rushed to her, the musicians began to play, the soldiers grabbed their guns, ready to fire blank volleys. The open coffin stood on the ground, Nadine's gaze rested on Turchin's face; concerned whether all is well now that her friend has allowed himself to lie on his back forever; the look is almost calm in its intensity, far from tears, from the manifestations of grief familiar to people. She knelt, supported by the arms of Horace Fergus and Elizabeth Howard, said something to Turchin, stroked him with her hand, kissed his forehead and said something to him again.

Vladimirov turned away, heard Virgie's cry, the rattle of ropes, the rustle of the earth and the loud voice of Christian: "He gave us brotherhood, the highest privilege of man!" The first clods of earth hit the coffin, then more and more muffled, softer: birds flew into the sky from rifle shots. Vladimirov no longer looked for Nadine, he was afraid that now he would find an old woman in her, just an old woman, and did not want this; the story of their life has come to the cemetery in Mound City, and from here it has no way.

INTER-CHAPTER TENTH

From a letter from Nikolai Vladimirov to his father.
London, July 1901

"I am sending you a medical record made when Turchin was admitted to the Anna shelter. I write only answers, where the question is clear by itself, and I highlight these answers - his own, Nadine's or

Tadeusz Drum's - with a light line. *Born 78 years ago. Father is Russian. Mother is Russian. Married. Engaged in journalistic work. Higher education. Never belonged to any church. No one knows anything about his family or relatives. Sanguine temperament. Didn't use opium, didn't drink. Previously smoked, recently smoked less. As far as is known, he did not suffer from head-*aches, *was not seriously ill. Previous seizures? - Did not have. Causes of the crisis? "Apparently, mainly due to general decrepitude and too hard work on books ... How did the current fit start? "A change was noticed in him about three weeks ago. He said that he suffered a lot, then suddenly grabbed his books and papers and lit a fire. I tried to burn them in the oven too...* What mea-sures should be taken? *"Measures should only be taken to keep him from destroying his own papers..."*

In this description, I did not find anything new for myself, but there was also a verdict on the work I started: Turchin is gone, but I have no way to Radom, to the old fire pit, to the burnt sheets. After Turchin's funeral, I did not meet his widow. Even before Mound City, I realized that looking for her was empty. With this weight on our hearts, we arrived at the port of New York; we were seen off by Virjie's relatives and some of the Turchin circle. And now Johnston caught up with us on the gangway, pulled me and Virgie to him, doused the smell of rum and took a package out of the briefcase. "Here is Madame's letter," he said. "It's a pity I couldn't see with at least one eye." "Did you just receive it?" I asked. "I have had it for over a week; but I was asked to give it to you at the very departure ...

"I rushed up the gangplank, stood next to Virgie; ungrateful, I waited impatiently for the steamer to sail and for me to be able to escape to my cabin. Soon you too will read Turchina's notebooks; what ruthlessness to everything, not excluding himself; what perseverance in the pursuit of a life goal, what pride, combined with a willingness to accept need and any blow of fate.

Chapter Thirty Seven

1

"I know where he and you stopped: in the newly begun Radom, in front of the church, there the road went up the hill, to hope, - then it lay broken, in pits and potholes. I will tell you the uncomfortable truth: you have become for him both hope and execution. How many times did he take up the pen, urging him to frankness, and what? - he printed other people's names, and tore paper, and broke when he tried to write his own. He needed a listener, a listener from Russia, a custodian of papers that tyrannized the brain, papers that he cherished and which he hated with an insane, destructive hatred. They imprinted his life, but they were not life itself, and the thirst to express this life and the inability to express it in words became his curse. Did it seem to you that, in an old-fashioned way, slowly, he wanders along the road of the past, where the dust has already settled, the corpses have been removed, the fallen bridges have been rebuilt, and the lie has disappeared into the air? Mistake: I found him broken after you and hated you for the hasty coldness of the listener.

He trusted the paper and your events, though, but not the soul. Is it possible? Are events possible without a soul? Is not thought its continuation, and even the highest expression of the soul? Perhaps, because what we call the soul is limitless. And often the one who gives her to people fearlessly, in the armour of thought, does not know how to quietly take her by the hand and lead her, undressed, suffering, dear confession. I thought about this: why do some novelists write themselves and their lives, while others run into someone else's and, looking around in shame, cross out every careless line of confession? Are the former proud of themselves and their lives, are the latter ashamed of it?

The answer is not here. She is in a special arrangement of soul and talent. He had nothing to hide and nothing to be ashamed of, but the thought that the world is important, and not he, someone's life is real, and not his, that no one owes him anything, but he owes everyone, this thought, which has become a feeling, and by instinct, kept the confession of the heart shut. And how you would have suffered if he had lived yet! Blue eyes, which have been looking out for the destitute all their lives in order to share bread and freedom with them, would become icy and hostile for you. Here is the right word - enmity, it would certainly settle between you. You demanded a story, but what could he tell? About how the millstones of memory turn? How does the black and light water of the past fall, causing them to flicker with the speed of thought, then gnash, grinding bones, slowing down to immobility? About flight into the sounds of the violin, which begging will not humiliate? About how disobedient in old age even birch logs are to an axe? About the fingers of an old woman, twisted after milking a cow in a winter barn? About the fingers that he warmed with his breath, kissed, hid in a warm, white beard ...

He wouldn't talk about it. He couldn't, but I can. I am accustomed to trust everything on paper, I make no exceptions either for the stench or for the bashful movements of the heart. It is true, he told you about my writings. He rushed about with them, tried to print, translated into Russian, hoped that the hour would come and in Russia they would race to print Nadezhda Lvova. But in matters of the spirit, victories happen even less frequently than on the battlefield... My time is wasted, the age rushed on, bringing mud and sand to everything that genius did not exalt above the chance of time. I tried to destroy my manuscripts. Ivan Vasilyevich saved them - why? What is their fate? One thing I know for sure: when he decided to burn the papers, he was guided not by madness, but by the order of will and mind. Life is done, and everything that was important in it expressed itself in actions. What is the use of papers, of their slow smouldering in other people's basements, by broken attic windows? What has not stood before life, then leaves.

He took his first steps in Radom without me. Turchin defended passion; even a handful of people in his eyes received the physiognomy of humanity; he commanded a regiment, a brigade, but had the fate of a company been assigned to him, he would have found in it the image of humanity. Such people keep the world, they are both company commanders and presidents. And I did not believe in the ideal Radom, at that time I lived differently. We had a women's war. Its own war of the North and South, the republic against slavery. In our army, the champions of women's freedoms converged; we demanded the right to vote for women. We were ridiculed not only by men, the black banner of voluntary slavery was also raised by the female army, led by the wives of General Sherman and Admiral Dahlgren. They frightened the townsfolk with the extremes of frilavisma (Movement for Free Love) [28], seduced them with the joys of home paradise, the blind happiness of motherhood, the sacred duty of a cook. It was then that Victoria Woodhole appeared in the army of women's freedom: she combined the blindness of spiritualism with the fanaticism of the American Joan of Arc. She believed that there is no final death for anything living, and beyond the threshold of bodily death there is the beginning of a new existence, free from the rough earthly shell, the life of the spirit, surpassing earthly life. In the early seventies, her patron spirit ordered her to go to Washington, DC, to the halls of Congress, with a special petition. There is no need to ask Congress for the right to vote for women, Woodhole said, because women already have this right, by the letter and meaning of the 14th amendment to the constitution. Woodhole became the victim of merciless slanders; she kept herself apart, in the dark vestments of spiritualism, in the sick pride of being chosen, but her idea seemed excellent. The Fourteenth Amendment stated that all persons - in the English text, of course, persons - born or naturalized in the United States, are citizens of the United States and the state in which they live, and no state has the right to make laws that reduce the benefits and privileges of citizens of the country. Are women not persons, and is not the right to vote the main privilege of citizens?

We achieved nothing in the inglorious war. I diverted you to explain why I went to Washington, Philadelphia or New York more often than to Radom.

And Radom lived. In burned-out Chicago, death roamed the streets and entered the shacks without knocking. The poor man despaired, and in Radom a man was waiting for him with bread in his hands and with an armful of free firewood. He was kind and generous and saw in everyone the possibility of a socialist. Was he right or was he deceived? An answer is needed. An answer was needed even hundreds of years ago, without an answer no one will take a step forward on earth even when the last monarchies fall. We both needed an answer, like the ground under our feet, like the main stone in the foundation of moral faith. And we always answered in full agreement: man is born equal and free, he is born for a just life. It is not the new-born's who are to blame for the enmity and inequality. If I had eased my earthly fate with the naive faith of Victoria Woodhole and appeared on earth in the form of a spirit in a hundred or two hundred years, even then I would not have changed our faith, even if I found the future earth in blood and pus, just as dirty and betrayed. It means that the work is not all over yet, I would say, sadly, and people are still wandering in the dark. Woe to the blind, but there is only one way to make them sighted: by healing the eyes. It is impossible to cut through their eye sockets in the forehead or in the back of the head, or cut them out in the chest, eyes will not grow there, they are once given to a person. Share fairly! is an eternal thought. Work hard and share fairly. Share the bread and wine, the work and the burden of the sword invoked by freedom. And freedom - especially hers! - divide fairly, reject the possibility of a slave. Which of the prophets of mankind wrote something different on their tablets? Even the one who came to strangle freedom, take away bread, turn work into humiliation, on his banner, though falsely, will write the same words: divide fairly! Born for equality, man forgot his taste, recognized the inequality of property, blood and skin colour. The possibility of a socialist has been stifled in him by greed, privilege, satiated servility, the lies of the priests, the humiliation of the spirit and the threats of the executioner—in some this possibility has already died, but in humanity it is alive, alive, alive, and that is the only way out, and there is no other way out.

Ivan Vasilyevich chose Radom as a playground for the struggle for justice. He was looking for people accustomed to work for the colony, who do not look at other people's necks with the secret thought of putting a yoke on them, throwing on the usurer's lasso, tightening the saddle so as to admire the beauties of the earth in the saddle. Two hostile forces gave Turchin the illusion of luck: the poverty of the colonists and the general's money! The poverty of the settlers kept them in outward equality, and he consoled himself with the thought that equality would become their essence. In one thing he was firm: his money did not give anyone a personal advantage, went to the general needs and fed everyone equally. It is to Turchin's money that Radom owes the success of the first two years; they lay down in the ground like manure in a ploughed field. Our money and the rest of Tadeusz Drum's savings, divided into a few, brought everyone an extra piece of bread, a saw, an axe, the opportunity to get a draft pair for a day or two. Need and cold brought people together even more closely; in the barracks there was also a common table for eight families, and in turn cooking. Radom did not know apostates. And only the priests did not hold on to the new parish. What drove them away? Turchin's godlessness? Financial need at the church, forgotten by the parishioners and the church hierarchy in Ashley? It was not Turchin's godlessness that was to blame, but the fact that he took the place of a shepherd in the hearts of the parishioners. The christening was far away, the people of Radom were in no hurry to die; and before the first confirmation one had to wait two years; so the priests spent their days and weeks.

Less crowded masses did not bring joy, the lessons of eloquence were wasted, the parishioners, exhausted by work, dozed off. The good fatalist priest Josef Muslevich disappeared from Radom. Priest Szulak, a Jesuit, came to Radom from Chicago with a self-made song "God Love

Poland", printed in large numbers on cardboard cards. Shulak called Turchin to war against the church, looked for the martyr's crown of thorns, Turchin did not respond, and Shulak, cursing us, returned to Chicago. Priest Kandid Kozlovsky, a former Capuchin and also a rebel of 1863, arrived in December 1874 from Cincinnati without thinking about leaving, but in the early spring of 1875, when the Radomites worked on the land from dark to dark, he read in the presence of Jan Kozelek, Mindak's tavern keeper and a bunch of old women preaching a heretical sermon about "the surest way to put an end to the church parish in Radom forever," and, at the risk of incurring the bishop's disgrace, left Radom on a dead March night. He was a glorious, lost man with tired, inquiring eyes; he was looking not for a field, but for a quiet haven for himself and did not find it in Radom.

3

It was then that the patience of the diocese ended, a bird of another flight flew to us. We saw a handsome gentleman, with pursed lips, an ambitious man, and next to him was the Madonna, his old housekeeper, Matilda Strizhevskaya. Here is his restless life in the blaze of national feeling, in the thirst for power and money. Theodor X. Giryk was born near the town of Marienwerden in West Prussia, early chose the path of serving God and joined the Prussian army as the youngest chaplain of the Austrian war of 1866. All his life two feelings fought in him, or rather, one feeling, but directed at two different subjects: love for the Poles and love for the Germans. Having loved the Poles, he, like a bad disease, hid his attraction to the Germans. Having given himself over to the Germans, he forbade himself to speak of the Poles except with pity and contemptuous pain. He came to the States as a German emigrant, Matilda Strizhevskaya, the wife of a Pole corporal, was born German; she preferred the sinful life of a housekeeper with Pan Theodore to the permitted pleasures with a corporal. But this is a feat: corporal Matilda could give birth to a bunch of children; devoting herself to a priest, she did not dare to think about motherhood.

Pan Theodore soon discovered how insignificant his career was among the millions of American Germans. They already had an extensive press and their prophets - Catholic, Protestant, revolutionary, from among the knights of 1848 and 1849. And one day, having fallen asleep as faithful German emigrants, the priest and his housekeeper woke up as Polish patriots. His patriotism turned out to be earnest and ardent, and Matilda shared obscurity with the priest, moved from parish to parish in the states of Illinois and Wisconsin.

Pan Theodor neglected millions of Germans for the sake of thousands of homeless Poles and did not lose. The first Polish newspapers, scattered churches, a hungry emigrant - everything was looking for a supreme leader. Soon the unknown priest changed his darned cassock for a solid church frock coat, and the provincial parish - for the church of St. Wojciech in Detroit, one of the main centres of Polish emigration. Even the distant homeland heard the voice of Pan Theodore. He announced in the newspapers that he saw the need to unite the Poles of all America, and who else, except for the church, could gather thousands of scattered, homeless, needy, and even desperate people scattered around the country? Together with priest Vincent Barczyński, he created the Association of Polish Roman Catholic Churches; but with resentment he discovered that this feat did not change his place in the church hierarchy. Then he turned his eyes to worldly affairs. He incinerated with an angry word, immigration agents, county and state authorities who settled gullible Poles in the wilds, extorted their last penny and left them to die in the wilds. He preached that the Polish emigration should come under the authority of an agency consecrated by a government patent; that every Pole, already on a ship deck in the Atlantic, should know in which county he will settle upon arrival in the States - having received his future place in a special depot of Pan Theodore's emigration agency, also created in Poland. The priest's plan was so good that the unfortunate emigrant would have fallen into a new slavery, having no freedom of choice, trusting the commission agents of Pan Theodore; from the lack of freedom of the monarchy, he would fall into a new lack of freedom and dependence. On the body of the republic, a district rash of the voluntary Polish Pale of Settlement would have poured out, where the church and its prophet, Pan Theodore, would reign.

But the authorities of the republic were adamant. Having taken care of the immigrants, Pan Theodore would have taken their money; Polish emigration was growing, the agents and commissioners of the Castle Garden had already adapted to fish out worn credit cards and coins of European minting from Polish frock coats and caftans - there was no reason to give this money to the priest.

Pan Theodore did not calculate his strength; the church gospel turned his head, the Catholic leaflets inspired pride. The ambitious probst went to Washington, besieged the White House and the Capitol, and returning to Detroit, found his place in the church of St. Wojciech. He was allowed to choose a new parish himself, and, to the surprise of everyone, Pan Theodore chose Radom with the village church of St. Michael the Archangel. Radom is his last redoubt, he has nowhere to go below, the enemies will not let him go above.

4

I told about Giryk so that you would know what enemy was revealed to us in the spring of 1875. In his arrival in southern Illinois there was Radom's confession that a mere backwater would not have attracted a glorious probst. He was looking for a strong community, a new land and a future. And the very name - Radom - originally Polish - attracted the priest; why not be a diocese of Radom? Ashley or Belleville are not Polish names.

The number of settlers reached four hundred. Farmers took cheap grain south to St. Louis and Cairo; winter crops came out from under the snow strong, thick, and the spring grain lay in good soil. The colonists got money, and with it the houses grew easier - other settlers, following the farm, lined up in Radom. Turchin's faith came true: the surrounding Irish, Yankees and Germans, at first Saturday guests of the Radom saloon with music, acquired plots and set about building.

In the evenings, without lighting a fire, we listened to the multilingual Radom at the window, and Turchin was glad that his paradise was completely successful. But the break was drawing near. It walked towards us on the short legs of a priest holding a strong torso.

Church of St. Michael the Archangel did not become the desert of Giryk. He often appeared on farms: on the neighbours - on foot, on the distant ones - in the saddle familiar from his youth. He asked about the economy, looked at papers, bills of sale, put aside the affairs of the Lord for the sake of worldly concerns. Only Tadeusz did not like Pan Theodore: his patent-leather boots, his hand clasped behind his back, the readiness of gray vain eyes to move from extreme concern with his interlocutor to cruel coldness. Turchin's lion's head, in a thick beard and graying hair over an immense forehead, Drum found normal, but the broad and intelligent face of a priest with a large forehead, his wavy chestnut hair covering his ears, his round, soft chin and tender cheeks irritated Drum. Perhaps our straightforward friend was oppressed by the role of Matilda Strizhevskaya under the priest: the noble Tadeusz professed his religion, every woman was originally holy to him and deserved a better fate.

One day the priest knocked on our door as soon as Tadeusz entered the house. Apologizing for the uninvited arrival, he said that he was glad to see Pan Teacher, he had business with Pan Tadeusz and Pan General, Radom's patron. It's time to build a school. Every good parishioner, even an illiterate peasant, who is taught to read and write above, will glorify the name and cause of the Lord in the Church. The diocese will give a certain amount for the school, and if you lay a foundation for a theological seminary, the diocese will release a lot of money ...

"There will be a secular school in Radom," Drum interrupted him. "Turchin donated forty acres for a school, the timber will arrive soon."

Turchin intervened in the conversation so that the hand of the giver would not hide behind his back.

"If the diocese gives at least a quarter for the school, we will cover the roof with iron. Please, sit down.

Giryk sat down without looking around, as if saddened.

"How can Radom think of a seminary without even having a spittle?" he said anxiously. "The homeless priests are leaving: and the servant of the Lord needs a roof over his head."

"Weak people left," Turchin confirmed. "The parishioners stayed, but the shepherds left."

"I will not leave." He drew a line between weak-faith pastors and himself. "Your barracks are empty, General."

Radom grew up, the innkeeper Mindak opened a house for visitors with cheap rooms and a boarding house.

"Take it!" exclaimed Turchin, pleased that the barracks would find a use. – "The diocese can afford it, I'll give it cheap."

He already saw the unfinished two-story school building ready, the large windows of the classrooms, the dawn of enlightenment in them, and he did not think that his former hostel of justice would turn into a pastorate, into the monastery of the church brotherhood of St. Michael the archangel.

"The house should be moved closer to the temple," said the priest.

And this suited Turchin: the station grew, a second warehouse and a fuel depot were built right next to the barracks, a high fence rose at its windows.

"That's how we did a nice job, like real merchants!"

With a lively gleam in his eyes, Turchin also called us to set aside tension, to give the guest courtesy.

"And your bearing is military, cavalry!" - Turchin cheered up when Pan Theodore got up, preparing to leave. "I knew the regimental chaplain, he sat in the saddle no worse than you."

"And it seems that they did not favour him? Especially Pan Tadeusz."

What is more cunning: to hide your secret knowledge of the interlocutor or to hint that you know everything?

"He was a brave priest: he changed his cassock for an officer's uniform."

"Our clothing often requires more courage than a uniform. And the chaplain, did he die in battle?

It seemed to me that the priest knew the answer."

"He returned to the church. He is in high ranks."

"You have done a lot for the local community," said the priest kindly. "For this you will be rewarded even in your lamentable unbelief."

"My faith is different: there is no place for God in it."

"God is omnipresent, and it is not for us to appoint a place for him."

Short-lived priests from Radom used to come to us in the hope of introducing the retired general to the Roman Catholic Church. They soon descended into good-natured grumblings or playful reproaches. Only the Jesuit Father Shulak left Radom convinced that we belonged to some dangerous Orthodox sect or a secret brotherhood plotting against the Holy See.

"There is a sure way to take root in Radom," Turchin said. "I advised the former priests: buy land."

"Our entire class is poor, and especially in a foreign land."

"The local conditions are comfortable; I will give you special benefits."

"Warm weeping, honest parishioners—this is my reward and all the benefits," the priest evaded.

On that they parted: this massive, broad-chested man on short legs was graceful in his own way, easy in evasiveness, in retreat, not an inch further than what was required.

5

Good luck settled in Radom for a long time. The summer was hot, but in the middle of the night rain often fell on arable land and meadows, falling with a quiet rustle, as if delayed in the sky by a caring hand so as not to disturb the short sleep of the workers. In the morning, the people of Radom, seeing the abundantly watered land, washed leaves, warm puddles on the road, and the growing ears rejoiced, and people of devout faith willingly agreed that they owed good luck and a new probst. Jan Kozelek was especially zealous; the former priests did not have an aura of learning in their eyes - with Kandid Kozlovsky, the one-eyed shoemaker even embarked on impudent disputes. Pan Theodore became Kozelek's idol; after a rivalry with several priests, he felt the need for humility. In addition, before the construction of the plebanium, the priest rented Kozelek's house; in May, the shoemaker moved his mattress and tools to the summer shed.

Turchin lived for a long time in Radom, and Michalsky lived in Chicago. The new settlers sent by him were all Poles. Without leaving Radom for a long time, one might have thought that all of Illinois was inhabited by Poles alone and that there was no one else to wait for in the colony, except for Catholic immigrants. From time to time an Irishman or Yankee also came - friends of Turchin or Drama, regimental associates, whom they managed to persuade to move by letters and promises of benefits.

One evening in early July, we met a family of new settlers from Mattoon at the station. Teddy Dawson came to Radom, conveyed greetings from Thomas's mother, took letters from us to her and a few banknotes. Now she and the family of her brother, a bankrupt farmer, have come to us. We loaded our handcart with our belongings and walked in the direction of our house, asking the Widow Morgan about Mattoon, grieving that the house where Thomas had grown up and where Lincoln had once been lodged had been bought by Illinois Central for demolition. The widow's face was still in tears when another cart appeared to meet us, with Kozelek and Jan Kowalski, and the priest and Matilda walked side by side. Plebania stood ready, with a high cross above the entrance. Our squat, warm house has changed: it looked sternly, with outward holiness and cold mountain peaks.

Pan Theodore greeted the new settlers in Polish and met with bewilderment. I explained who they were, the priest inquired in English about their faith and heard that they were all Protestants. He wished them God's help and went on his way, not waiting for Kowalski and Kozelek to push the overloaded cart.

The next morning, a loud conversation broke out between Turchin and the priest.

"Why are you interfering with our cause, Pan General?"

"What case are you talking about?"

"Give us, even in a foreign land, to converge in your brotherhood, even here do not oppress us.

... The priest spoke pitifully, with a request, like a conquered Jew with a Roman procurator, like a petty gentry with a governor of a white king; in his humiliation, like a poisoned blade in a sheath, there was an insulting impertinence.

"I knew many Poles," Turchin replied. "They loved their homeland and died in the Carpathians for the freedom of the Magyars."

"But Pan General served Paskevich!" the priest interrupted him.

"It was not General Turchin who served Paskevich, but the young lieutenant Turchaninov," I intervened. "This man atoned for his mistake with his whole life, but how did you atone for your sins, Prussian chaplain?"

The priest politely bowed to me: "It's good when people meet who know everything about each other!"

"So it wasn't a Polish woman who came to Radom with you," I didn't let him dissuade him with a trifle, "but a charming German woman ... and is this a sin?"

"Pani Matilda is a servant and a good Catholic. Blacks are also hired as servants, Pani Nadine."

"Don't you dare call her Nadine, you... church dude!" Turchin was furious. "She is

Mrs. Turchina, Nadezhda Alekseevna Turchina, that's who she is for you, and not Nadine."

"Excuse me," the priest bowed humbly. — "Excuse me, ladies, we Poles are not used to patronymics."

"I'm not only a servant, but I'd rather take a black one as a brother than you!" Turchin roared.

"I will eat from the same bowl with him!"

"Why is the Polish blood so guilty, Pan General?"

"What a Pole you are! You are a tiny ecclesiastical Napoleon, ambitious, hungry for a career. Our former priests were nice people... but you have more of the devil."

A strange smile descended from the eyes to the chin on the gentle face of the priest: as if he knew in advance what Turchin would say.

"Godless lips cannot offend me." I noticed how his muscles trembled, straining the black cloth, and I felt hatred and thirst for violence in him.

"Who promised you that from Radom you would be allowed to make a county where you would be a police officer, and a priest, an inmate and a marshal of the nobility!" Tadeusz Drum alone has won over more hearts to Poland than a hundred seekers like you.

"Tadeusz Drum is a bad Pole."

"He is a true Pole!" exclaimed Turchin. "With a subtle soul, noble, brave, bitter conscience of the century! Is it possible to be more Polish than Tadeusz!"

"Born a Pole - a Pole. But, having lost the faith of his fathers, he is already half Pole."

"And Dudzik? What about Vojtech Malinowski? What about Jan Kowalski or the Gajewski brothers? They want to live in peace with neighbours of any blood."

"Protestant Yankees set fire to God's house," the priest recalled.

"It was set on fire by blind people like you, fanatics not of God, but of the church order!" A few years will pass, and Polish girls will enter the houses of Germans or Irish as wives, and young men will find brides there, a Catholic will get along with a Protestant, and what do you want?

"To protect the Poles. Keep them pure in faith and nationality."

"Why did you choose the worst place on earth for this!" Turchin was angrily perplexed.

"A country where a nation is being formed, absorbing not only thousands of Poles, but millions of other Europeans. You, like a blind mole, are digging under the old earth, not knowing that there is another life above... - And he turned to me with a bitter smile: - These are truly poor fellows who will believe the prophets who have lost their minds!"

When Turchin spoke about what secretly frightened even the priest himself, Pan Theodore lost his self-control. He shouted to Turchin that he saw self-interest and calculation behind his every act, that Turchin deceived the colonists when selling land, drew up bills of sale and land surveys in favour of the railway company to the detriment of farmers, that he was a secret agent of the Russian ambassador and the Antichrist, and his whole goal was to corrupt Poles with the guise of goodness, desecrate the faith and holy Polish blood, mix it with alien, wild and strong blood. The priest's mouth twitched, a veil fell over his eyes, they went out, as before an epileptic seizure, he groped in the air with his hands, as if looking for support - the disease opened up in him.

I rushed to the kitchen, scooped up water, Turchin turned to the window so as not to humiliate the priest with a look when he came to his senses. Pan Theodore did not take a cup of water, gray eyes with an enlarged pupil looked irreconcilable, he hated Turchin and our habit of having not only a name, but also a patronymic, as if this was not a common habit for our people, but a privilege of the Russian nobility. He came out of the fit so quickly that I doubted if he was sick? — but the priest had been ill for a long time and seriously. Then we learned what a salvation Matilda Strizhevskaya was for him, his beloved woman and nurse during sleepless nights, when Pan Theodore rushed between light and darkness, between Prussia and Poland, comprehending the truth with an extraordinary mind and being afraid to follow it.

6

Turchin looked miserable. He was not afraid of the priest, but bitterness penetrated him from a presentiment of what kind of war he would have to wage, how low and dirty its weapons would be. We have seen the possibility of a brotherhood of man; let it be flashes, illumination, brevity of a spark, as in the experience among retorts and flasks. These moments are accessible to the eye of the observer, but not every heart rejoices completely. And not everyone can see their connection with the life of mankind - here my vision was inferior to Turchin. He was not the destroyer of a nation, of any nation, neither in thought nor in deed; I don't want even the shadow of such a suspicion to cross your mind. But he did not find on earth such blood that would deserve to rise above other blood, and only in this, the original equality of people, he saw almost the divine justice of nature. He was the enemy of everyone who, like Pan Theodore, brought blood to the marketplace and called out prices that were favourable for some, and humiliating for others, and even equal to the sentence of death. His look of equality opened the way for friendly contact, made the earth a safe home, - every other promised war. That is why he, a Russian, a Russian in everything—in love, in anger, in battle, in friendship, in generosity, in stubbornness— not only called himself a citizen of the American Republic, but was one, and had, like few, the right to speak of her: our republic. He yearned for Russia. In the middle of winter, when short-lived Radom blizzards were singing, he sometimes rushed out of bed to the window, hoping to see snowdrifts, Russian....

I'm talking about friends. But Turchin also found enemies in the republic. Host of enemies! Turchin never said this; but I believe that he went to America and for the enemies. Strange, would Turchin lack them in Russia - and me along with him - if he challenged them? However, it is neither strange nor funny, if we remember that natures, passionate to the point of recklessness, can only handle an open battle, and in pretense they are not worth a penny.

Now a new enemy has appeared before us, in a priest's frock coat, but with the same badge of a black rebellion. What could we fear? Church curse? The intrigues of the diocese in Ashley? But the diocese did not touch Turchin when the priests fled from Radom, and Pan Theodore settled here, received from Turchin a house for weeping and free land, ordered St. Rosary, nimble sister Tutu - what will the diocese charge us for? And the church is not so strong in the States to decide the fate of strangers.

We expected empty volleys from the church and weeping over logs that had now become the enemy's fortress. How naive we were!

The priest did not touch Turchin in the church, being an image of tolerance and peacefulness. Pan Theodore knew how to uproot oaks; he set to work from the roots, cut them down, tore them with tunnels, and even gnawed with his teeth, sparing no incisors.

The company called Turchin to Chicago. In the presence of Michalski, Hansom advised the general not to interfere with the ethnic homogeneity of the colony and began to gently reproach the general that he, a thoughtless philanthropist, settled the Mattoon poor in Radom, giving them money to buy land, and there was talk of usurious interest in Illinois Central ("people are envious , Turchin, they do not believe in noble motives..."). He knows the general's generosity and worries whether Turchin will suffer bankruptcy and poverty - after all, the general has already made a loan in Chicago. "Bankruptcies befall money people, Mr. Hansom," Turchin said. "I may be in need, I am ready for this." "But you took the money against the security of the house and land in Radom." "Isn't this land my property?" "You are not a simple farmer, you are an agent of the company, nothing should be done behind her back. You seek special benefits for those you know personally, this also casts a shadow: is there a hidden benefit behind your patronage? - "The people of Mattoon are poor, on the verge of despair..." - "I remember this family," said Hansom disgustedly. "We must avoid the beggars, Turchin!" - "I think you have well fenced off from them, Hansom, but I prefer to remake the beggars into happy citizens of the republic." "They're also Protestants." "And you are a Protestant; this does not prevent you from being one of the directors of a company where there are many Catholics. "There's my church in town for me, but how should they go to Ashley or Nashville? Let Protestants settle with Protestants, and Catholics with Catholics..."

At the end of September, on the day of St. Michael the Archangel, a new railway agent arrived in Radom. Until now, this position was filled by Turchin, it provided the funds we needed - Turchin now rarely managed to sell the land, and Michalsky received financial independence. And before Christmas, a notary settled in Radom; previously he managed Josef Krefta's insurance office. The notary had not yet shown up, and a letter had already arrived from Illinois Central that no bill of sale would be valid without a notary's seal, and during the winter of 1875-1876 all old bills of sale must be confirmed by him. He showed up at our place immediately upon arrival: a young cheeky fat man, efficient and intelligent, with a good-natured chuckle and small palms, which he zealously, in a strange agitation, rubbed one against the other. He made a claim to friendship and frankness, but he did not receive friendship and became unbearable.

The lights in the windows of the parsonage went out early and the day began like a peasant early there, - Sister Tuta was the first to fly out. From her, unkind rumours spread through Radom, they harmed us in the eyes of the local inhabitants. Upon learning that Turchin had left, one day Christina, the wife of Jan Kovalsky, came to me with her son at her breast. She admitted with tears that they speak badly of us, but she does not believe that the general and I are not husband and wife, and that God punished us with orphanhood for our sins. "Don't cry," I reassured her, "for a quarter of a century we have loved each other, this is almost your entire life on earth, and for twenty years we have been husband and wife. We don't recognize the church, it won't even get us dead." Kovalskaya did not understand: the wedding is so glorious, there was no better minute in the life of each of them - angels hover over their heads, temple silver and gold burn, and the faces around are cheerful & kind - what's wrong with that? "Mr. Nadin, the wedding is good: I take you, Christina, you will be my married wife, in joy and sorrow, in poverty and wealth, in health and illness, in youth and old age ..." - "We live like that, Christina. We were connected by a feeling. Are you and Jan holding one wedding? - "We live well, in harmony ..." - she was embarrassed. "And in love! Don't be afraid of this word. Without love we would be wild beasts; but what about animals, because they love in their own way, and human love drives everything, and the future is in it, in your children born in love. Kovalskaya looked scared, and I came to her aid: "I know what you want to ask. Nature did not give us children, not God. I got married late: at the age of thirty. And then we left Russia, wandered without a home, then four years in the saddle, and such terrible deaths that if a thousandth of them are reflected in the foetus, then even then his life would be unhappy. And so motherhood is gone. - Still seeing bewilderment on her face, I continued: - Here is Pan Theodore, a young, strong man, but he will end his life alone, without a wife, without children. You will say that God called him to a feat, this is his reward. But the world is big, there are not only church concerns in it. We try to give everything to people, and take small things for ourselves, without which we cannot live. - I wanted to bare my soul; I should have once said that to another woman. "I was exhausted by unborn children ... What a suffering it is when the body and soul hurt, when everything is thirsty, demands a birth, wants to hear his cry ... "I don't know who was more unhappy at that moment: me, a belated, already obsolete pain, or kind Christina Kovalskaya. As if spellbound, she approached me, held out her son and immediately took it away, pressed it to her chest and held it out again uncertainly, fearing to offend me and fearing to give her son into sinful hands. So she touched me and I hugged them both. "But you said: "married," she was looking for a way out and hope for me. "So you are husband and wife before the law?" I couldn't explain everything to her. How many enlightened people on both sides of the Atlantic grimace at the words "civil marriage", as if a stench hit their noses; what to demand from a young mother, accustomed from childhood to the indisputability of the church and the priest.

8

With the arrival of notary Joshua Ford, I did not leave Turchin alone: he too easily switched from laughter to clenched fists. The most insignificant mistakes were revealed in the old merchants; without exposing the forest land, one cannot fully see its space, small ravines, mounds hidden in the hazel, the borders of two possessions. The notary put the suspect papers aside; on the left - mistakes, allegedly at a loss to Illinois

Central, on the right - to the detriment of farmers. So that you know what a miserable game Joshua Ford played, I will say that Turchin was summoned in three cases as a defendant in the world court - there were three cold & wicked! - and what were they looking for from a man who gave the colonists thousands? Here are the shameful numbers: 5 dollars 54 cents, 3 dollars 48 cents, and ashamed to say about the third, how the justice of the county agreed to summon its priests for the sake of these 75 cents! Even if we wanted to pay off our claims, not allowing us to go to trial, Tadeusz Drum would not allow us to do this; followed by trials, verdicts in favour of the plaintiffs, Turchin's appeals to the District Court of Washington County, and our three new defeats, and writ of execution in the hands of the offenders.

Leaving the court, Tadeusz Drum - he represented Turchin in court - invited the winning plaintiffs to his forest for a full settlement, promising money and some other gifts, and if they turned up to the general, he added, then there they would be met by the gun of Drum himself : for such a pleasure, he is ready for hard labour. The plaintiffs got cold feet, and Joshua Ford came to us; in his briefcase were warrants of execution bought from farmers, a paltry lawsuit by Illinois Central against its land agent for the errors of old merchants, and, alas, Turchin's Chicago bill of which Hansom spoke. Josef Krefta handed over the bill to the notary: the moneylender bought both the bill and the mortgage from his Chicago counterpart. The bill expired in April, the $300 borrowed threatened us with ruin: the loss of our house and land.

The winter dragged on, the snow fell sparingly, not covering the ground; frozen, it loudly responded to the step of the colonists and the hooves of the horses. In the midst of the cruel clarity of this winter, under the winds that roamed more freely around Radom, under the crimson sun, no matter how you turned, poisoned arrows fell into Turchin. Conscientious people, not knowing how to intercede for us, began to walk past our house less often. Friends came: Jan Kowalski with his wife, both carpenters, the Gaevski brothers, Jan Nowak, Mikhail Witta. And they, opening the gate, going up to our porch, looked around at the windows of the weeping. It happened that Turchin was stopped on the street or in the forest by a timid farmer; taking off his hat or cap, losing his tone, he would show his papers, wondering if there was a mistake in them, pointing to some places circled with a notary's pen. The Catholic Gazette hit me in the back; a quarter of a century has passed since that day, but we have not pulled out an arrow from a living body, the poisonous tip has remained at the very heart. The newspaper sympathetically wrote that *"General Turchin, who did so much for the Polish colony, is now in financial difficulty, but there is nothing shameful in the general's bankruptcy, no gesheft (work) or impurity, but only the result of the indiscretion and generosity of Pan Turchin bordering on extravagance."* The newspaper even called to help Turchin, warning, however, that the fate of the Polish settlement required the cold-blooded intervention of wealthy people. *"Pan Turchin will always find bread and shelter in Radom,"* the newspaper concluded.

Michalsky is his own man in the Catholic Gazette, but Turchin, after a long, furious silence, said that it was not him, Michalsky could not fall like that. Indeed, a day later, Michalsky, bypassing the weeping, rushed to us so that we could see his innocent eyes. "So that's what he is! Here is his hand!" Michalsky exclaimed, not daring to name the priest aloud. Our enemies have resorted to this poison more than once: condoling, attacking the government for forgetting the merits of the general, they casually reported that Turchin is in poverty and lives on charity funds. Nothing tore apart Turchin's brain like this lie; let everyone who touched us remember - there was need and hunger, eating away everything inside, and momentary despair, but every loaf of bread, even its crust, was earned by labour, and not put into our hands.

You will say how little the arrows of cowards and clerical hooks mean to one who went to the buckshot! Error! Error! It is with them that the noble is lost; assuming pride in the enemy, he does not foresee a bite, a dirty spit, he wants to break the sticky cobweb at once, and now he is already sitting tighter than before, his ears are covered with mud, his eyes are spat out by weeping creatures. What was it like for me to see him unhappy, to watch how kindness leaves blue eyes. I also had my own eyes, but the happiness of love lies in the fact that the eyes of the beloved are yours; you see life in them, and if you are destined to lose your eyes, then first you give yours.

During these months, Turchin became addicted to the violin. The draftsman's pencil and the engraver's engraver once fed us. And this gift - music - healed the heart; it never occurred to us that the violin, carved three centuries ago by a master from the shores of Lake Salo, would soon give us bread. In the spring, the windows of the house were thrown wide open until the darkness of the night, and passers-by heard our music; I say - ours, because I often echoed him on the piano. But it happened that he improvised, the violin followed him to the Don, to the southern steppes of Russia, to the parquet floors of Warsaw and Krakow, to Tennessee and Alabama, to the black huts, where the blacks, having come together, sing their psalms.

9

One day in April, when Turchin's bow was moving from one military fire to another, I noticed the silhouette of a motionless gentleman in front of the fence. The sun was setting behind him: a lean man on thin legs, a top hat in one hand, and in the other a business Yankee bag. He seemed to be preparing to take off, his right leg was set aside, his busy hands were raised. "What else, a literate man appeared for our souls!" I thought with annoyance. The evening sun slipped over the roof of the warehouse, and I realized that I was looking at a black man: a head in curls, a face in a short beard. He sank lightly on his knees, dropped his top hat and bag onto the path, and folded his arms across his chest.

It was our Abraham. He was among the few whom Lincoln's proclamation lifted above disenfranchisement and poverty. Mind and erudition opened the way for him; helped by Frederick Douglas and the white circle in Boston. Abraham found the position of secretary of the tax agent.

Frantisek Gajewski galloped into the forest after Tadeusz. For a long time there had not been such fun at our table; three veterans sat behind

453

him - no, four, let me stay in their company - two Russians, a Pole and a Negro. Drum invited Abraham to Radom; but the black secretary—by manners the first gentleman at the table—smiled, revealing his infantile clean teeth, and shook his head—he had already chosen his path and did not live for himself alone. He has business with us: there was a book in the bag - a collection of military articles published in New York and Boston on the tenth anniversary of the military victory and delayed in publication. Embossed binding, excellent paper, engravings and vignettes; galloping cavalrymen, tall wagons, overturned guns, bayonet fighting, rebel ships sinking from Monitor bombs ... "Only pictures are good here," Abraham noted, "they do not lie. I will cut them out, and we will burn the rest..."

The soldiers had not yet taken off their uniforms when the Chicago publisher Clark printed the first volume of a work compiled by T. M. Eddy, editor of the North Western Christian Advocate. In the summer of 1865, the title of the book sounded like a proud dedication: *"An outline of the civil and military history of the state during the years of the war to preserve the Union, and the history of operations in which the soldiers of Illinois covered themselves with glory, stories of distinguished officers, a list of glorious dead, the movement of medical and Christian services, with portraits of famous people engraved on steel".*

The book also found a place for the 19th Illinois Regiment, and the engraver's chisel transferred to a steel plate and a portrait of John Basil Turchin in the unbuttoned uniform of a brigadier general. But every year the South returned to the hands of those who rebelled and evaded the bullet. They began to write little about the war, as if sobered up after the massacre made in a drunken fumes. *"There are real generals here, but there are also the losers of the war,"* Abraham continued. *"What ruined by military stupidity is now tinted with hired feathers. However, Buell himself wrote his lie..."* Turchin had already come across Buell's pages; Buell gave the whole catastrophe of the army in Alabama to two officers - Turchin and Mitchel. As if there was no annulment of the sentence and a presidential patent for the rank of brigadier general, and then the dismissal of the commander, but there was a military genius

Don Carlos Buell, born to save the Union, and next to him officers - the ruins of his cause. Abraham said that there are people who are ready to lift the gauntlet; only some papers and documents are needed. And Turchin jumped up from the table, laughed, pointing to the book, and spoke merrily:

'Losers of the war!' Well said, "Abraham, thank you for coming and saying so nicely." He affectionately ruffled his gray rings, looked into his brown eyes. – "We should have Bingham here, good Bingham-Napoleon, he would advise how to deal with enemies in a Christian way, so that not a single hair falls from their heads." "We didn't know that he wasn't talking about distant Buell, but about tomorrow's business. - Here, Nadine, our true friends!" he exclaimed, pointing to Tadeusz and Abraham. Brothers in humanity! He slammed his fist into his own hand. And the enemies are still alive! Suddenly it was revealed that they were alive, and there, - he pointed to the book, - and here! His finger pointed out the window. - They are alive, and we give them our houses ... I need you, Abraham, and you, Tadeusz; the day after tomorrow we go to Mission Hill. Agree on drummers? he finished enigmatically.

Buell brought him back to the battlefield. Turchin pulled his feet out of the swamp of merchants, bills and mortgages.

10

The colonists, summoned by Turchin and the priest, came to the plebanium. The parishioners entered the hall appointed for meetings of church brotherhoods, and people of other faiths crowded there, some crowded in the square, stood at the open windows. At the entrance to the parsonage, two gentlemen stopped with sheets of paper in their hands: Tadeusz and Abraham. The paper is divided by a vertical line, and the names of Turchin and Pan Teodor are written at the top of each half.

Looking around at the people sitting on the benches, I felt how great their confusion was. Behind Turchin is the heart, the kinship of the forest pioneers, on other faces it is read, on many? But Pan Theodore, the first priest who lived with them for a year, was a servant of God, a friend; he visited every farm, held every bill of sale in his hands.

Turchin raised his hand.

"Pan Theodore will speak first; within these walls - the first word to the shepherd. Today you will say goodbye to one of us, and to whom - you decide. Pan Tadeusz and an old black friend of mine are standing at the door; everyone, leaving here, will put his name on paper: under Pan Theodore, whoever wants him, or under my name, whoever is with me. If it pleases Pan Theodore, let not mine, but his friends take the papers."

"I have no chosen ones," said the priest humbly.

The priest confidently waited for the denouement; had it been his will, he would have silently let the parishioners go, if only to make sure as soon as possible that the people were with him, with the holy church. But the speech was prepared, Pan Theodore knew how to keep in his head the thought and the word, and the prudent question of the catcher, after which he silently looked from the pulpit at the parishioners. This time he spoke with special skill; a dull voice complained to the parish, was perplexed, sought understanding. The priest asked himself: does he want Pan Turchin to leave Radom? - and after a pause, he answered: no, he doesn't want to! Turchin is ardent, sometimes unfair, but the servant of the Lord wants everyone to disperse as brothers from this evening and everyone to give a hand to yesterday's enemy.

The priest disposed of people with calm speech. They looked at Turchin, would he go to the world?

Pan Theodore mourned over our unbelief, but did not blame the general for the expulsion of the former priests; on the contrary, he mentioned that Turchin did not interfere with the construction of the church and spitting. But life without confession is a sin, just as a worldly marriage without a wedding is sinful.

"Don't touch their love!" shouted Tadeusz. "Or I'll shake your bed."

Great was the temptation for the priest to go into battle with the pagan, the keeper of the manor grave, but he restrained himself.

"Faith is at the foundation of everything," continued the priest, "but today the parishioners have come together in order to decide worldly affairs. One of them stands close to faith, belongs to it, like a body to the soul, this is a national matter; the other is daily bread, land, houses, prosperity." - And, without taking the lists in hand, looking into the hall and choosing one by one the colonists, the priest reminded that the land agent cared little about the future of the farmers, the interests of Illinois Central for him stood above the needs of the farmers. "The general is a generous man," said the priest, "no one will suspect that the farmers' dollars were stuck behind the lining of his coat, but the company demanded service, and Turchin served. Let the loss happen even in five, and even in three dollars, and this money was obtained with sweat, with blood; even if the mistakes are small, they can result in a great disaster - a deed of sale without sufficient guarantee can be declared invalid in a court of law, and the Pole will lose his land. We don't want to believe in Pan Turchin's bankruptcy," the priest remarked. "Radom used to see him in good spirits, but even the newspapers are already writing about overdue bills. " The priest went from farm to farm and everywhere he found a flaw, a hole, a possible mistake. He spoke with reproach about the war veterans, who got the land for free, unmeasured, so they are kept by the forest princelings.

"They paid for the land with blood," Turchin remarked.

"Those who donated blood do not need land: God took them to himself."

"Or the devil!" shouted the notary; he was drunk. – "These guys are willingly cleaned up by the devil."

"What is good for a veteran with a gun in his hands," continued Pan Theodore, "is not good for a peaceful Pole. He needs an honest fence, so that the land passes faithfully from father to son. Why did it happen, the priest asked, that even a worthy person failed?" He mournfully looked around the hall, waiting for an answer from people, did not wait and answered himself: Because the Polish cause is alien to Turchin. The best part of the priest's speech began. He remembered Turchin in passing, in a hint, so that he would not forget that the general was

Russian, but he knew Poland, he lived in Poland, however, in Russian garrisons, with Tsarevich Alexander. This thread he pulled inconspicuously, he wove the national banner in front of the parishioners, and tears welled up in more than one farmer. And, seeing the parishioners captivated by his word, Pan Theodore spoke about the Polish nest in America, about the future fertile state, which would set an example, if not for the world, then for the whole country.

Not a word, not a whisper in the hall, when he finished and prayerfully closed his eyes.

As soon as Pan Theodore stepped back from the pulpit, blinded by the fire of the sun that he had kindled, all eyes turned to Turchin. Few looked at him with hope. Others - gloomily, as if for the first time they saw us. Jan Kozelek, although with one eye, burned through the motionless figure of the general.

"Did any of you think that he was keeping me in Radom?" Turchin asked without calling, cherishing the silence achieved by the priest. "It was not the seekers of easy bread who came together here; violence drove you across the ocean, need pushed you out of Chicago and Detroit. What keeps me in Radom? - The parishioners were silent; Turchin was Radom for them, they received it before the earth, before the merchants, do you ever wonder why this is above your field, and not another sky? — Church? I did not kneel in it, I helped you lay down its walls and walked away. Earth? Soon you will take the third harvest, and I did not plow my own land. Do you keep income? They could, could be, Josef Krefta would have made big money here, as the Yankees say, and I am a beggar. So your shepherd just announced: Pan Turchin is bankrupt! You know where my money went, but the probst doesn't even need to know; let him think that you always had white bread and meat for dinner.

Listening to his home conversation, the parishioners hoped that Turchin also wanted peace.

Why am I here? The crisis is over, the company is building roads, laying second tracks, that's where my money is. And I left Chicago, I'm with you - why? You have nothing to think about if everything that we talked about is forgotten; but is it possible for me not to think, with such a stubborn head! Smiling, he tilted his head slightly. - Leaving Radom, I will not extend the hand of a creditor to you; you have no debt, no papers, no receipts. But what is there? Yes, I dreamed of a brotherhood. Not a separate Polish brotherhood, not a church brotherhood, but a brotherhood in humanity: in front of him, no money has a price ...

"And you speak Polish with us!" shouted Kozelek. "You are a soul catcher!"

"I want to be heard, that's why I speak your glorious language. My friend Abraham will forgive me for speaking incomprehensibly, he knows that in any language I will tell the truth. - And Turchin explained to Abraham: "I speak Polish, this is a good language for the truth."

"It's not words that lie, it's thoughts, General!" Abraham said cheerfully.

"You are used to John Turchin taking care of your affairs, and Pani Nadine heals ailments; so a man gets used to trees that give him shade, but at the first need he cuts them down with an axe. Yes, I am Russian, but in the opinion of your pastor, a bad Russian. Not because I served in the army of Paskevich or came to Warsaw with the headquarters of the Russian corps, but solely because I am not busy with Russian business. If I had advanced on Poland in 1963 with an army, even then I would have been a good Russian for Pan Theodore, an enemy, but a Russian, in my work, in my blood. Only she is worth something in the eyes of the priest. Give him a separate Polish county. Well, just like the Irish will take a separate land for themselves, there are millions of them; and the Germans too; and what, in the republic that accepted us, there is no longer America, but only rags, proud voivodeships with ambitious people on the throne! Today priests and Joshua Ford have found errors in our bills of sale, and tomorrow they will find errors in state borders. And here it's not up to the justice of the peace, the guns will judge: war! massacre! No, I don't want to go to this paradise."

"And I say no!" exclaimed Tadeusz. "No, wicked shepherd!

On this day, the priest showed his strength; confident that he owns the hearts of the parishioners, he looked at Drama with forgiving sadness.

The ploughman, Pan Theodore will tell you, before throwing grain into the ground, he will weed out not only the tares. Take away both frail grain and someone else's, not from this earth and not from this sun, if you want a good harvest. And for grain these words are true. But to transfer such truths to people is blasphemy. Remember, I asked you about the usurer Kreft: do you want him? You said - he was a money changer, a horse dealer, such Jesus expelled from the temple. And I rejoiced, not because you rejected the Pole; you have driven away the impurity. And what does your priest want, you heard; he spoke well today, now it is not difficult for you to choose one of us. Only one more thing: this year I was asked a lot of questions; for each bill. The priest asked, and more the notary, and I answered all the questions. Isn't it true, Pan Theodore?

" Truly so."

"And you confirm, Joshua Ford?"

"Your soul is wide open, your coat is unbuttoned and your tie is on one side," the notary cheered.

"Turchin was nicknamed the wild Cossack, and he humbly answered questions. Let Pan Theodore, taking God as judge, answer three questions."

"I'm ready, Pan Turchin!"

"The Catholic Gazette is far away, in your Detroit, Pan Theodore: who told them about my bankruptcy?

"If I prayed for your well-being, they couldn't hear me in Detroit."

"Tell the truth, Pan Theodore; truth relieves the soul."

The priest proudly closed his lips, moving away from the conversation.

"I'm not bankrupt yet," Turchin said. "True, I asked to postpone the bill, but the first term will come in three days."

Pan Giryk! Michalsky's voice trembled. "When I was leaving for Detroit, you handed over letters with me to the Provost at St. Wojciech and Priest Vincent Barczyński. Maybe, in your anxiety for Rad, you wrote something to them?"

"The soul of a shepherd is always in fear for the parishioners," Giryk evaded. "Detroit priests won't carry letters to the newspaper. I vouch for them."

"God is a guarantee to Pan Theodore, and he is to the Detroit priests!" Turchin concluded cheerfully. "Who gave my bill to Joshua Ford?"

The priest was offended, and many in the hall thought that it would be better to ask this question to the notary.

"Pan Theodore doesn't know," said Turchin. "It's a pity! And you, venerable Joshua Ford?"

"I wish I didn't know!" - I'll see how you play your violin in three days."

"Here is a worthy man," Turchin encouraged him. "Who gave you the bill?"

"Josef Krefta!" I didn't borrow from him.

"And this man made two hundred merchants!" the notary was amazed. – "In your opinion, the one who lent you money cannot sell the bill? What is it, do you have such orders in Russia?"

"And we have usurers - Christ-sellers!" Turchin reassured him. "And we have bills going around. That's where the world stands."

"So Krefta bought the bill. For whom Krefta will take, he will not let him out alive."

The parishioners were relieved of their hearts: the priest remained on the side-lines.

"And now the last thing, Pan Theodore. How many acres of land did you buy from Krefta for yourself? And why don't you tell people: soon I will go for the plough with you."

Earth, earth! The simple matter of life, it most of all occupied the minds of farmers. They wanted to know who the new settler was, how much land he had bought, and if there was any danger to them?

The Church does not forbid us to own land.

"But why secretly, Pan Theodore?"

"I did not have time to inform about the purchase: a little time has passed."

"The land was purchased in November last year."

The eyes of the parishioners flickered from one to the other; Turchin's words fell on the cup of truth, bent it down.

"Is it really possible to put a piece of land as a reproach to the shepherd!" — Excerpt left the priest.

Turchin laughed with spiritual relief. He learned about the purchase from the side, from the draftsman, who made a copy of the survey map.

"From now on, your probst is the first gentry in Radom—four hundred and twenty acres of land!" So take one of us with your bare hands: here I am, a bankrupt, and here is Pan Theodore, and not alone, but with Krefta, with a notary, a glorious company, they will cope with any gesheft.

"Bismarck! the thunderous voice of Dudzik was heard."

It was perhaps the biggest curse among Poznań and especially Silesian Poles. We watched as the parishioners moved towards the doors, showing their backs to both the priest and Turchin, turning away from the uncleanness into which they had been plunged.

11

And the city of the Sun did not stand in the place of forest cuttings. When the Franciscan fathers, Leon Brandys and Desiderius Liss, brought church bells to Radom, they glorified in three tones not the commune, not the hostel of justice, but the village streets smashed by carts, the Krefta steam mill, the new brick spitting, oak shutters on the windows and card tables at the Mindaka establishment.

The priest did not collect even a third of the signatures. The parishioners bowed to Turchin; columns of names stretched out on both sheets below his name. The priest made his way to the porch, to Abraham and Tadeusz, intimidating the parishioners with a stern look. Convinced that the parish had leaned towards the general, the priest shouted about the betrayal of the faith, about the blind moles, whom fate had brought into the light of day in vain, and that the Radomites were a depraved mob, ready to stone their prophet. "Hejže on Soplice! Hejže on Soplice!" he shouted in rage, recalling the gentry betrayed by the mob from Mickiewicz's poem. Pan Theodor left Radom and the bosom of the Polish church. Matilda took him to Wisconsin, to a village with a German name Berlin, and at the end of the summer they returned to our district, to their farm, 14 miles northeast of Radom. Here he lived in friendship with the German Catholics. Ambition consumed him, illness knocked him down; in the autumn of 1878 he died before reaching the age of 42. Matilda did not give his body to the priests from Ashley, the stubborn wife of a Prussian army corporal woke up in her. She put Pan Theodore in an oak coffin, brought him to our church on a cart and buried him at the Radom cemetery. The priest in Radom was then her compatriot, a German who spoke poor Polish, the Franciscan father Marek Tanel; he laid the foundation for the domination of the Franciscan order in our parish.

The Priest Giryk left, Turchin remained. We redeemed the bill from Joshua Ford on time: Tadeusz collected the money. Abraham left with a light heart: the general allowed him to light the redeemed bill and from this fire to light his pipe.

Well, did Turchin's unbelief win?

No. Next to the church brotherhood of St. Michael the Archangel and the Sisters of the Mother of God, St. Other brotherhoods also arose - St. Francis and St. Casimir; a bell tower was attached to the church, and the ringing floated over the surrounding forests.

So the priest and the diocese won?

No, faith in Radom has not increased. Pan Theodore lost, but Turchin did not win either. From the talk we returned home joyful, and the next day, looking around, we did not find either the captured guns or the reclaimed land under our feet. The desert... Were the parishioners afraid of God's punishment, betraying the shepherd and seeing on Mindak's wagon trunks and suitcases of a priest, instead of beer barrels, sacks and boxes? No, there was no fear. The colonists lived, rejoiced at Saturday evening, to drink a mug of beer at Mindak, smoke a cigar, talk about their old voivodeship, about a few dollars sent overseas to their relatives, about whether the Poles were real Kashubians, and certainly about the accursed Bismarck. They lived, as Fergus and his compositor Charles had predicted, with their worries, indifferent both to the blood-bitten lips of Pan Theodore and to the mountain peaks of the socialist Turchin. They gave him their vote and a small penny to redeem the bill, could you expect a greater sacrifice from them!

Having defeated the priest in the tournament of honour, we nevertheless woke up in the morning on the ashes of the spirit. The hostel of justice slipped away, melted away like the mist of this warm April morning, leaving odourless and colourless moisture on the seeking palms. Radom lived, roosters called to each other from dawn, bells put on cows rang back in the forest, at the appointed hour a thick, three-tone bell ringing floated around. "The century is passing by! Turchin told me as we escorted Abraham to the train and Tadeusz Drum said that Ashley's school trustees demanded that the anatomical model not be brought to school. "A century is passing by," repeated Turchin, "and on the soles, like dust, like graveyard clay, the accursed last century sticks. We have not achieved here and what was in the regiment. But it was, it was!" I stopped him on Varshavskaya Street, took off his hat, knowing that passers-by could also see us, ran my hand over his cold forehead, across his cheek, from temple to chin, and laughed with a laugh that contained both careless tenderness, and love, and all the inseparability of our past and future years. "Why have I never had the feeling of a fleeting life? - he asked. "It's as if several lives have been lived, and each one did not pass suddenly, each one was dense, and long. And do you feel it?" I couldn't lie and shook my head. "How so? he wondered. "Why do you have one life and I have several?" he asked. "All that I have experienced from the Carpathians to Abraham, who has just left, is you. It happened, and I lived for a minute, but still in my heart - you."

"It is in you that my one life lies, no matter how long it lasts, one and one. I'm your serf, Vanya, - I laughed again." "Alexander let go of the serfs in Russia, and before me, across the ocean, his arms were short, and I remained a serf girl, Nadeyka Turchaninova ..."

12

How to live?

We are beggars: Turchin did not bow to Hansom, the cash registers of Illinois Central slammed shut. Radom sank for us to the ordinary, to an accidental dwelling. Everything happened - extreme need, plots that humiliate the republic, but not him; never, not for an hour, did he lose his independence. Now you know a lot, look around and think: could we have been intimidated and with what? Losses that would surpass past losses? Neither man nor god invented them. Fear? Killing fire? Everything was, everything, except dishonour. Having lost your homeland for the sake of a great ideal and hardened your heart with the torment of a lifetime, is it possible to be afraid of losing money or a house? Chicago became the second city of the Western Hemisphere, a stream of business people and workers seethed along the new streets, in the stone banks. War veterans, admirers of Turchin, were everywhere, but the city did not belong to them. Chicago lived not in memory, but in the present day. What could he give us now: service in Illinois Central? Harassment of a pension from a forgetful Congress, money without which one cannot live in a big city? A slow transition to the Chicago poor working class?

"The workers of the railway company still remember you," Fergus, Turchina called. "You found your language with them." "I worked with them, tried not to be robbed by the administration: that's all my language." - "Nothing else is needed; just support their fight. And you can write, write about the labour issue ... "Fergus began to praise Turchin's military pamphlets, said that he remembers them. "What you say shows that I am right: everyone should do their own thing. I decided to write only my own, what I know, which is my life. Such is my pen: it will not write someone else's; no matter how much you dip it in the inkwell, I'll pull it out dry."

It's easy to say: write! We need old papers, reports from the commanders of both sides, written immediately after the battle, with a still hot pulse, we need dispositions and military maps, quartermaster reports and orders from commanding officials. Not a dispute with

Buell led Turchin's pen; the republic and the new-born nation were to recognize themselves in the mirror of his books. But how to write without having your own bread? Others make their writings a means of life; but we never had this joy and relief. Scribbled sheets were not fed; like chicks in a nest, they themselves opened their hungry beaks.

So we have become farmers, and a non-trading farmer is a sufferer, crucified on the cross of his needs. Shoes, clothes, salt and sugar, kerosene, ten papers, iron feathers and tobacco will not grow on the farm. Sometimes Turchin travelled to Chicago to lecture on past battles, especially memorable for Illinois. He lectured in the Panorama of the Battle of Mission Ridge and in the old concert hall, on the corner of State Street and Randolph Street, which is now Marshall Field's general store. On the podium, Turchin forgot everything, like a child, like a simple-minded settler in Kansas or Utah. He experienced everything all over again, again led the brigade to storm the Missionary Ridge and, having reached the summit, wiped the sweat from his forehead and smiled at General Willet, who, having caught up with Turchin, congratulated him on his success. I remember how it happened at the first lecture. "Just at that moment," Turchin said, "a heated fight flared up on my left, and I apologized:" Damn it, Willett, I have to go to my brigade! "No sooner said than done: Turchin flashed his eyes, turned, rushed to the left and nearly fell off the stage. He did not feel that he was in Chicago, that he was lecturing, and not pursuing General Bragg's army, retreating to the old Chicago field.

Lectures gave crumbs: for tickets to Washington and back. As long as the War Records Office was headed by a worthy officer, Colonel Robert N. Scott, Turchin could receive copies of official papers. Years pass, and more and more bookish people write about the war, hired pens or bilious ambitious people: they find comfort in rearranging tin soldiers, rolling toy guns across the pages of books. Turchin was looking for truth and soon became a rare expert on the subject. Increasingly, letters came from Chicago, Springfield, Philadelphia and Washington, their authors were looking for details from Turchin, a correct assessment of events, reconciled the numbers of losses or the number of prisoners. "Why doesn't the general go for letters himself?" - the postmaster asked Misha Florikh, who was running after our mail. "Pan General writes," Misha boasted. "And Mrs. Turchina?" This question made it difficult for the boy, he tried to remember what I was doing. "Pani general's wife is milking the cow..." So we lived in the imagination of those who know how hard it is to plow the land and harvest grain. Chickamauga was coming to an end, Missionary Ridge was paper mounds on every table when Robert N. Scott died, and after his funeral, Turchin was barred from the archives. The brigadier general, one-armed, in iron glasses and black robes, the new head of the archives, removed Turchin from his papers. Turchin went at him with his chest, clenching his fists; but these people, without their own passion, are not ashamed of a stranger. He waited too long for a place and, having finally received an exemplary cemetery of papers, did not want to let the living there. "I know every cupboard, every folder..." Turchin was angry. "The late Scott allowed many to rummage through the papers." "That's what they're here for." "It's not time yet. Many want to base their ambition on my papers, I will not allow this. Since when are they yours, Mr. All-to-my-self-take?" "My name is..." and with loud boredom he announced his full name, rank and military merit. "My name alone gives me that right."

He seemed to be just old, dumb and boring. But we learned that others had gained access to the archive, they freely entered there, and a broken sergeant stood in front of Turchin, demanding permission from the military department. "Mr. Turchin," the one-armed man said when Turchin reappeared in his office, "I was told that you do not receive a military pension?" - "Let the cripples or rich idlers get it; I feed with my own hands." The dim, close-set eyes above the bony nose lit up at last with a feeling - alas, hateful. "But the War Department and Congress denied you a pension?" "I didn't please them, and I didn't ask for it." - "And everyone says that your general-ship is doubtful. As if you owe the general's patent to protection ... The late president joked, and when he missed it, you were already somewhere in the army, and there was no time for you. Turchin jumped up to him, and the one-armed one stood up: dead cloth, dull, neat seams, two rows of buttons, mortal boredom, everything is like on a dead man, dried and set up like an anatomical model of Drama! Turchin stepped back and said through his teeth: "I would call you, General, and kill you, but I doubt whether you shoot well with your left hand? Be quiet! Real generals are made by the war, not by ministers and presidents: they are only clerks in the war," - thundered Turchin. "I'll be in the archive, if only so that you don't burn the papers of the North!" I'll be in the archive, and you - beware!

Turchin wrote to the War Department and Ulysses Grant and obtained access to the archive. Under Scott, he entered here as if he were in his own house, now he wandered along the corridors and rooms, as if through someone else's fortress, along the trench, where the unbeaten enemy still sat down. The requested folders strangely disappeared, they were not found on the spot - it turns out that someone asked for these papers before Turchin, or damage to the numbered sheet was noticed in the file, and it was sent for correspondence, for gluing and repair. Turchin was told that such and such cases had been sent to the military department.

One might have thought that archival folders were delivered by vans to the authorities, and yet the order required that even the Minister of War himself appear in the archive.

13

And Turchin decided to visit an important person. I don't write who this person is; the republic owes him, and he is not all bad. This is not James Garfield, but a man who also knew what Turchin was like in the war.

Turchin appeared to him on July 4, the American Independence Day, in a dark alpaca troika, with a violin at hand - the owner was at home and immediately went into the living room. America knew well this face framed by coarse hair, reminiscent of the prickly yawn of a sea animal, his lips, folded imperiously and implacably in all portraits, an angry and pityless look from under his eyebrows. When Turchin spoke without taking off his hat, the general recognized him, but did not give a look:

"I was informed that an Italian musician had come, but are you an American?" - "I have an Italian violin, and in any case the instrument is more important than a person ..." He opened the case, showed the dark wood of the violin by Gasparo da Salo. "I am old, and this violin is two hundred years older than me," "Do you want to sell it?" Turchin bared his graying head. "In my whole life I have not sold anything, and when I bought it, it was certainly at a loss." - "And I, to confess, would not have bought it, and would not have taken it as a gift." Turchin gave him, at choice, the opportunity to recognize his officer or play a performance, and he chose the latter. "But you wanted an Italian musician and went out to him." - "It is a duty of hospitality, especially on a day like today," the general replied. "Actually, I would rather do without foreigners; the army knew it." - "They say you, like no one else, owe it to the blacks!" Turchin objected. "This is greatly exaggerated, they were taught on the plantations to do the job properly, that's all." - "Now it seems our army is clean again?" Turchin asked. "Yes, we don't need anyone in particular..." Even the general has already entered into an evil game. "In peacetime," Turchin said, "you can leave only drums, shako and adjutants in the army, but, God forbid, war, and where is there to monitor blood," - "If you don't sell a violin," the general said dryly, "Then what did you come with?" - "I could teach music to your children." The general laughed. "They grew up."

"I would teach my grandchildren: every time needs honest music!" "It seems to me that the violin is not an American instrument," the general said with irritation. "That's right, let's give it to the Italians; the French - a guillotine, we will take a drum and a cheque book. And you listen to how the violin speaks with the voice of our Republic!"

The beard had already pressed the violin, a full arm bent into hands, the bow slid over the strings and, without preparation, extracted the music of the blacks from the violin. Playing without notes, Turchin lowered his eyelids, saw nothing and heard nothing but music. This time he woke up because someone held his elbow. Turchin opened his eyes - a gray-haired, red-faced old man was holding out to him a hat with a five-dollar bill on the bottom, the general was not in the living room. "Here, if you please," the old man said enviously and whispered theatrically: "My God! Ivan Vasilyevich! What are you?.." The old eyes watered, the flabby eyelids quickly turned red; Saburov spoke in a whisper and in French, biting his lip apprehensively when a Russian word escaped. Decrepitude approached him; when Turchin shook out the banknote on the carpet and Saburov bent down to pick it up, his hand was trembling. "I recognize the Russian soul and rejoice like a brother ... I'm fed up with this prison, but what can you do," he said quickly, following Turchin, managing to close the doors. "They are fleeing from fortresses, from iron bars, I myself am in the soul of Monte Cristo, but how to escape from satiety, from satin captivity ..." "I have lost the habit of French, Saburov!" Turchin said loudly. "You leave me without a piece of bread! I am a Frenchman here, they do not need another me; French and not even Saburov. An immortal nature, a lizard that changes colour, he himself already believed that he was a Frenchman, a natural Frenchman. "What are you, his butler? Lackey? "Why are you trampling?! I'm an educated person, I also took other courses at the Sorbonne... - Lying about the Sorbonne, he believed that too. - I'm a tutor! Gouverneur-s!" They were approaching the gates of the manor. Saburov grew bolder, spoke more cheekily: "I am not a servant here, sir, I am a member of the family; remember how it used to be in Russia, French tutor! Now these boors have adopted our fashion - let them get a Frenchman! - He laughed. "Why didn't you ask the general for yourself? Turchin asked. "This man is like a Turkish pasha, he can do anything, and you have done so much for our army..." Saburov blocked the wrought-iron gate with iron foliage. "Whoever remembers the old, that eye out! I saw you and froze: I hid behind the curtain, and you, my dear, on the violin. Take the money!" - "This is lackey money, they got into the lackey's pocket!" "That's true," Saburov smiled. - Gouverneur-s - the same lackey. He was still holding the gate with his hand. Why didn't you ask the general, you ask? In a war, a general is scared, they will kill him, but for what? If I happened to die in the Caucasus, if you please, I understand everything, for the king and the fatherland, but here? They even have it written on the banner in Latin: "E Pluribus Unum" - one of many, - they took from the ancients, they didn't come up with their own words: the republic of merchants! "Are you thinking of pleasing me with swearing, Saburov?"

"I would like to go to Russia ... Will they give me a place? Will houses be valued? Though a ruin, but to return ... Oh, Lord, after all, ruins have their own beauty, and they adorn the earth ... " Turchin stopped. "Do you want the truth? Then listen: you are not a ruin, you are a corpse. The corpse of a leper is one of those that are thrown into pits, on logs, poured with pitch and burned, burned, burned! "And who are you, my dear," Saburov also broke loose and squealed after him, risking his place: "A counterfeiter! Thug! And you will die, you are not needed either here or there! .. "

"Why did you go to the general with a violin?" I asked after listening to the story, and Turchin answered easily: "He always did not love me. Ulysses Grant respected, but this one did not like. Turchin was not there; if he had said something unkind to me, I would have given him a piece of red goods. And he swerved, swerved, isn't that nice!" Poor Vanya! I am writing this now, leaving him in Mound City to wait for me: poor Vanya! How difficult it is to live openly, to always give the first shot to another, to put on in battle not chain mail, but a thin, white shirt of honesty. Poor Vanya! - I say now, when he does not hear me, but even now I hasten to add next: good, the best of all whom my eyes have seen, a wise madman, the happiest of all the unfortunates of the century ...

"The curse of Saburov," he continued, "is not the end of my Washington day. The old man's cries did not stain me; not for Saburov to stand between us and Russia. Do you 57 remember what Herzen told me: "It is clear that the Russian people are destined to die for the freedom of others." True, true, but not all; having taken the right to fight for someone else's freedom, we also suffer bitterly from our abandoned land. And I wanted to calm my heart, find my comrades in the brigade, and with a ticket in my pocket, I went along unfamiliar, new streets towards the shipyards, to the house where Barney O'Mullen lived. I recognized his wife, she did not immediately. Their adult daughter called out from the room the address of the saloon where they could look for their father. I pointed out that I had left a book for Barney the night Lincoln was assassinated; can i get it back? They deliberated about something, quarrelled, then the mother brought out a book wrapped in paper. "I recognized you, you were with a beautiful lady." "This is my wife," I said, "she is alive and as good as ever." - "You have a violin in your hands," the poor woman sighed, "it means that you can feed your wife even without a gunshot."

14

There was no Irishman in the saloon; they gave the address of the beer hall, where Barney went with a drunken gang. It was getting dark, the sky above the capital lit up here and there with lights, as if not a hundred years ago, but the country had just thrown off the yoke of the British. Turchin found Barney in the company of two gunners from the Gekker regiment, the former batman of General Baird and a sailor from the Monitor - he was on the supply of cores when the Monitor exchanged fire with the Merimak. There was also an old man with a turkey crop and bulging eyes; he swore he was the last to leave Fort Sumter when Major Andersen threatened him with a pistol; that the banner of the fort was carried not by Andersen in a suitcase, but by him - on his chest, under his uniform. "Aby did not dare to support our fort," the old man wept, "to send ten ships with cannonballs, with flour and bacon; Abie was still smearing lard on the asses of the damned southerners, so we lost the war. The old man accepted the surrender of the fort as a final defeat in the war; the victories of four years did not assuage his grief.

Turchin came to life, he was sitting at a table with friends. A tall glass of beer stood in front of him, the foam subsided, he did not take a glass in his hands, and this amused the veterans more than the fact that an unfamiliar fat man in a black suit was a general, and even from Russia. The owner of the beer hall came up, replaced the untouched glass with a new one, with a head of foam. "Do me a favour," he asked, "take a sip. This is a good beer, I'm a White House supplier, and the ladies drink it there. At the word "ladies" Barney tore off his chair, he shouted that he demanded a drink for the mother-confessor, for madam, and whoever refuses, let him go out the door with him and they will talk there. Then Turchin dipped his mustache in beer, opened the violin case, raised the instrument and said: "Here is my wine, guys; old, seasoned, there is nothing like it in the prince's cellars. Before playing, he showed Barney his book, it was in a violin case: "I took it, Barney." — "Okay; I really didn't have time to look into it," "That's a quarter of a century, Barney!" - "All business, general, business; oh, how many things a beggar loafer has to do.

The crowd was noisy, not everyone wanted to listen; the violin voice is sharp, it does not leave the soul calm. But, starting to play, Turchin did not hear anything around him; and the room was silent. People were arriving, walking from the street and from the kitchen, the German family was crowding in the back doors: daughters, sons-in-law, children - that evening they too received a pardon from an early bed. Turchin could not firmly tell me what he played that evening. He began with Alar, but briefly, so that the violin would declare itself, and then it was not someone's music, not the notes he remembered, but what was and remained his life: a sobbing Magyar song, and a polonaise that interrupted the tears of the Magyars, and the voice of the Don steppe, and then closer and closer, in the steepness of his life, to "John Brown", to "America" - this second anthem of the republic - to bivouac songs and soul-stirring psalms.

How many times then did the old man with a puffy face sway slightly in front of people in the saloon or in the same beer halls of the Germans; how many times an indifferent crowd met him and demanded his music - and he yielded, cunningly, yielded for the sake of appearance, so that a person's ear would not be completely closed before his violin. He played something entertaining or sentimental, and now a thin string, an insignificant hair, holds them with a lasso, and he takes his own, and laughs at them, forcing them to listen to what they, as usual, do not listen to.

But that evening, July 4, he was surrounded by friends. When he lowered the bow and raised his face from the violin, there was a standing ovation. The German innkeeper made his way to him, threw two dollars on the table and said: "The guys are joking, they call you a general, and you are a musician ..." - "This is not a joke: I am a brigadier general" - "It's a pity! The German reached for the silver coins. "I wanted to invite you to play with me." Turchin was ahead of him and took the coins. "This is a worthy payment for me," he said. "Today I played a little play in one house, only two minutes, and they paid me five dollars for them. I gave them to the lackey, and I'll take these.

15

Now you know how the violin began to feed Turchin. Not a day or a month, but ten years, while the fingers were enough for both the bow and the pen. When he had to choose, he chose the need and the pen.

The Chickamauga is out. A book is brought from the printing house, and it seems that you are holding in your hands a projectile that will fly - and more than once - around the earth, make the crowds raise their heads from the mud, stir every bell in the bell tower, echo in almost every house. But this is the fate of few books, and not only its creator is happy - he is almost a god - but also the one who reads it.

It's done - Chickamauga is out! Fergus wrote in the preface that its author, General Turchin, *is one of the most widely educated military men in our country, that the book in which he describes one of the most memorable and important battles of our last war should attract the attention of all enlightened people.* Alas, enlightened people went their own way, and the author of Chickamauga fed on music. He was called to the best halls of Boston and Philadelphia, not to mention Chicago - he refused, and soon the impresario left the original alone, enrolling him in the category of tavern musicians. He was not afraid of concert lights; tail-coats, evening dresses, delivered from Paris in the morning, he was not afraid to falsify, playing a virtuoso piece. The reason was different: we have chosen our audience forever. Everything in the republic that, in appearance and essence, was drawn to monarchical Europe, to the nobility, albeit a plebeian, check nobility, we forever threw out of our everyday life.

I write - we, having the right to do so. We were one life, and travelled together, and where they found at least a piano broken by tavern hands, I sat down at the keys and echoed Turchin's improvisations. If such a duet offends you: the general with the violin and Princess Lvova at the unwashed keys, blame yourself - life has taught us to laugh in the face of the nouveau riche, and not be afraid of any work. From trips we brought a little money and spent it with discretion, so that the interest of the stomach did not distract us from work. On better days, I served bread and butter, bacon, beans, and coffee; but we moved without regret from good bread to bran bread, from sugar to cheap molasses, from coffee or tea to boiling water. Only complete hunger could drive us out of the Radom nest to earn money. And then we walked to the station, in our best dress, carrying a violin and a light bag, we walked through the village, hand in hand, and the Radom people believed that we were going to visit someone, bored with village life. Outwardly, everything was going well. I, as before, received the newspaper "Revolution", now free of charge; good Mrs. Stanton, publisher of the newspaper, did not send it to me alone from her bounties. We left Radom cheerful, as if in a conspiracy, light from hidden hunger, and returned with purchases, it happened that Turchin could even treat the oncoming cigarette with Virginia tobacco. Even Tadeusz Drum did not know when we ran out of meat, when we switched to bread with beans, and then to bread alone with hot Radom water.

You ask, why did the Radomites not come to our aid? They owed Turchin and remembered it; the poor man keeps even small debts in his head. I don't blame them: Turchin himself blocked their way. Years passed, and the priest's lie about the general living on alms to charity did not die, it is repeated now. Before I had time to bury Turchin, I also read about myself in the Chicago Record that I actually live only on charitable funds. Shortly after leaving Radom, Pan Teodor Turchin read the same words about himself, and not in the Catholic Gazette, but in the serious one, in the St. Louis Dispatch. You recognized Turchin and you can imagine how he met his debtors who came to us with dollars. Nothing helped, neither the pleas of Christina Kowalska, nor Dudzik's scolding, nor Michalsky's persuasions. Someone sent 15 dollars by mail, Turchin refused to receive it, calling it a trap, dirty money. Judge us with a worldly, quick judgment; he will not awaken the heart with belated regret; we have chosen our own life, let other conscientious people try to understand us. There were no exceptions for the wealthy Elizabeth Howard, nor for the generous Johnston, nor for the Ferguses, nor for Tadeusz Drum; and here's what I'll tell you, and you try to hold back a smile - money, estates separated these good people, and we, when they happened to be with us, equalized and made their brotherhood possible. Isn't that better than a good hunk of bacon or a holiday turkey!

You saw Drum at the funeral. He knows how to wear his only dress, knows how to take care of it, but he is as poor as we are. Here's how it happened: remember this too - perhaps through his act you will better understand us. At 75, Turchin demanded a pension; he wrote that he expected to go to the forefathers before his strength failed him, but fate played a cruel joke on him, and his father and mother, not knowing the thrifty frugality of the republic, supplied him with too strong a heart. Turchin's request was rejected. Only two years later, thanks to the insistence of two senators, Congress passed a private resolution to pay Turchin a $50 pension!

Perhaps you will also think: how awkward their life was. Then you did not understand our existence, and Turchin was mistaken in you. We ourselves laid down our lives, and not only in our youth, when many people can do it, but also by picking up a stick to make walking easier.

Now, after Turchin's death, my pension is also denied. My war has already been recognized, no one reproaches me for medical assistance to the wounded. Why not give such a glorious heroine something to eat? And here's why: listen, man of the new century, the voice of the last century. You can't give me money, this will offend the morality of the republic: after all, Mrs. Turchina is not the wife of a general before God, my request was rejected, not recognizing our marriage in Russia. That's how nice it turned out! It turns out that we are alive, and that distant Russian impudence of ours, a worldly civil marriage without a wedding, is also alive. Even now, in my thoughts, I am ready to fall in front of the unfortunate father, hug his knees, say that I myself suffered, breaking his heart, but let him forgive me from the grave, the sacrifice was not in vain if it infuriated the Orthodox aide-de-camp Prince Lvov, and priest Pan Theodore, and the Tartuffes of our republic. How good it is to realize that you are not dead to them, but still an adversary, albeit a dying one, because they are more afraid of a dead lion than a living dog.

16

So, the plot is over, not life, but what is subject to words. I did not say one thing: when Turchin was denied a pension, I decided to remove from him at least care for me. Of the few things that made up our life, only love and a book remained, which Turchin completed. I don't mean a description of the Battle of Missionary Ridge - this book was ready, it surpassed Chickamauga in style and passion, it was half the size, it lacked only maps and an index of names. Turchin pushed Missionary Ridge aside for a while and devoted himself to a great work, a complete outline of the war for freedom. I decided that our union, having endured half a century, would also endure separation, and I left light, as if to visit Elizabeth Howard, but the alms house became my haven - do not be afraid of this word! - an alms house secretly arranged for me by Mrs. Clara Barton, a lady benefactor, whose star rose during the war, in the management of hospitals. I understood that Turchin would write to Mrs. Howard and, having received no answer, would rush to search, and I wrote to him every day, arranging it so that my letters would come from different cities. I claimed that I had fallen into an unthinkable social whirlpool, that my old friends from Philadelphia, from the paramedic school, and good friends of Victoria Woodhole, and sister-in-law Lucy Stone, and even the publisher of the Revolution, Mrs. Stanton, vied with each other to call me. My frequent travels freed Turchin from writing to me; in mine, I was glad that he could devote every minute to work, not wasting even on short lines of writing. And all this time I lived in an arranged barracks, listening to the instructions of Miss Adams, an employee, that the rich need the poor just as much as the poor need the rich, that she will unite us into one family, as the Lord united people, and in gratitude for her deeds, the mayor's office put to the establishment of a special policeman, guardian of law and order.

A lot was burned in the fire set up by Turchin in the spring of this year, and my letters were also burned, about a hundred in number, a letter on each of those days that I spent in a decent prison, founded according to the method of Miss Adams, until Turchin burst in and took me away by the hand like a girl. I regret that these letters do not exist, but meanwhile what are they? Each of them is a page of an imaginary life, sketches of generous houses, a new landscape, kindness and wit of my imaginary hostesses and a page of my confessions, love, gratitude to the person who gave me his life. I wrote every day, and if my self-imposed exile lasted for years, I would not get bored writing to him even then. Perhaps you will smile or frown here too, comparing such a confession with the image of old people; Your will, I warned you that my relationship with paper is short, I do not lie to it. If you smile at the whole past century, finding it sentimental, I will answer that I bless this century, as the century of explorers of freedom and great faith. And I repeat what I once wrote to Herzen in London: freedom in practice is a great happiness!

I will not write about Turchin's books. What survived - survived, and what was put on fire is gone forever. I'm still sorting through the papers and again marvel at the power of his thought. I do not recognize pity, and I asked you to transfer this package to the gangway of the ship so that you would not be born with the temptation of pity and generous Russian charity. The need helps to work - what I managed to save, I will have time to sort it out, but I don't need more, I have no other things to do, except for those that are impossible - to bow to the graves of my father and the old people of Ivan Vasilyevich. After all, I will soon be to him, yes - to him, although I know that there is darkness and there is no other life, and yet to him, in order to equal him in death, as we were equal in life.

For half a century in America, we have formed the image of an emigrant ship. Not the one that crosses the Atlantic, but another, under a white canvas, with a team falling from exhaustion. There are old people and children in it, women in demolition, everything is in it, just like on the ark, whose name is Earth, humanity. And when it snows, the strong ones go with shovels and spades, go ahead of the wagon, clearing the way for the horses.

Union Army blacksmith at the front line, Petersburg, 1864.

Epilogue

"Radom (State of Illinois). - The widow of General D. B. Turchin, who in the civil war followed her husband inseparably, having performed many heroic acts herself, now actually lives only on charitable funds. General Turchin recently died and was unable to provide for her future, so now she is in the care of her friends and friends of one who was one of the most famous soldiers of Illinois.

All the efforts that have been made over the past few years to obtain a pension for Ms. Turchin have failed, since the War Ministry has rejected these requests, based on the failure to comply with some formality regarding the marriage of the General and Mrs. Turchin in Russia.

"Chicago Record", January 1902

"Had the political slogan of the early 1850s, 'America for the Americans', prevailed among us, it is more than likely that the result, long before this second decade of the 20th century, would have been 'America for the British' or some other foreign power, because in this case, after the civil war, we would not have come out as one whole - "E Pluribus Unum", but "E Pluribus Duum", [29] , if not more, and we would have become easy prey for one of those who were just waiting for the denouement.

By the way, Uncle Joe Cannon, in the Saturday Evening Post of May 3, 1913, jokingly says that "in the early fifties we had the American Movement, which was also called the Know-Nothing Movement, because none of its members could brag that he knows what it is."

So, thousands of people whom this "patriotic" movement would prefer not to let on our doorstep, to our shores, bravely fought for America, which became their foster mother, as for the only country where they knew the joy of freedom. And our country should be infinitely grateful to these fighters not born in our country. And one of the greatest heroes among them was "that terrible

Cossack," as he was called on both sides of the Mason-Dixon line [30], Ivan Vasilyevich Turchaninov, or John Basil Turchin, under which name he entered American history.

At the National Cemetery in Mound City, Illinois, on a marble slab, which does not particularly stand out among many other soldier's graves, this unusual inscription is carved[31]:

John, B. Turchin
Brig, Gen'l, U.S.A.
Dec. June 24, 1822 - June 18, 1901

Nadine
His wife
Nov. July 26, 1826 - July 17, 1904

It is not known whether the dates of their birth, carved on the plate, correspond to our chronology or the Gregorian calendar, which is still accepted in Russia both in secular and church affairs.

Madame preferred that the general be buried in Chicago until it was found that she could lie down next to him. It is difficult to give an exhaustive portrait of Turchina in order to describe her character, mind, refined culture, which distinguished her all her life: both when she lived in prosperity, and when she endured all sorts of hardships with great dignity and lived in poverty.

It is possible that if she were the wife of some American general, as illustrious a hero as her husband, then she would be sung in songs, and another Place would be assigned to her in our history, so that in the memory of every schoolchild in our country her name would probably be imprinted second, only after Martha Washington, in a number of heroic American women.

"Illinois Central Magazine"
September 1914, Volume 3, No. 3, 3

FINIS

NOTES

Notes

1—Hereinafter, the discharge is replaced by bold (note by layout designer). OUT

2—All documents and original texts are typed in italics in the novel. OUT

3—In the name of the Father, and the Son, and the Holy Spirit! (lat.)

4—City Hall of Paris.

5—Corporate spirit (French).

6—All mine I carry with me (lat.).

7—An order to present the arrested person to the court to consider the legality of the arrest.

8—"Cholera" (French).

9—L. Stone (1813-1893) - a prominent figure in the struggle for the equality of women in America and against slavery.

10—Mr. Bell does not advise me to marry this lady because she has seven children. Well, what is it? Mr. Bell cannot love for me. I myself must love, and I love her!

11—"Jesus made the blind see, Jesus healed the lame, Jesus restored hearing to the deaf. Come in, good Jesus!"

12—J. B. Turchin. military rambles. Chicago, 1865, pp. 30–31.

13—Illinois Central Railroad.

14—"The Girl I Left"

15—"One, two, three, four, five, six, seven - tiger!"

16—White plebeians, mob,

17—Out of many, one (lat.).

18—More royalist than the king himself (French).

19—Johnston refers to Scott as a colonel, as the commander of the regiment. In May 1862 Scott was a lieutenant colonel. — Approx. author.

20—That is, officers - employees of the commissariats, as the departments of military formations that were in charge of supply were called.

21—"The Sinfulness of Dancing" (English).

22—Illinois Central Magazine, September 1914, Volume 3, No. 3, article "General John Basil Turchin and Nadine, his wife."

23—"Copperheads" is the nickname of northerners, supporters of a compromise with slave owners, and their agents.

24—"Ten dollars a month! Three of them for clothes! I'm going to Washington. Fight for Lincoln's daughters!" (English, distorted).

25—Daring and Suffering. Boston.

26—Take God to help, and the devil into a fist! (Polish)

27—Death and wife are predetermined (Polish).

28—Movement for free love, freedom from divorce, civil marriage, etc. (from free love - free love).

29—Out of many, two.

30—Line of demarcation between free and slave states.

31—John B. Turchin. Brig, General of the US Volunteer Army, 24 Dec. 1822 - June 18, 1901. Nadine, his wife, - Nov. 26. 1826 - July 17, 1904

Lincoln on the battlefield, 1862.

John Basil Turchin (January 30, 1822 - June 19, 1901)
Russian hero of the US Civil War

(born Ivan Vassilievich Turchaninov) was an American briga-dier-general, a famed hero of the Civil War, whose exploits are written down into American history....

Ivan Turchaninov was born in the Don region, into the family with long-reaching military traditions. Therefore, the boy was enrolled into the First Cadet Corps in St. Petersburg, and later into the famous court" Mikhailovsky artillery school.

Due to his extraordinary military abilities, he was accepted into the Imperial Guard. And in 1852 he also graduated from the Nikolay Staff Academy in St. Petersburg, where he got acquainted with Prince Alexander, the future Emperor Alexander II. In 1853 he went to the Crimea War, and fought bravely for three years, and later served in the Baltics and in Poland.

During his travels across the country and his service, Turchin became a staunch supporter of abolitionism, which he considered a hindrance to Russia's development, and in 1853 even started a secret correspondence with A. Herzen. His dissatisfaction with the existing order grew. And so, in 1856, while serving in Poland, where he had been sent to for his bravery in the Crimean campaign and where he had successfully reached the high position of the head of staff for a whole infantry corps, he suddenly left - basically escaped - for Chicago, USA, together with his young wife Nadezhda (born Lvova).

In America, Turchaninov, having taken a new name – John Basil Turchin – failed to adapt and went bankrupt. Even more, he was disap-pointed with the country: in his letters to Herzen, he wrote that US is plagued by just as many problems as Russia.

But he also noted that "...one thing I am grateful to America form, is that it has helped me get rid of my snobbish mentality, took me down to a level of a mere mortal. No labour is too dirty for me now".

Turchin graduated from an engineering school, and Nadezhda received medical education, and so soon the couple returned to wealthy life and to the high society. He even got acquainted with a certain businessman George McClellan, and a certain lawyer named Abraham Lincoln!

With the breakout of the Civil War (June 1861), Turchin enrolled into the Federal Army, receiving the rank of a colonel at once, and taking over the 19th Illinois infantry regiment.

It was after this decision, the news of which reached his original homeland and caused a small scandal, that the Tsar signed an order expelling Turchin from the ranks of the Russian Army, and even making a personal note "An American officer cannot be a Russian of officer as well!", also taking away the of officer's commission Turchin had held for all those years.

He distinguished himself in a series of large battles, but his star hour came in 1863: during the battle of Chickamauga (September 19-20, 1863), the lines of the North were breached, and the federal troops turned to run. Turchin personally led his brigade into a counter-attack, broke through the South positions, finding himself deep behind enemy lines, then turned around and fought back to the North army! This exploit was entered into American history as Turchin's raid behind enemy lines.

During the battle at Chattanooga (November 24-25, 1863), Turchin's brigade found itself under heavy cannon re at the foothills of the Missionary ridge. He personally led his men into a charge up the ridge, withstanding everything including bombs with fuses lit, being thrown down at them by the defending artillerists, and knocking out the enemy guns. The casualties reached 282 killed, but the victory at the Missionary ridge changed the balance in the whole war radically into the North's favour.

An interesting fact from Turchin's biography: having fallen ill, he put his wife Nadine, a nurse at the hospital, as the regiment's commander, and she, surprisingly, was quite successful at the position.

The commander was personally acquainted with three US Presidents: Lincoln, Grant and Garfield. And his soldiers made an ironic song about him, called *"Turchin's got your mule"*, which became his brigade's battle anthem.

After a heart attack in October 1864, Turchin left the military service, returned to Chicago where he worked as a patent engineer, and also helped settle immigrants coming to the US from Europe. It was him who in 1873 founded the Radom Polish commune in Illinois.

He also wrote memoirs and historic novels about the Civil War – works that would later be featured in the libraries of all the military school of America: *"The Battle at The Missionary Ridge"* and *"Experiences and Impressions of The Civil War"!*

At the very end of his life, Turchin wrote to Emperor Alexander II a plea to grant him the right to return to Russia but was denied.

The former hero died in poverty in 1901, aged 79, having lived his last years on a meager military pension. He was buried with military honours on a military cemetery in Mount City (Illinois), at the expense of the local authorities.

Sources:

The voice of Russia, Find a grave, A Journey Through Slavic Culture.

The Battle of Missionary Ridge.

www.ingramcontent.com/pod-product-compliance
Lightning Source LLC
Chambersburg PA
CBHW031212050726
47495CB00017B/241